Advances in Intelligent Systems and Computing

Volume 1085

The series "Advances in Intelligent Systems and Computing" contains publications on theory, applications, and design methods of Intelligent Systems and Intelligent Computing. Virtually all disciplines such as engineering, natural sciences, computer and information science, ICT, economics, business, e-commerce, environment, healthcare, life science are covered. The list of topics spans all the areas of modern intelligent systems and computing such as: computational intelligence, soft computing including neural networks, fuzzy systems, evolutionary computing and the fusion of these paradigms, social intelligence, ambient intelligence, computational neuroscience, artificial life, virtual worlds and society, cognitive science and systems, Perception and Vision, DNA and immune based systems, self-organizing and adaptive systems, e-Learning and teaching, human-centered and human-centric computing, recommender systems, intelligent control, robotics and mechatronics including human-machine teaming, knowledge-based paradigms, learning paradigms, machine ethics, intelligent data analysis, knowledge management, intelligent agents, intelligent decision making and support, intelligent network security, trust management, interactive entertainment, Web intelligence and multimedia.

The publications within "Advances in Intelligent Systems and Computing" are primarily proceedings of important conferences, symposia and congresses. They cover significant recent developments in the field, both of a foundational and applicable character. An important characteristic feature of the series is the short publication time and world-wide distribution. This permits a rapid and broad dissemination of research results.

** **Indexing: The books of this series are submitted to ISI Proceedings, EI-Compendex, DBLP, SCOPUS, Google Scholar and Springerlink** **

More information about this series at http://www.springer.com/series/11156

Sonali Agarwal · Shekhar Verma ·
Dharma P. Agrawal
Editors

Machine Intelligence and Signal Processing

Proceedings of International Conference, MISP 2019

 Springer

Editors
Sonali Agarwal
Department of Information Technology
Indian Institute of Information
Technology Allahabad
Allahabad, India

Shekhar Verma
Department of Information Technology
Indian Institute of Information
Technology Allahabad
Allahabad, India

Dharma P. Agrawal
Department of Electrical Engineering
and Computing Science
University of Cincinnati
Cincinnati, OH, USA

ISSN 2194-5357 ISSN 2194-5365 (electronic)
Advances in Intelligent Systems and Computing
ISBN 978-981-15-1365-7 ISBN 978-981-15-1366-4 (eBook)
https://doi.org/10.1007/978-981-15-1366-4

This Springer imprint is published by the registered company Springer Nature Singapore Pte Ltd.
The registered company address is: 152 Beach Road, #21-01/04 Gateway East, Singapore 189721, Singapore

Organization

Patron

Prof. P. Nagabhushan, IIIT Allahabad

General Chair(s)

Prof. Shekhar Verma, IIIT Allahabad
Prof. S. Chakravarthy, University of Texas

General Co-chair

Dr. Shirshu Verma, IIIT Allahabad

Organizing Chair(s)

Dr. Sonali Agarwal, IIIT Allahabad
Dr. Pavan Chakraborty, IIIT Allahabad
Dr. Vrijendra Singh, IIIT Allahabad

Workshop Chair(s)

Dr. K. P. Singh, IIIT Allahabad
Dr. Manish Kumar, IIIT Allahabad

Local Organizing Committee

Dr. Pritish Varadwaj, IIIT Allahabad
Dr. Satish Singh, IIIT Allahabad
Dr. Rahul Kala, IIIT Allahabad
Dr. Triloki Pant, IIIT Allahabad
Dr. Mohammed Javed, IIIT Allahabad
Dr. S. Venketesan, IIIT Allahabad

Program Committee

Prof. T. Lahiri, IIIT Allahabad
Prof. O. P. Vyas, IIIT Allahabad
Prof. Anupam Agarwal, IIIT Allahabad
Prof. R. S. Yadav, MNNIT Allahabad
Prof. Ashish Ghosh, ISI Kolkata
Prof. R. B. Pachori, IIT Indore
Dr. M. Tanveer, IIT Indore
Dr. Vijay Kumar Chaurasia, IIIT Allahabad
Dr. Neetesh Purohit, IIIT Allahabad
Dr. N. S. Rajput, IIT BHU, Varanasi
Dr. P. Ramaswamy, University of Kent, UK
Dr. Mohammad Syafrullah, UBL, Indonesia
Dr. Shafaatunnur Hasan, UTM, Malaysia
Dr. Muneendra Ojha, IIIT Naya Raipur
Dr. Ajay Singh Raghuvanshi, NIT Raipur
Dr. Partha Pratim Roy, IIT Roorkee
Dr. Mohammad Najim, MMU Mullana
Dr. Sumit Kushwaha, KNIT Sultanpur, India

International Advisory Committee

Prof. G. C. Nandi, IIIT Allahabad
Prof. Srikanta Tirthapura, Iowa State University, USA
Dr. A. R. Alaei, Griffith University, Australia
Dr. V. Janeja, University of Maryland, USA
A. B. M. Shawkat Ali, University of Fiji
Prof. Dharma P. Agarwal, University of Cincinnati, USA
Prof. D. Kotzinos, University of Paris-Seine
Olivier Colliot, Aramis Lab

P. N. Suganthan, NTU, Singapore
Prof. G. Prasad, Ulster University, UK
Prof. A. Krishna, Curtin University, Australia
Prof. Gaurav Sharma, University of Rochester, NY
Prof. Sharma Chakraborty, UTA, USA

Preface

This volume of proceedings provides an opportunity for readers to engage with a selection of refereed papers that were presented at the Second International Conference on Machine Intelligence and Signal Processing (MISP-2019). MISP aims to bring together researchers in related fields to explore and discuss various aspects of data mining, artificial intelligence, optimization, machine learning methods/algorithms, signal processing theory and methods, and their applications. This edition was hosted by Indian Institute of Information Technology, Allahabad, Prayagraj, Uttar Pradesh, during September 7–10, 2019. All submissions were evaluated based on their significance, novelty, and technical quality. A double-blind review process was conducted to ensure that the author names and affiliations were unknown to technical program committee. This proceeding contains 36 research papers selected for presentation at the conference. Our sincere thanks go to all the authors for their interest in the conference and to the members of the review committee for their insightful and careful reviews.

We are grateful to General Chairs for their support and express our most sincere thanks to all keynote speakers for sharing their expertise and knowledge with us. Our sincere thanks to Prof. P. Nagabhushan, Director, IIITA, for his valuable suggestions and encouragement. We would like to thank the organizing committee and many other volunteers who worked behind the scenes to ensure the success of this conference. We are very thankful to the EasyChair conference system for providing support for submission and review process. Finally, we would like to acknowledge Springer for active cooperation and timely production of the proceedings.

Allahabad, India   Dr. Sonali Agarwal
Allahabad, India   Prof. Shekhar Verma
Cincinnati, USA   Prof. Dharma P. Agrawal

Contents

About the Editors

Dr. Sonali Agarwal is Associate Professor at Indian Institute of Information Technology Allahabad, Prayagraj, U.P., India. Her main research areas are data mining, software engineering and machine learning. She has published several research papers in academic journals and international proceedings. She has served as a guest editor for special issues in several journals, program committee member for many conferences and chaired several conferences. She also organized several national/international workshops and conferences. She is a senior member of IEEE and a member of IEEE CIS and ACM.

Prof. Shekhar Verma is a Professor at Indian Institute of Information Technology Allahabad, Prayagraj, U.P., India. His main research areas are wireless sensor network, cryptography, cloud computing and machine learning. He has published several papers in academic journals and international proceedings. He has served as a guest editor for special issues in several journals, program committee member for many conferences and chaired several conferences. He also organized several national/international workshops and conferences.

Dharma P. Agrawal has been the OBR Distinguished Professor at the University of Cincinnati, since August 1998. He is co-author of textbooks Introduction to Wireless and Mobile System (4th edition), and Ad hoc and Sensor Networks (2nd edition), co-edited Encyclopedia on Ad Hoc and Ubiquitous Computing, and Embedded Sensor Systems. He is the Fellow of IEEE, ACM, AAAS, NAI, IACSIT, and WIF. He is a Golden Core member of the IEEE-CS and recipient of the IEEE Third Millennium Medal. He has published over 684 articles, 42 keynote speeches, 67 intensive courses, 8 patents and 25 invention disclosures, supervised 75 Ph.D. dissertations and led UCBT Bluetooth package. He has been on the editorial boards of IEEE Transactions on Computers, IEEE Computer, Journal of High Speed Computing, JPDC, and is serving IJCN, JECE, IJSIA, IJDSN, IJAHUC, IJAHSWN, JDSN, and IJWMC and founding EIC of the Central European Journal of Computer Science. He is Editor-in-Chief for the Journal of Sensor and Actuator Networks. His research interests include applications of sensor networks in monitoring Parkinsons disease patients and neurosis, applications of sensor networks in monitoring fitness of athletes personnel wellness, applications of sensor networks in monitoring firefighters physical condition in action, efficient secured communication in Sensor networks, secured group communication in Vehicular Networks, use of Femto cells in LTE technology and interference issues, heterogeneous wireless networks, and resource allocation and security in mesh networks for 4G technology.

Ring Partition-Based Fingerprint Indexing Algorithm

Manas Jyoti Gogoi, Amit Kumar Trivedi, Dalton Meitei Thounaojam, Aniruddha Bhattarchajee, Rahul Debnath and Kaushik Borah

Abstract The shear number of fingerprints in a modern database makes exhaustive search, a computationally an expensive process. It is in this context a new indexing mechanism is proposed to speed up the process of identification of fingerprints. In this model, concentric circles are made around the core of a fingerprint, and by grouping the fingerprints according to the number of minutiae points in each of these rings, we can select the best prospective fingerprints for a particular query fingerprint, thus greatly reducing the number of fingerprints in which exhaustive search is to be employed and increasing the speed of Automated Fingerprint Identification Systems (AFIS). The proposed model was tested on different datasets of FVC2000 database, and the results show that the model achieves high CIP with a low penetration rate and our model was able to significantly speed up the process.

Keywords Fingerprint · Indexing · Minutiae points · Core · Clustering · Rings

1 Introduction

Nowadays, the information and communication technology plays very important roll in day-to-day life, but it also poses the problem of person authentication and information theft. The classical technique based on password and token is no more reliable. The biometric-based security and authentication is more reliable. Particularly, the fingerprint biometric has gained much acceptance. The higher acceptance of fingerprint biometric can be attributed to its uniqueness, persistence, and easy to use. There are two general ways in which fingerprint-based biometric systems are used: verification, which is a 1-1 matching process and identification and is a more complex 1-N matching process. The three most common approaches to solve the identification problem are: (1) exhaustively search the entire database and look for

M. J. Gogoi · A. K. Trivedi (✉) · D. M. Thounaojam · A. Bhattarchajee · R. Debnath · K. Borah
Computer Vision Lab, Department of Computer Science and Engineering, National Institute of Technology Silchar, Assam 788010, India
e-mail: rivedi19@gmail.com

© Springer Nature Singapore Pte Ltd. 2020
S. Agarwal et al. (eds.), *Machine Intelligence and Signal Processing*,
Advances in Intelligent Systems and Computing 1085,
https://doi.org/10.1007/978-981-15-1366-4_1

a match which is like repeating the verification procedure for every database finger-print, (2) fingerprint classification which divides the fingerprints into certain classes based on some observed traits of fingerprints, and (3) fingerprint indexing which will select some specific traits of each fingerprint and store it as key along with the corresponding fingerprint that can be used to retrieved it later on. As the size of the database increases, the first approach becomes impractical due to its very high search time. The second approach shows some chance of reducing the search time. But the different classifications used in this method are taken from the Galton–Henry clas-sification system which categorizes the fingerprints into five classes: right loop (R), left loop (L), whorl (W), arch (A), and tented arch (T). However, this classification fails on two grounds: (1) The classes are relatively low in number, and (2) the classi-fication is skewed and the distribution is uneven among the five classes (31.7% right loop, 33.8% left loop, 27.9% whorl, 3.7% arch, and 2.9% tented arch) [17, 24]. Some researchers have proposed other classification systems based on other features [2, 8, 9, 19]. However, those methods also failed on the two above-mentioned parameters. The third approach of fingerprint indexing is supposed to help in reducing the search list to decrease the matching time.

2   Background and Related Works

The main purpose of indexing is to significantly reduce the number of candidate fingerprints to be matched. Over the years, researchers have come up with lots of indexing mechanisms, but all these have two main components: (1) feature extraction and (2) index space creation and candidate retrieval. The fingerprint features that are used by most of the fingerprint indexing technique can be grouped into: (1) Level 1 features: the ridge flow and frequency, (2) Level 2 features: the minutiae and the direction of a minutia, and (3) Level 3 features: sweat pores and outer contours (edges) of the ridges.

Based on the class of feature used and the index space, the various indexing mechanisms can be broadly grouped into four categories.

2.1   Global Features Based Mechanisms

The features used in this method are usually ridge frequency, orientation field, and singular points. An unsupervised learning algorithm (such as k-means) is used for the generation of index space [18, 26]. Many researchers are working on deriving other distinct features about a fingerprint by evaluating various parameters using Delaunay triangulation from the fingerprint minutiae points [22]. Jiang et al. used orientation field and dominant ridge distance features for indexing [18]. They proposed a new distance measure to remove the inconsistency of orientation differences and the proposed regional weighting scheme to visibly improve fingerprint retrieval performance.

2.2 Local Feature-Based Mechanisms

In this method, minutiae points and local ridge information are used to extract some translation invariant or rotation invariant features which are then used as the basis for index space creation. Features such as minutiae location, direction, triplets obtained from minutiae, local ridge-line orientation, and ridge density are recognized as local feature. To increase the searching speed in large database, Cappelli et al. proposed a hash-based indexing [6]. They designed a "Locality-Sensitive Hashing (LSH)" using "Minutiae Cylinder-Code (MCC)." The MCC is very efficient and effective for transforming minutiae representation into transformation-invariant fixed-length binary vectors [5]. The indexing technique was tested with six different databases, namely: NIST DB4, NIST DB4 (Natural), NIST DB14, FVC2000 DB2, FVC2000 DB3, and FVC2002 DB1, and better result was reported in respect of penetration rate and computation time.

Feng et al. [10] combined minutiae and its surrounding ridge pattern to form a local substructure. A binary relation was defined on this substructure which are invariant for indexing. The experimental result on FVC2002 shows promising results.

Biswas et al. [3] reported that the utility of local ridge-line curvature together with minutiae local geometric structure can be used to increase the efficiency of fingerprint indexing.

2.3 Sweat Pore-Based Mechanisms

This method uses the features extracted from the Level 3 details of a fingerprint such as information obtained from sweat pores, incipient ridges and dots. During the index space creation, the index values are computed from the extracted features. They normally use Level 3 features for their feature set.

2.4 Other Mechanisms

Reduced scale-invariant feature transformation (SIFT) uses the positions and orientation of minutiae and special points for feature extraction. A symmetrical filter-based method generates a fixed-length feature vector which is used to create the index space [25]. Li et al. used the response of three filters to map the orientation image.

For different orientation, they get different response and use it for indexing. The filters were divided into two one-dimensional filters to make the feature extraction faster [21]. This paper is organized as follows.

In this paper, the core point of the fingerprint is detected at first, and then taking this core point as center, concentric circles/rings are made. The location of minutiae points and their count inside each ring and position relative to the rings are used for feature extraction and thus to create the index space. The use of concentric rings around the core makes the features invariant to rotation.

3   Proposed Model

The indexing of the fingerprint database is carried out offline. The retrieval of candidate list is done in online mode. The basic operations for both the indexing and retrieval are same. In indexing, the index value is calculated to store the fingerprint template in database and updation of index table. The different phases of the proposed model (see Fig. 1) are organized as follows.

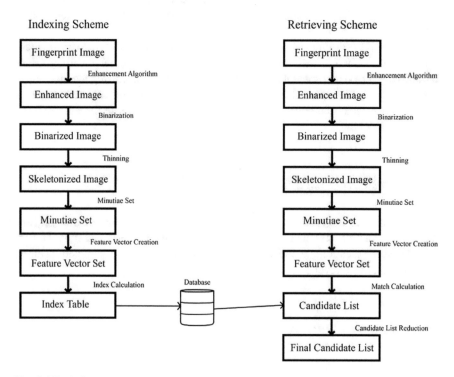

Fig. 1 Block diagram

3.1 Preprocessing

The preprocessing stage is composed of three stages: Image enhancement, binarization, and skeletonization [11].

Image Enhancement: The image enhancement algorithm (See Fig. 2) [13] is made up of five stages:

1. The first stage in enhancement is normalization where image pixel intensity values are normalized to have the desired mean and variance.
2. In the second stage, the orientation image is computed from the normalized image.
3. In the third stage, the frequency image is determined from the normalized image and the orientation image [14].
4. The fourth stage classifies each block in normalized image into recoverable and non-recoverable regions, finally giving a region mask as output.
5. Finally, in the fifth stage, an array of Gabor filters [7, 15], which is set up according to the local ridge orientation and ridge frequency, is used on the normalized image to get the final enhanced image as output [20].

Binarization: After the enhanced image is obtained from the previous step, the ridges in the image are represented by black pixels, while the valleys or furrows in the image are represented by white pixels. This results in an image where ridges appear black and the valleys appear white. This process is known as binarization [11].

Skeletonization: This stage eliminates the redundant pixels from the ridges till the ridges are just one pixel wide.

(a)   **(b)**

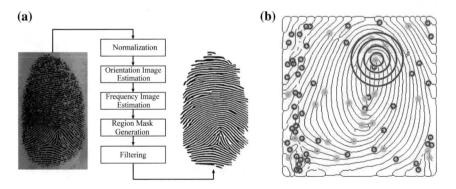

Fig. 2 **a** Flow Chart of fingerprint image enhancement. **b** Minutiae extraction and core detection

3.2   Minutiae Extraction

The minutiae points are obtained from the skeletonized image. Minutiae points are discontinuities in fingerprint ridge lines. A single fingerprint image has many minutiae points, which collectively makes the fingerprint unique. The location, direction, and type of minutiae points are generally used as features to differentiate fingerprints from each other. Generally, the number of minutiae in a fingerprint varies from 30 to 80, which also depends on the quality of the fingerprint image. In the proposed model, two different types of minutiae points are used: bifurcation and termination minutiae. Bifurcation is a point in the fingerprint where a given ridge line gets divided into two separate ridges. On the other hand, termination is a point where a certain ridge line terminates, or ends. The ith element, m_i of a minutiae feature vector, $M = \{m_1, m_2, m_3, \ldots, m_n\}$, is a triplet of (x_i, y_i, θ_i), where x_i and y_i are coordinates of the minutiae in the fingerprint image and θ_i is its direction.

3.3   Core Detection

The core point of the fingerprint is obtained by complex filtering. The core is the point in the fingerprint image where maximum curvature of the fingerprint ridge occurs [1, 4]. The core is a type of singular point. If two or more core points are detected, generally the topmost one is selected as the core. This point remains unaffected by the effects of rotation and translation which in turn makes it robust to use as a feature. Each fingerprint has a core point (x_c, y_c), where x_c and y_c are its respective x- and y-coordinates.

3.4   Ring Partition

The fingerprint image is partitioned into rings by using a ring partition algorithm. Four concentric circles are made with the core point (x_c, y_c) as the center. Since core point is unaffected by rotation, the circles are also unaffected by the effects of rotation. Now, the number of minutiae points is counted in each ring. This gives us a 4-tuple $N = \{n_1, n_2, n_3, n_4\}$, where $n_1, n_2, n_3,$ and n_4 are numbers of minutiae in the rings 1, 2, 3, and 4, respectively. This tuple is the final feature vector which is used for primary indexing.

3.5   Primary Hashing

Let n be the number of minutiae in a ring of the fingerprint. For the index space creation, we convert n to $\{(n/2) + 1\}$ and then by using $\{(n/2) + 1, i\}$ (where i

belongs to $(0, 4]$ representing the ring number) as the key, we append the fingerprint id to the index space. The values were converted to incorporate tolerance in the hashing process, so as to overlook minute differences in minutiae count of the fingerprints that might occur due to various reasons. As a consequence of this primary hash, the index space so created has dimensions $\{((m/2) + 1) \times 4\}$, where m is the maximum number of minutiae in a ring of a fingerprint in the database and the second dimension corresponds to the number of rings chosen.

3.6 Indexing Stage

For each image in the database, the image is first pre-processed, and next, the minutiae in the image is extracted and the core of the fingerprint is determined. The core determined is manually verified, and if it is falsely detected, the fingerprint image is rejected. Also, if the core detected is too close to any of the edges of the image (i.e., if it is within a certain threshold for the database), the fingerprint is rejected. The minutiae in the fingerprint image are now partitioned into concentric rings with the core as the center which gives the feature vector N for the image. Using the four values in the feature vector N, the fingerprint is hashed to its corresponding position in the index space. Therefore, two images are kept in the same bin if any corresponding element in its feature vectors is equal. The primary hash is done to reduce the number of heavy computations, as only the images which correspond to the same bin, as that of the query image is selected for further evaluation.

Algorithm 1 Indexing Scheme

Input   : Fingerprints, (I_i), i=1,2,...,N ; N is the number of fingerprints in the database
Output : Hash table, H_d.

1: **for** i=1 **to** N **do**
2:    $P_i \leftarrow$ **Preprocess**(I_i)
3:    $M_i \leftarrow$ **ExtractMinutiae**(P_i)
4: $(X_{ci}, Y_{ci}) \leftarrow$ **CoreDetection**(P_i)
5:    $R_i \leftarrow$ **RingPartition**(I_i, X_{ci}, Y_{ci})
6: $MaxRad_i \leftarrow$ **MaxRadius**(R_i)
7:    **if** $MaxRad_i \geq th_1$ **then**
8:        **RejectImage**(I_i)
9:    **else**
10:       $Table1(i) \leftarrow$ **DistanceCore** (K, M_i, X_{ci}, Y_{ci})
11:       **for** $j = 1$ **to** $NumberOfRings(R_i)$ **do**
12:           $Table2_{ij} \leftarrow$ **CountMinutia**(R_{ij})
13:       **end for**
14:       $Table2con(i,:) \leftarrow (Table2(i,:)) /2 + 1$
15:       **for** $j = 1$ **to** NumberOfRings(R_i) **do**
16:           $bins(Table2con(i,j),j) \leftarrow \{bins(Table2con(i,j),j),i\}$
17:       **end for**
18:    **end if**
19: **end for**
20: **return** $H_d \leftarrow bins(Table2con)$

3.7 Retrieval Stage

In the retrieval stage, the query fingerprint is processed and based on the number of minutiae in each of the four rings, the fingerprint images are retrieved. Those fingerprints ids that have been retrieved at least two times based on the minutiae count of the four rings are shortlisted. A fingerprint id can be retrieved for a maximum of four times or a minimum of zero times. Next, for all the shortlisted images, the distance of all the minutiae points from the core is calculated and sorted in ascending order $D_i = \{d_{i1}, d_{i2}, ..., d_{in}\}$, where d_{ij} is the distance of the jth minutiae from the core point of the i_{th} fingerprint. The first ten distances from the distance vectors of the shortlisted images are chosen for comparison using the LCS algorithm [12] with the first ten values of the distance vector of the query fingerprint. The first ten values of the distance vector are chosen so as to take into consideration all the minutiae points inside the rings. The LCS scores of the comparison of shortlisted fingerprints with the query fingerprint is sorted in decreasing order, and the top k fingerprint ids are selected as final output of the system (k is a threshold, which is the size of the candidate list given as output by the system). The longest common subsequence between the distance vector of the query image Dt and that of the shortlisted image D_i is taken as a measure of similarity between the two fingerprints, and we store these normalized longest common subsequence values in a sorted vector. If the value of any of the top three LCS scores of the comparison is less than the threshold $c,$ then the query fingerprint is rejected by the system and the result is a no-match found, otherwise the top k fingerprint id is given as the final output of the system.

Algorithm 2: Retrieving Scheme

Algorithm 2 Retrieving Scheme

Input : Query fingerprint, I and Hash Table H_d
Output: Shortlisted list of fingerprints , F .
1: $P \leftarrow$ **Preprocess(**I**)**
2: $M \leftarrow$ **ExtractMinutiae(**P **)**
3:(X_c, Y_c) $\leftarrow$ **CoreDetection(**P **)**
4: $R \leftarrow$ **RingPartition(**I, X_c, Y_c**)**
5: **if** $MaxRad \geq th_1$ **then**
6:      **RejectImage(**I**)**
7: **else**
8:      $Table3 \leftarrow$ **CalDistCore** (K, M, X_c, Y_c)
9:      **for** $j = 1$ to $Number\ of\ Rings\ (R)$ **do**
10:         $Table4(j) \leftarrow$ **CountNoOfMinutia(**R_j**)**
11:      **end for**
12:      **for** $j = 1$ to $Number\ of\ Rings\ (R)$ **do**
13:         $ShortlistedImages(j) \leftarrow bins(Table4(j),j)$
14:      **end for**
15:      **for** $j = 1$ to $Number\ of\ Images(ShortlistedImages)$ **do**
16:         **if** $LCS(Table3,\ Table1(j)) \geq th_2$ **then**
17:             $SelectedImages(j) \leftarrow ShortlistedImages(j)$
18:         **end if**
19:      **end for**
20: **end if**
21: **return** $SelectedImages$

---

4   Experimental Results and Discussion

The proposed model was implemented in MATLAB. Various performance evaluations were measured based on the tests performed on the FVC2000 [23] datasets
which are particularly tricky and difficult to study and experiment on. The model
was tested on the four different datasets of the FVC2000 database.

In accordance with the general convention, two metrics [16, 27] have been defined
for performance evaluation of the proposed indexing model:

1. CIP: The CIP or Correct Index Power is defined as $(N_{ci}/N_i) * 100$, where N_{ci}
    is the number of correctly indexed/identified images and N_i is the total number
    of images that should have been accepted from the database. So, in a way, CIP
    gives us a measure about the hit rate of the proposed model. The higher the CIP,
    the more images the model has been able to correctly identify from the database.
2. CRP: The CRP or Correct Reject Power is defined as $(N_{cr}/N_r) * 100$, where N_{cr}
    is the number of correctly rejected images and N_r is the total number of images
    that should have been rejected from the database. The higher the CRP, the more
    images the model has been able to correctly reject from the database.

The proposed model was tested on various datasets and evaluated on the measures
mentioned above.

4.1   Results for the Dataset FVC2000-DB1

The feature extraction and index space creation of the proposed model was done on
the basis of the fingerprints in the database DB1. For testing, a new dataset was made
with some images taken from the DB1 (the database on which it was trained); in
addition to these images, some images were also added to the dataset which where
part of other dataset (i.e., the fingerprints the model was not trained on) and the
objective of the proposed model was to correctly identify the fingerprints from DB1
and reject the fingerprints of other datasets. In the sorted shortlisted set of fingerprints,
if the LCS value of any of the top three fingerprints were less than 0.67, the query
fingerprint is rejected and the output is a no-match otherwise a match. This threshold
which is termed as c is obtained empirically. The performance of the model was
measured by varying the penetration rate from roughly 5–27%, and the best CIP
achieved was 89% with penetration rate 27%. The CRP attained by the system was
70%. The results obtained for this database are shown in Fig. 3.

4.2   Results for Database NIST-DB4B

Similarly, the feature extraction and index space creation of the proposed model were
done on the basis of the fingerprints in the database DB4B. For testing, a new database

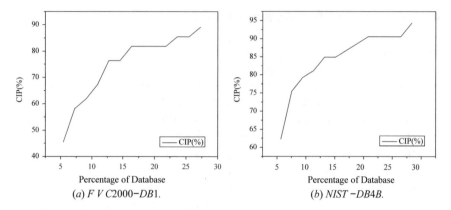

(a) F V C2000−DB1.   *(b) NIST −DB4B.*

Fig. 3 CIP

was made with some images taken from the DB4B (the database on which it was trained); in addition to these images, some images were also added to the database which where part of other databases (i.e., the fingerprints the model was not trained on), and the objective of the proposed model was to correctly identify the fingerprints from DB4B and reject the other fingerprints. Like the previous dataset in the sorted shortlisted set of fingerprints, if the LCS value of any of the top three fingerprints is less than the threshold 0.69, the fingerprint is rejected as a no-match otherwise a match. In the same way, the threshold c is determined experimentally as in dataset DB1. The performance of the model was measured by varying the penetration rate from roughly 5–28%, and the best CIP achieved was 94% with penetration rate 28%. The CRP attained by the system was 70%. The results obtained for this database are shown in Fig. 3.

It is evident from the graphs that the model was able to achieve a good balance between CIP and CRP while keeping the number of searched images to a minimum. Most of the fingerprints used in this case were particularly of bad quality, so the proposed model is also robust to fingerprint captured with low-resolution sensors. Thus, from the above two sets of results, we can assert that the proposed model is accurate, fast, and robust.

5 Conclusions and Future Work

Fingerprint is the widely used biometric for verification and security of a system. In this paper, a fingerprint indexing system is proposed. The proposed system uses a ring partition concept where the number of minutiae in each ring is used for indexing. To reduce the size of the candidate list obtained from the indexing algorithm, a matching algorithm using LCS is used. For the experimentation of the proposed systems, standard database FVC2000 is used. From the experiments, it is observed that the system achieved good results. Even though the system had a drawback that it could not be applied on the datasets DB2 and DB3, still the results are good for the other datasets. Thus, the system can be implemented to reduce the overall time required in searching the identity of a person. Although the datasets were very challenging, the systems proved to be accurate as well as efficient in such an environment.

References

1. Bahgat, G., Khalil, A., Kader, N.A., Mashali, S.: Fast and accurate algorithm for core point detection in fingerprint images. Egypt. Inform. J. **14**(1), 15–25 (2013)
2. Biswas, S., Ratha, N.K., Aggarwal, G., Connell, J.: Exploring ridge curvature for fingerprint indexing. In: 2008 IEEE Second International Conference on Biometrics: Theory, Applications and Systems, pp. 1–6 (Sept 2008). https://doi.org/10.1109/BTAS.2008.4699384
3. Biswas, S., Ratha, N.K., Aggarwal, G., Connell, J.: Exploring ridge curvature for fingerprint indexing. In: 2nd IEEE International Conference on Biometrics: Theory, Applications and Systems 2008 (BTAS 2008), pp. 1–6. IEEE (2008)
4. Cao, G., Mao, Z., Sun, Q.S.: Core-point detection based on edge mapping in fingerprint images. J. Electron. Imaging **18**(1), 013013 (2009)
5. Cappelli, R., Ferrara, M., Maltoni, D.: Minutia cylinder-code: a new representation and matching technique for fingerprint recognition. IEEE Trans. Pattern Anal. Mach. Intell. **32**(12), 2128–2141 (2010). https://doi.org/10.1109/TPAMI.2010.52
6. Cappelli, R., Ferrara, M., Maltoni, D.: Fingerprint indexing based on minutia cylinder-code. IEEE Trans. Pattern Anal. Mach. Intell. **33**(5), 1051–1057 (2011)
7. Daugman, J.G.: Uncertainty relation for resolution in space, spatial frequency, and orientation optimized by two-dimensional visual cortical filters. JOSA A **2**(7), 1160–1169 (1985)
8. de Boer, J., de Boer, J., Bazen, A., Gerez, S.: Indexing Fingerprint Databases Based on Multiple Features, pp. 300–306. Technology Foundation (STW) (2001)
9. Deblonde, A.: Fingerprint indexing through sparse decomposition of ridge flow patches. In: 2014 IEEE Symposium on Computational Intelligence in Biometrics and Identity Management (CIBIM), pp. 202–208 (2014). https://doi.org/10.1109/CIBIM.2014.7015464
10. Feng, J., Cai, A.: Fingerprint indexing using ridge invariants. In: 18th International Conference on Pattern Recognition (ICPR 2006), vol. 4, pp. 433–436. IEEE (2006)
11. Greenberg, S., Aladjem, M., Kogan, D., Dimitrov, I.: Fingerprint image enhancement using filtering techniques. In: 15th International Conference on Pattern Recognition, vol. 3, pp. 322–325. IEEE (2000)
12. Hirschberg, D.S.: Algorithms for the longest common subsequence problem. J. ACM (JACM) **24**(4), 664–675 (1977)
13. Hong, L., Wan, Y., Jain, A.: Fingerprint image enhancement: algorithm and performance evaluation. IEEE Trans. Pattern Anal. Mach. Intell. **20**(8), 777–789 (1998)
14. Jain, A., Hong, L., Bolle, R.: On-line fingerprint verification. IEEE Trans. Pattern Anal. Mach. Intell. **19**(4), 302–314 (1997)
15. Jain, A.K., Farrokhnia, F.: Unsupervised texture segmentation using gabor filters. Pattern Recogn. **24**(12), 1167–1186 (1991)
16. Jain, A.K., Hong, L., Pankanti, S., Bolle, R.: An identity-authentication system using fingerprints. Proc. IEEE **85**(9), 1365–1388 (1997)
17. Jain, A.K., Ross, A.A., Nandakumar, K.: Introduction to Biometrics. Springer Science & Business Media, Berlin (2011)
18. Jiang, X., Liu, M., Kot, A.C.: Fingerprint retrieval for identification. IEEE Trans. Inf. Forensics Secur. **1**(4), 532–542 (2006). https://doi.org/10.1109/TIFS.2006.885021

19. Kawagoe, M., Tojo, A.: Fingerprint pattern classification. Pattern Recogn. **17**(3), 295–303 (1984)
20. Laine, A., Fan, J.: Texture classification by wavelet packet signatures. IEEE Trans. Pattern Anal. Mach. Intell. **15**(11), 1186–1191 (1993)
21. Li, J., Yau, W.Y., Wang, H.: Fingerprint indexing based on symmetrical measurement. In: 18th International Conference on Pattern Recognition (ICPR 2006), vol. 1, pp. 1038–1041. IEEE (2006)
22. Liang, X., Asano, T., Bishnu, A.: Distorted fingerprint indexing using minutia detail and delaunay triangle. In: 2006 3rd International Symposium on Voronoi Diagrams in Science and Engineering, pp. 217–223 (July 2006). https://doi.org/10.1109/ISVD.2006.42
23. Maio, D., Maltoni, D., Cappelli, R., Wayman, J.L., Jain, A.K.: Fvc 2000: fingerprint verification competition. IEEE Trans. Pattern Anal. Mach. Intell. **24**(3), 402–412 (2002). https://doi.org/10.1109/34.990140
24. Maltoni, D., Maio, D., Jain, A., Prabhakar, S.: Handbook of fingerprint recognition. Springer Science & Business Media (2009)
25. Peralta, D., Galar, M., Triguero, I., Paternain, D., Garc´ıa, S., Barrenechea, E., Benítez, J.M., Bustince, H., Herrera, F.: A survey on fingerprint minutiae-based local matching for verification and identification: Taxon. Experim. Eval. Inform. Sci. **315**, 67–87 (2015)
26. Wang, Y., Hu, J., Phillips, D.: A fingerprint orientation model based on 2d fourier expansion (FOMFE) and its application to singular-point detection and fingerprint indexing. IEEE Trans. Pattern Anal. Mach. Intell. **29**(4), 573–585 (2007). https://doi.org/10.1109/TPAMI.2007.1003
27. Wayman, J., Jain, A., Maltoni, D., Maio, D.: An introduction to biometric authentication systems. In: Biometric Systems, pp. 1–20. Springer, Berlin (2005)

A Novel Approach for Music Recommendation System Using Matrix Factorization Technique

G. S. Ananth and K. Raghuveer

Abstract A recommender system provides personalized content to its users to handle the ever expanding information overload, thus improving customer relationship management. Music is a subject used widely across the world in different aspects of life. Music recommender systems help users to listen to the right choice of music. With the advent of mobile devices and Internet, access to different music resources is easily available. In this paper, we provide music recommendations to the Million Song Dataset using the TuriCreate's core ML library and with a focus on two methods of collaborative filtering techniques: user-based and item-based recommendations. Results are deduced exploring numerous metrics to measure the similarity of users and items such as cosine metric, Pearson correlation, latent matrix factorization and others. A comparison of different evaluation metrics is carried out to check for the effectiveness of the recommender system.

Keywords Collaborative filtering · CoreML · Cosine coefficient · Jaccard coefficient · Matrix factorization · Million Songs Dataset · Precision · Recall · TuriCreate

1 Introduction

In this modern world of technology and digitization, with multiple products available in multiple domains, we humans tend to get confused to purchase the right product as per our requirement. The word 'domain' may be swapped with the word 'subject.'

This paper intends to use the word domain. So, the domains may be of various examples starting from books, music and movies to even complex ones like consumer

G. S. Ananth (✉)
Department of MCA, NIE, Mysuru, India
e-mail: ananth.gouri@nie.ac.in

K. Raghuveer
Department of ISE, NIE, Mysuru, India
e-mail: raghu@nie.ac.in

© Springer Nature Singapore Pte Ltd. 2020
S. Agarwal et al. (eds.), *Machine Intelligence and Signal Processing*,
Advances in Intelligent Systems and Computing 1085,
https://doi.org/10.1007/978-981-15-1366-4_2

products and people. In most cases, we seek for help from our coworkers based on their experiences or rely upon purchasing the best and the most popular product.

This is where the role of recommendation systems or simply called recommender systems (RS) comes into picture. A RS takes an input of one particular domain at a time and provides an output of the best recommended products of that domain. These recommended products are also in simple terms called as recommendations. Recommendations are playing a vital role in this current era and helping people or consumers avoid the problem of confusion.

1.1 Introduction to Recommender Systems

Recommender systems are broadly classified into two major types as collaborative filtering recommender systems and content recommender systems. A third category hybrid recommender system is a logical combination of collaborative and content-based systems.

A collaborative filtering (CF) recommender system is one of the most widely implemented and used system. It works on the logic of aggregation of ratings given to a particular product, recognizes the common differences between users on the basis of the ratings provided and finally provides a list of recommended products based on inter-user comparisons.

Whereas in a content-based recommender system, the user gets recommendations based on the features of a product category which many a times also act as the keywords of the system. Though the user would have provided ratings, the recommendations are generated on the basis of the user's interest dependent upon the features present in those ratings.

Hybrid recommender systems usually combine the overall functionality of these above two types into a single system. It uses the ratings of a product combined with its features to provide a different class of recommendations to the user. In fact, the use of hybrid techniques of generating recommendations avoids problems like cold start, sparseness of data and to some extent resolved scalability issues as well.

Overall recommendation systems also act as a class of systems dealing with information overload, that is, systems segregating data through the form of information retrieval or information filtering. The retrieval is a dissemination between providing relevant versus irrelevant data.

These days RS is also categorized into other types as knowledge-based RS, demographic RS, model RS, context-aware or context awareness RS, social network analysis-based RS and group recommendation systems among others which will be little discussed at this context.

But with the sudden boom of old school technology called artificial intelligence (AI), a different class of RS is emergent over the years called computational intelligence (CI)-based RS. Generating recommendations from the CI RS involves techniques revolving around AI that includes Bayesian classification, artificial neural networks, various clustering mechanisms and fuzzy set logic. Also, most of these computational intelligence techniques are a subset of machine learning algorithms.

2 Music Recommender Systems

The music recommender system (MRS), as the name suggests, generates music recommendations. Music seems to be the part and parcel of life for many people. People listen to music when they feel bored or feel happy. They listen to music when they have free time or even to spend their time. At the same time, there are people who are not just consumers of music but also generate or produce music. They are called 'musicians' at the high level. A commonality between people who generate music versus people who consume is—there are circumstances when they may need to search and look for music of their likings. There are as well scenarios where we need to suggest and recommend music to our friends. A system like music RS does the same job of providing recommendations and suggestions for its users.

2.1 Types of Music Recommender Systems

Music recommender systems are of various types based on different self learning questions like

1. Is the music recommended of one single genre?
2. Does the RS suggest just songs or does the user of the system need a playlist of the songs?
3. Is music recommended based on a single musician or is it based on the demographic locations of the musician?
4. Is music recommended dependent upon a specific dataset or is it generic in nature?

3 Related Work and Literature Survey

Recommender systems are a technique evolved almost from the 1970s. The same continues for music RS. With the web growing so fast and thanks to latest sophisticated hardware and software components, watching movies and listening to complete discographies of music are not a tough task. The speed of Internet playing a major role as well, downloading music, movies and related content from legal sites is a child's play. But at the same time with so much of data at the disposal, it is a tough task to decide the resource. It results in frustration to users while deciding the proper content.

Music is an e-resource, and the usage of recommendations generated from MRS is called as utilization of services. Overall, these e-resources could also refer to other sub-categories of data like videos, music and many a times to documents. The system that generates the recommendation to this domain is also called e-resource service recommender systems.

There is enough work done and implemented based on MRS like Flycasting, Smart Radio and RACOFI [1]. In most of these systems, rating a resource of music is an explicit task. Hence, most of these MRS work on the principle of CF.

4   Proposed Work and Details

This paper proposes music recommendations using CF techniques for a dataset called Million Song Dataset (MSD).

This work built two models using different techniques. The non-personalized model, in future called as NPM, is a basic recommender system which did not provide any recommendations using any ML techniques, but provided predictions using the total number of times a song was listened to parameter. The personalized model with its different variants using ML provided recommendations that were based on the techniques that included few of the common memory-based CF models using representations like neighborhood-based CF with various correlations like Jaccard coefficient and cosine coefficient. The recommender algorithms with item/user-based CF techniques were the popularity recommender algorithm, item similarity algorithm with variants and matrix factorization algorithm as detailed in Table 1.

To overcome few of the problems faced in CF techniques, like the large and vast size of the datasets, sparsity of the rating matrix, the matrix factorization algorithm is widely used and becoming popular in recent times [2]. The matrix factorization algorithm decomposes a rating sparsity matrix into latent factors of users and items. A user can rate few music songs that he likes, and as well does not give any rating for the songs that he does not listen or he may not like. The predominant idea behind a matrix factorization technique is to fit in those slots of the matrix where there is no rating at all. In most cases, it is important to observe that the number of such latent features or factors should be less than the number of users or items.

Table 1 Proposed different models and variants

Sl. No.	Model type	Type variant
1	Non-personalized	Maximum listens
2	Personalized (ML applied)	Item similarity with the Jaccard coefficient
		Item similarity with cosine coefficient
		Matrix factorization recommender

Matrix factorization algorithms make use of different variants like singular value decomposition (SVD), principal component analysis (PCA) and probabilistic matrix factorization (PMF). The matrix factorization in this proposed work predicts a song recommendation based on the scores obtained through ranking. The TuriCreate-based matrix factorization algorithm's model makes use of different solvers like stochastic gradient descent and alternate least squares to improve the scores. The rating score for prediction is obtained as weights combined of user and item latent factors along with other side data. The side data is a set of optional fields including bias, side features and the related pairwise combinations.

The usage of matrix factorization models gained popularity after the Netflix challenge. In fact, the first spot winners of the challenge won by the team Yehuda Koren et al. were able to score 8.43% better than Netflix evaluation itself. Then, with joining the team called BigChaos, they were able to further improve the RMSE score to 9.46% [2]. Significantly, enough number of dimensions and parameters were matrix decomposed providing space to newer challenges.

4.1 Machine Learning, Role of Datasets in Machine Learning and the Million Song Dataset

Machine learning (ML) is a class of different algorithms and models that the computers are made to use to improve the performance of a specific task progressively. These machine learning algorithms make use of a model of sample data called the 'trained data' to either predict or provide decisions without being actually programmed.

There are in-numerous applications evolving through the usage of machine learning in the field of computer Science. They include filtering of emails, network detection of intruders, prediction applications and facial recognition applications among others.

Datasets are becoming an integral part in the field of machine learning. Through the usage of datasets with ML, we humans are able to produce advance results in various fields. Again, these datasets are classified into various types based on the type as follows: image data, text data, sound data, signal and sensor data.

Introduced by Thierry, of LabROSA and Brian, Paul of The Echo Nest in the year 2011, the Million Song Dataset is a freely available collection of audio features and metadata for a million songs [3] and is a part of sound data.

The overall size of the dataset is about 280 GB and available for complete access across few universities in the USA, and there are available subsets of about 1 and 2% of the original size for download and called as subsets of MSD. There are also few different clusters of data of the MSD available for download to evaluate for music information retrieval (MIR).

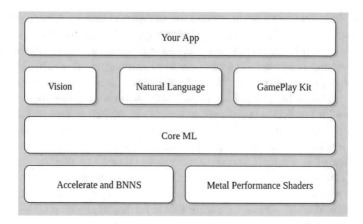

Fig. 1 CoreML architecture

Overall, the introduction of the MSD acted as a platform for music and MIR researchers to use machine learning in particular to not just produce recommendations, but as well work with the few of the different features provided with the data.

For the proposed work, we used ML techniques using a free and open-source tool set called TuriCreate with a subset of MSD to produce music predictions. Turi was initially a machine learning startup in Seattle, known for its product GraphLab that simplifies common machine learning tasks. In 2016, Apple acquired Turi and later in December 2017 made TuriCreate as an open-source project.

GraphLab is a framework that is proprietary, but is free to use for one year as an academic license. TuriCreate is built on top of the CoreML framework. It has ready-to-build models for different ML tasks like recommendations, classifications, object detections, classifiers and regression [4].

Figure 1 depicts the CoreML architecture. The location of 'Your app," colored in purple color is the space where most of the applications run.

5   Results

A recommender system provides recommendations to users. It is through these systems that personalized content is delivered in today's technological world.

The next few images show the recommendations provided by both the NPM and personalized models following the same order of Table 1 (Figs. 2 and 3).

As we can observe, all the models provide different set of recommendations to the users.

```
# Top Songs with most total lisens:
+---------------------+-------+-----------------------------+
|       songID        | plays |            title            |
+---------------------+-------+-----------------------------+
| SOBONKR12A58A7A7E0  | 35432 |       You're The One        |
| SOAUWYT12A81C206F1  | 33179 |            Undo             |
| SOSXLTC12AF72A7F54  | 24359 |           Revelry           |
| SOFRQTD12A81C233C0  | 19454 |        Sehr kosmisch        |
| SOEGIYH12A6D4FC0E3  | 17115 | Horn Concerto No. 4 in E f... |
| SOAXGDH12A8C13F8A1  | 14279 | Dog Days Are Over (Radio Edit) |
| SONYKOW12AB01849C9  | 12392 |           Secrets           |
| SOVDSJC12A58A7A271  | 11610 |       Ain't Misbehavin      |
| SOUFTBI12AB0183F65  | 10794 |           Invalid           |
| SOHTKMO12AB01843B0  | 10515 | Catch You Baby (Steve Pitr... |
+---------------------+-------+-----------------------------+

+----------------------------+----------------------------+------+
|          release           |         artistName         | year |
+----------------------------+----------------------------+------+
|     If There Was A Way     |       Dwight Yoakam        | 1990 |
|      Vespertine Live       |           Björk            | 2001 |
|      Only By The Night     |       Kings Of Leon        | 2008 |
|     Musik von Harmonia     |          Harmonia          |  0   |
| Mozart - Eine kleine Nacht... | Barry Tuckwell/Academy of ... |  0   |
| Now That's What I Call Mus... | Florence + The Machine     |  0   |
|         Waking Up          |        OneRepublic         | 2009 |
|         Summertime         |         Sam Cooke          |  0   |
|       Fermi Paradox        |          Tub Ring          | 2002 |
|      Catch You Baby        |       Lonnie Gordon        |  0   |
+----------------------------+----------------------------+------+
[163828 rows x 6 columns]
```

Fig. 2 Recommendations of max listens NPM model

```
+-----------------+------+-----------------+-------+---------------------+
|   artistName    | rank |     release     | score |       songID        |
+-----------------+------+-----------------+-------+---------------------+
|     Yölintu     |  3   | Kaikki kohdallaan | 307.0 | SOPUONM12A81C203D7  |
|     Bracket     |  6   | Novelty Forever | 185.0 | SOKGUUL12A6D4FE599  |
|       Us3       |  2   |    Questions    | 537.0 | SOFVLYV12A8C145D8F  |
|     Bracket     |  6   | Novelty Forever | 185.0 | SOKGUUL12A6D4FE599  |
|       Us3       |  2   |    Questions    | 537.0 | SOFVLYV12A8C145D8F  |
|       Us3       |  2   |    Questions    | 537.0 | SOFVLYV12A8C145D8F  |
| Dananananaykroyd |  10  |   Hey Everyone  | 150.0 | SOGLBYS12AB0184353  |
|     Yölintu     |  3   | Kaikki kohdallaan | 307.0 | SOPUONM12A81C203D7  |
|     Bracket     |  6   | Novelty Forever | 185.0 | SOKGUUL12A6D4FE599  |
|     Bracket     |  6   | Novelty Forever | 185.0 | SOKGUUL12A6D4FE599  |
+-----------------+------+-----------------+-------+---------------------+

+-------------------+---------------------------------+------+
|       title       |             userID              | year |
+-------------------+---------------------------------+------+
|   Syntymäpäivä    | 64f74807362a33dbd40eb289ea...   | 1997 |
| Back To Allentown | a25a3ec830e3c36f85f9bf34e7...   | 1997 |
| Believe In Yourself | 5eece595b8a9f96e423ac34d79... | 2004 |
| Back To Allentown | 1efed18670ad14cc09d6ce382e...   | 1997 |
| Believe In Yourself | 3800f917c4e5bbfb0cec44c13c... | 2004 |
| Believe In Yourself | d7fe2e86d59ae0162c7cc2d10e... | 2004 |
|    Watch This!    | 239e1e008953dcd31aec705d67...   | 2009 |
|   Syntymäpäivä    | 29a249836bd9014d201810d46b...   | 1997 |
| Back To Allentown | dbf0b10dd68b4a248acb2c0b80...   | 1997 |
| Back To Allentown | 66014bf518cbc370d31086313d...   | 1997 |
+-------------------+---------------------------------+------+
[1100000 rows x 8 columns]
```

Fig. 3 Recommendations of the popularity model

The NPM model's recommendations are based on the max listens–plays field. The songs with the highest 'plays' count are listed. The popularity model works on the concept of the score. Songs with higher scores are listed first.

userID	songID	score	rank
c6cd84996b43c5962606138445...	SOWCTHS12A8C139A97	0.015151510636011759	1
550a3b591a250974e5dac2db20...	SONPBNC12A8C13AB1F	0.022727272727272728	1
1ad2bd26a77b31fd84d4cc62b2...	SOULHNH12A8C13B2CB	0.01081196665763855	1
2f8451db01d393665f650c05ff...	SOEXGMS12A6D4FCE2D	0.038461538461538464	1
e988930c3c139727b870fb3b3e...	SONNNEH12AB01827DE	0.012128353118896484	1
437476f813dd891044341f56bf...	SONPBNC12A8C13AB1F	0.015625	1
cad15e2b38bcf176f457a2ec2e...	SONNNEH12AB01827DE	0.019914650917053224	1
a199fefde2d118ccf4a5bbcbe6...	SOGDUWA12A8C137022	0.05555555555555555	1
6f96db0287e30d586e486078b3...	SOKEMEM12A8C133D92	0.03922615945339203	1
ae85ed0b9ba0daac60a5d1b5a6...	SOMYJOZ12A58A7D16F	0.025689221918582916	1

title	release	artistName
Dope Man	The Very Best Of Hip Hop M...	N.W.A.
My Yoke Is Heavy	Don't Be Scared	Daniel Johnston
La negrita	Re	Café Tacvba
Blood Ties	Balls of Fury	Randy Edelman
Lithium	Nevermind	Nirvana
My Yoke Is Heavy	Don't Be Scared	Daniel Johnston
Lithium	Nevermind	Nirvana
Devil In The Belfry	The Scarecrow	Avantasia
Con la Casa en Órden	El Rock De Mi Vida	Guasones
Make Up Your Mind	Theory of a Deadman [Speci...	Theory Of A Deadman

Fig. 4 Recommendations of the item similarity with the Jaccard coefficient model

userID	songID	score	rank
fd50c4007b68a3737fe052d5a4...	SOAUWYT12A81C206F1	0.12883153557777405	1
fd50c4007b68a3737fe052d5a4...	SOSXLTC12AF72A7F54	0.11161501208941142	2
fd50c4007b68a3737fe052d5a4...	SOUFTBI12AB0183F65	0.09241822361946106	3
fd50c4007b68a3737fe052d5a4...	SOOFYTN12A6D4F9B35	0.08921386798222859	4
fd50c4007b68a3737fe052d5a4...	SOUNZHU12A8AE47481	0.08225348591804569	5
fd50c4007b68a3737fe052d5a4...	SOVDSJC12A58A7A271	0.07987940311431885	6
fd50c4007b68a3737fe052d5a4...	SOPUCYA12A8C13A694	0.07870896657307942	7
fd50c4007b68a3737fe052d5a4...	SONHWUN12AC468C014	0.07627292474110921	8
fd50c4007b68a3737fe052d5a4...	SOHZMFE12A6D4FB412	0.0739159882068634	9
fd50c4007b68a3737fe052d5a4...	SOBOUPA12A6D4F81F1	0.07256762186686198	10

[1100000 rows x 4 columns]
Note: Only the head of the SFrame is printed.
You can use print_rows(num_rows=m, num_columns=n) to print more rows and columns.

userID	songID	score	rank
556057ceddeb4e47b5bf813794...	SOZWUOD12A6D4FB294	0.06415002875857884	1
624d3d7cd65bdd4892f1c15bd8...	SOJZQCK12A6D4F8E7B	0.07071068286895751	1
f6809f15f3dc170bffab04820b...	SOWJFCJ12A8C135D19	0.05547002553939819	1
031d0b638cf92a3260e7cdb78e...	SOWJFCJ12A8C135D19	0.10101525272641863	1
9d1aa4dbabaec00771cdf24137...	SOTNVRG12A8AE46240	0.07875239849090576	1
eaa55c6f5737cd251968923be2...	SOJZQCK12A6D4F8E7B	0.03140371579390306	1
b8d76ac698f39dfc87fa58d23e...	SOTNVRG12A8AE46240	0.056968102852503456	1
21457438da8c965fe0bd8c4084...	SOYJDYT12AB0186F0A	0.05540444254875183 4	1
b5ce32cced565b78af611de349...	SOJZQCK12A6D4F8E7B	0.03140371579390306	1
0f52936b553092bc7e9d5f9851...	SOTNVRG12A8AE46240	0.056968102852503456	1

title	release
Got It Right Here	Da Real Live Thing
Blink	The Destroyed Room
Guilt Within Your Head	Enter: The Conquering Chicken
Guilt Within Your Head	Enter: The Conquering Chicken
Showdown (album version)	In Silico
Blink	The Destroyed Room
Showdown (album version)	In Silico
The Last Song	Deja Vu: The TFK Anthology
Blink	The Destroyed Room
Showdown (album version)	In Silico

Fig. 5 Recommendations of the item similarity with cosine coefficient

```
+--------------------------------+-------------------------+----------------------+--------+
|             userID             |         songID          |        score         | rank |
+--------------------------------+-------------------------+----------------------+--------+
| 7af7fbbce195a5fdb5dc890c96...  | SOXTUWG12AB018A2E2       | 97.08676689290424    |   1    |
| 2ef2bff7b5139b6dce1b965cc5...  | SORJSQI12A6701D62D       | 36.63021507286449    |   1    |
| a1854b008044d34009e8c33f5c...  | SOXTUWG12AB018A2E2       | 68.06145312809367    |   1    |
| 45c6127b67a1d13a2c103dafae...  | SORJSQI12A6701D62D       | 61.54068079018016    |   1    |
| 687ec011f9066902fca8397526...  | SOXTUWG12AB018A2E2       | 71.42428507828136    |   1    |
| fcfa0d1b1574b87d22907a3b56...  | SORJSQI12A6701D62D       | 44.22126235985179    |   1    |
| 923aa605311651809b13df8b64...  | SOPMHHE12AB01845F6       | 51.29095127605815    |   1    |
| b5820dad2cad6f8a0204776a9c...  | SOXTUWG12AB018A2E2       | 47.244608903163865   |   1    |
| 46dbe6d754f3b5a8adde6fa086...  | SOXTUWG12AB018A2E2       | 111.72361054443736   |   1    |
| 73211e4fbc1046547694a07788...  | SOGKNQF12AB017FD67       | 39.88988151573558    |   1    |
+--------------------------------+-------------------------+----------------------+--------+

+--------------------------------+-------------------------+----------------------+
|             title              |         release         |      artistName      |
+--------------------------------+-------------------------+----------------------+
| Drop The Hammer (Album Ver...  |   A Search For Reason    |       Kilgore        |
| Say Hello (Angello & Ingro...  |        Say Hello        |      Deep Dish       |
| Drop The Hammer (Album Ver...  |   A Search For Reason    |       Kilgore        |
| Say Hello (Angello & Ingro...  |        Say Hello        |      Deep Dish       |
| Drop The Hammer (Album Ver...  |   A Search For Reason    |       Kilgore        |
| Say Hello (Angello & Ingro...  |        Say Hello        |      Deep Dish       |
|  What Is a Young Girl Made of  | Intro: The Beach Boys - EP |   The Beach Boys   |
| Drop The Hammer (Album Ver...  |   A Search For Reason    |       Kilgore        |
| Drop The Hammer (Album Ver...  |   A Search For Reason    |       Kilgore        |
|           Swan Song            |    Bomb In A Birdcage    |    A Fine Frenzy     |
+--------------------------------+-------------------------+----------------------+

+------+
| year |
+------+
| 1998 |
| 2005 |
| 1998 |
| 2005 |
| 1998 |
```

Fig. 6 Recommendations of the matrix factorization model

The item-based similarity models predict songs with a higher rank value. Figures 4, 5 and 6 show the music recommendations based on different coefficients with the ranking that have a value of one.

6 Evaluation Metrics for Recommendation Algorithms

To know the importance of evaluation of a RS, we need to also understand the overall workflow of RS. Figure 7 depicts the generic workflow of a recommender system.

Evaluating a recommender system or the algorithm used in the recommender system is not an easy task. This could be due to various reasons.

The algorithm used in the RS might not work in the same manner for all the datasets used for evaluation. The parameters used for the metrics might play a different role depending upon the data.

The second reason is that the RS evaluation could fail due to the motive of the result produced by the algorithm. The algorithm may produce different results—wherein it fails to match to the expected results.

The last reason is the challenging task of using what parameters of a metric that could lead to the better evaluation of a RS.

Overall, most of the evaluation metrics for a RS was duly conveyed in the seminal paper described by Herlocker [5].

Fig. 7 Workflow of a RS

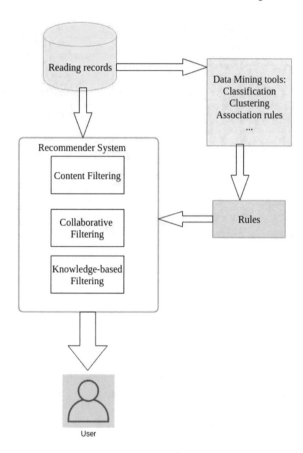

6.1 Introduction to Evaluation of RS

Most of the RS are evaluated on the fundamental concept of two parameters: precision and recall. In fact, precision and recall are classical evaluation metrics for most retrieval tasks.

Precision is a parameter that tells the total number of predictions that are relevant.

Recall tells the total number of relevant predictions that are recommended.

6.2 Evaluation Metrics for Music RS

We discuss a few metrics that can be used related to our proposed work of music recommendations as below.

Based on the rating of the predictions or recommendations provided by a RS, we have mean absolute error (MAE), mean squared error (MSE) and root mean squared error (RMSE) as the major metrics. Of the MAE and MSE, RMSE is found to be a better metric.

Most of the papers written earlier used MAE, but these days we find a lot more papers using only RMSE or some means of R-squared metric to be the default modeling metric [5].

In fact, RMSE was the major metric used in the Netflix challenge as well [4].

6.3 Defining MAE and RMSE Modeling Metrics

MAE is the measure of average magnitude of the errors in a result set of predictions; whereas, RMSE is the square root of the average differences between prediction and actual observation. The formula for calculating is given below:

$$\text{MAE} = 1/n \sum_{i=1}^{n} |x_i - y_i| \tag{1}$$

$$\text{RMSE} = \sqrt{\left(1/n \sum_{i=1}^{n} (x_i - y_i)^2 \right)} \tag{2}$$

Table 2 shows the RMSE values at the first run for the proposed models.

Figure 8 depicts through a graph plotting these Table 2 RMSE values of personalized models.

Table 2 RMSE values of proposed models

Sl. No.	Model Type	Type Variant	Evaluation Metric	
			RMSE (I run)	Precision_Recall (Yes/No)
1	Non-personalized	Maximum listens	NA	NA
		Popularity recommender	7.35	Yes
2	Personalized (ML applied)	Item similarity with the Jaccard coefficient	6.78	Yes
		Item similarity with cosine coefficient	6.78	Yes
		Matrix factorization recommender	7.25	No

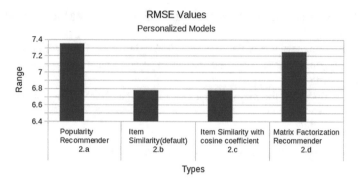

Fig. 8 RMSE values of personalized models

7 Comparison and Conclusion

Through the inference of the above results and the evaluation of metrics through Table 2, we could induce that the personalized model's variant—matrix factorization of the proposed work was a better model of getting expected recommendations. Although the RMSE value of popularity recommender was slightly higher in few of the runs, we need to note that since only 58% of the overall 1 M MSD songs were only listened once by the users, and the matrix factorization model was a better model due to the deviation in ratings.

8 Future Work

This paper was a platform to start the use of the large-scale datasets and also to understand the complex problems involved in music information retrieval. Some of the MIR tasks that can be carried out in future would be to use as follows:

1. Other ML techniques like Natural Language Processing, Word and Keyword extraction.
2. Make use of other data sources like Free Music Archive or Webscope provided Yahoo Music sources.
3. Recommend songs to users based on their mood. This can be also called as sentiment music extraction.

9 Challenges in Music Information Retrieval

Music information retrieval is not an easy task due to some of the below mentioned challenges:

- Datasets are extremely vast and extensive. The download of the complete instance of MSD is far from possible. Special request access should be sent to procure data.
- After the procurement of such data, again processing the data on a single machine is improbable and a great challenge. Most of the datasets can be regarded as Big Data. Hence, processing behavior changes from one data to another.
- The data present in the MSD and in other sound data will be in unsupervised mode. So, to manipulate with labels like tags will not be easy. Creating labels to the data is also an extensive procedure and might not be possible.
- Few datasets like MSD and also Free Music Archive dataset contain too many numbers of fields. In a way, this is a boon for performing research, but as well can be considered a disadvantage.
- Copyright and other infringement issues are also a major concern. A user should be extremely careful before using a dataset. The license and freedom of each data source should be read before in use.

References

1. Pazzani, M., Billsus, D.: Learning and revising user profiles: the identification of interesting web sites. Mach. Learn. **27**(3), 313–331 (1997)
2. Bokde, D., Girase, S., Mukhopadhay, D.: Matrix factorization model in collaborative filtering algorithms—a survey. Procedia Comput. Sci. 136–146 (2015)
3. Bertin-Mahieux, T., Ellis, D.P.W., Whitman, B., Lamere, P.: The Million Song dataset. In: ISMIR (2011)
4. Garcia, C., Rona, L.: The Netflix Challenge (2018)
5. Herlocker, J.: Evaluating collaborative filtering recommender systems. ACM Trans. Inform. Syst. **22**(1), 5–53 (2004)

Generation of Image Captions Using VGG and ResNet CNN Models Cascaded with RNN Approach

Madhuri Bhalekar, Shubham Sureka, Shaunak Joshi and Mangesh Bedekar

Abstract Recent advancements in technology have made available a variety of image capturing devices, ranging from handheld mobiles to space-grade rovers. This has generated a tremendous visual data, which has made a necessity to organize and understand this visual data. Thus, there is a need to caption thousands of such images. This has resulted in tremendous research in computer vision and deep learning. Inspired by such recent works, we present an image caption generating system that uses convolutional neural network (CNN) for extracting the feature embedding of an image and feed that as an input to long short-term memory cells that generates a caption. We are using two pre-trained CNN models on ImageNet, VGG16 and ResNet-101. Both the models were tested and compared on Flickr8K dataset. Experimental results on this dataset demonstrate that the proposed architecture is as efficient as the state-of-the-art multi-label classification models. Experimental results on public benchmark dataset demonstrate that the proposed architecture performs as efficiently as the state-of-the-art image captioning model.

Keywords Computer vision · CNN · Deep learning · ImageNet · VGG16 · ResNet-101 · LSTM · Flickr8k

M. Bhalekar (✉) · S. Sureka · S. Joshi · M. Bedekar
School of Computer Engineering and Technology, MIT World Peace University, Pune, Maharashtra, India
e-mail: madhuri.bhalekar@mitwpu.edu.in

S. Sureka
e-mail: shubham.sureka@mitpune.ac.in

S. Joshi
e-mail: shaunak.joshi@mitpune.ac.in

M. Bedekar
e-mail: mangesh.bedekar@mitwpu.edu.in

© Springer Nature Singapore Pte Ltd. 2020
S. Agarwal et al. (eds.), *Machine Intelligence and Signal Processing*,
Advances in Intelligent Systems and Computing 1085,
https://doi.org/10.1007/978-981-15-1366-4_3

27

1   Introduction

In the field of computer vision and natural language processing, language generation is an important challenge. While this seems to be easy for human beings, it is quite difficult for machines because it requires the machine to understand the image content and express their relationship in a natural language. Obtaining sentence-based description for images is becoming important and has many applications such as navigation for the blind people, subject and its context identification for robotic system, question answering system on videos for public and personal security and surveillance. Also, the image captioning system should be capable of capturing implicit semantic information of an image and generate human-understandable sentences. Due to the enormous challenging nature of this process, there has been a recent need of research interest in the image caption generation problem. Advancements in the field of computer vision have happened as a result of conceptualization of neural networks and their consequent implementation by training them on large image classification datasets (e.g. ImageNet [3]) and image captioning datasets (e.g. Flickr8k [9]) thus improving the correctness of caption generation significantly. The major factor behind this is that deep learning has introduced the nature of intuitive learning which was absent in any computer vision technique used before.

The main inspiration of our work comes from recent advancement in the field of machine learning and artificial intelligence. Over the last decade, it has been proven that CNN is capable of producing a rich description of the input image by embedding it to a fixed-length vector, which can be applied in variety of different systems. In this work, we propose two CNN models, VGG16 [17] and Resnet-101 [7] that are pre-trained on ImageNet [3] dataset, which acts as an encoder and along with its pre-trained weights they are used to encode an image to its feature set. The CNN stack starts with conv layer, responsible for applying mask over a portion of the image repeatedly to obtain a matrix of extracted features of the entire image. This is followed by applying rectified linear unit (ReLU) and incorporating MaxPool layer to extract relevant features describing the context of the image. Then, the image encoding is computed by the fully connected layer depending upon the distinct weights of activation. A long short-term memory (LSTM) network takes image encoding as an input and generates the next most probable words which are equal to the beam size given to us as an output. We have also employed an RMSprop optimizer for computing a new learning rate to change the step size and minimize the cost function such that corresponding weights are suitable for generating captions. With the current training on the Flickr8K dataset, running test on the 1000 test images results in 0.5143 BLEU-1 score while the human performance averages 0.70 which significantly yields better performance than state-of-the-art approaches.

2   Related Work

Throughout the research conducted, we came across many papers related to implementations of image caption generation. Redmon et al. [16] uses a simple to construct method that throws away complex pipelines used by state-of-the-art methods. Multiple bounding boxes in an image with a confidence score to each bounding boxes. The objects are identified with the one having the highest class probability. In contrast to above, a multimodal recurrent neural network for classification and labelling of images is used by Mao et al. [22] consisting of language model, image model and multimodal part. Language model learns dense embeddings for each word in the dictionary and stores the semantic context in recurrent layers. The image part contains a deep CNN which extracts image features, and then the multimodal part connects the language model and the deep CNN. Concretely, high-level feature extraction and detection using state-of-the-art classifier on deep CNN have produced efficient results [20] where it has been used for object detection and relaxed multiple instance SVM for classification. Here, SVM plots feature in n-dimensional space and then classifies using hyperplane.

Guillaumin et al. [6] using semi-supervised learning classifies an image and then associated keyword, with labelled and unlabelled images. Thus, both images and their associated keywords are learned, using that, unlabelled images are scored. Bidirectional mapping between images and their sentence-based descriptions with the help of recurrent neural network (RNN) [21] attempts to dynamically build a visual representation of the scene as a caption is being generated or read. The representation automatically learns to remember long-term visual concepts. Compared to above approaches passing an entire image input, instead only certain regions deemed relevant are given as input by Jia et al. [10] by providing a framework for implementing RCNN, Softmax function for classification and regression for bounding boxes, and by using the concept of selective regions from the given input to pool dominant features.

Vision deep CNN and language generating RNN are uniquely combined to generate natural sentences given the training image as proposed by Vinyals et al. [21]. The goal of model is to maximize the likelihood of predicted description sentence with the target description sentence given the training. Xu et al. [23] proposes an attention-based model that automatically learns to describe the content of images. For that it uses backpropagation techniques and also by maximizing variational lower bound, it is able to learn to gaze on salient objects. Object recognition can be further enhanced by advancing scene understanding [14], with an introduction of new dataset and by pipelining the annotation as labelling the image, instance spotting and instance segmentation.

Captioning can also be done by extracting grammatical semantics from the regions of the images [19]. The system trains on images and corresponding captions and learns to extract nouns, verbs and adjectives from regions in the image. These detected words then guide a language model to generate text that reads well and includes the detected words. Fang et al. [4] considers the concept of learning like a child, a model has been trained with large visual concepts and proposes a method to enlarge its word dictionary. This avoids extensive retraining and saves a considerable amount of computing power.

A new framework of encoding–decoding models was introduced by Kiros et al. [13]. For encoding, a joint image-sentence, embedding is used where sentences are encoded using long short-term memory cells. Whereas for decoding, a new neural language is introduced called the structure-content neural language model (SC-NLM). It guides the model in word structure generation phase by providing a soft template to avoid grammatical mistakes. Another approach which uses deep convolutional neural network along with long short-term memory (LSTM) network, a semantic representation of an image was generated and then decoded by Soh [18] also R. Kiros et al. [12] provides the multimodal neural language model.

A similar concept that includes bounding box annotations [16] but combines unsupervised learning has been proposed by Dai et al. [1]. Rather than moving the masks pixel by pixel in a convolutional network, an unsupervised method is used which is trained to look for semantic features of the images directly, exploiting bounding boxes to supervise convolutional networks for semantic segmentation. Lastly, Soh [18] uses the concept of binary mask. Pixel with value one contains an object else none. This binary mask is applied in both vertical as well as horizontal fashion until the localization of the object is as precise as possible. After a detailed research on aforementioned as well as a few other works like in [4, 11], our model was inspired by Show and Tell: A Neural Image Caption Generator [21] for implementation.

3 Proposed System Architecture and Model Framework

3.1 Basic Architectural Design

Our image caption generator has a state-of-the-art architectural approach concatenating a high-level image feature extractor with a natural language generator.

Image feature extractor takes an entire image as input and undergoes a series of complex mathematical operations and then it learns every aspect of the input image where it extracts the present prominent entities, contrasting objects, sharp features in the foreground as well as background and jointly learned relations between all the image features. This is represented and stored in the form of an extremely crucial data structure, i.e. standard sized vector that serves as one of the main components of the architecture.

As shown in Fig. 1 natural language generator takes the image embeddings generated by image feature extractor and start-of-sequence (SOS) token denoted by '<START>' as input to predict the next word successively based on the information it has until it gets the terminating token.

Fig. 1 Image caption generator

3.2 Image Feature Extraction Model

We are considering two pre-trained models VGG16 and ResNet-101 for image feature extraction.

3.2.1 VGG16

We have used a pre-trained VGG16 [17] convolutional neural network on ImageNet as our feature extractor model depicted by Fig. 2. The CNN is in the form of repeated blocks of conv-ReLU architecture, each terminating with a maxpool layer which selects features with the high significance relative to other features in the image. The input image is an RGB image and is resized to size of $224 \times 224 \times 3$. The image is convolved by a filter mask of size 3×3 with stride of 1 in the row-wise and column-wise direction. The convolution operation computes dot product for every iteration throughout the image volume followed by a ReLU activation function to weigh all the features to a particular degree according to their relevance. The weight parameters used, are learnt during the training process of the neural network. Initially, 64 such filters extract the low-level features from the image. Similarly, another such conv layer is repeated to extract features which are better among those obtained from the activation of the features which were extracted in the previous convolutional layer.

In order to pool, to remove the noise and filter out the irrelevant features maxpool operation is performed, where a mask of size 2×2 is convolved over the previous layer with a stride of 2 in the row-wise and column-wise direction. This stack of extraction and sub-sampling is repeated several times with 128, 256 and 512 filters successively. Then, the final maxpool layer ($7 \times 7 \times 512$) is flattened in the form of

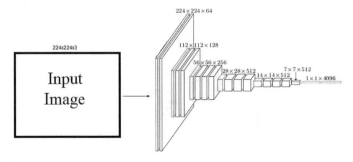

Fig. 2 VGG16 architecture [5]

a single vector. At last a fully connected layer ($1 \times 1 \times 4096$) is formed with each unit connected to every unit of the flattened layer having the joint representation of the learned features. In our architecture, we pop the final softmax layer ($1 \times 1 \times 1000$) from the stack which predicts the class of the input image and as an alternative we use the fully connected layer that contains the image embeddings required for further computations to generate captions.

3.2.2 ResNet-101

VGG16 is a shallow network with less number of layers, as you go deeper the feature extraction gets improved but it introduces vanishing gradient problem. In order to address this problem, we used residual networks-101(ResNet) as shown in Fig. 3, it provides a solution by adding the identity mapping of the previously occurring intermediate layers to the present layer hence performing better feature extraction than its shallower counterparts. The operation by which the resultant layer is generated is called as the **residual function**. This layer is represented as $H(x)$, where x is the layer taken from the shallower network and $F(x)$ is the output learned from the current layer. $H(x)$ is given by

$$H(x) = F(x) + x. \tag{1}$$

The residual networks are composed of sequential structures called **residual blocks**. A residual block is basically defined as

$$y = F\left(x, \{W_j\}\right) + x \tag{2}$$

where x is the layer copied from the shallower network. In the above equation, y is defined as the output of the element-wise addition of $F(x, \{W_j\})$ and x where $F(x, \{W_j\})$ is the residual function and its output represents the residual mapping to be learned, whereas x represents the layer copied directly from the shallower network. The residual block where $F(x, \{W_j\})$ and x have the same dimensions is called as an **identity block** as shown in Fig. 4.

If x has different dimensions than $F(x, \{W_j\})$ then the equation is represented as

$$Y = F\left(x, \{W_j\}\right) + W_s x \tag{3}$$

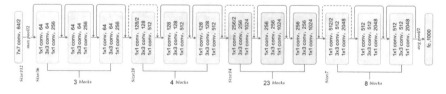

Fig. 3 ResNet-101 architecture [2]

Fig. 4 Identity block

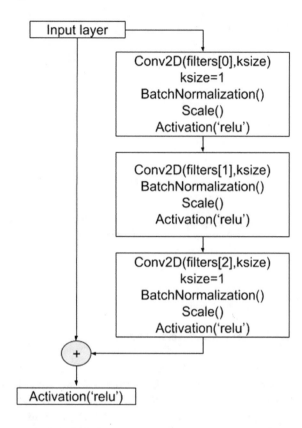

where $W_s x$ is a linear projection of x obtained in order to match the dimensions with $F(x, \{W_i\})$ for element-wise addition and W_s is a square matrix. This type of residual block is called as a **conv block** shown in Fig. 5. Finally, after each residual block a nonlinear activation function is performed.

3.3 Natural Language Generation Model

We have used LSTM [8] model for generation of natural language that describes the image. As shown in Fig. 1 we pass the fully connected layer to the hidden LSTM layer of our model. Consider a sentence $S_i = W_0 \ W_1 \ W_2 \ \ldots \ W_{j-1}$ where $j \leq n$ and n is the length of the longest caption in our dataset. LSTM uses probabilistic language model where a word W_k at a position k $(k \leq n)$ is predicted by $W_k = \max(p(W_j | W_0 \ W_1 \ \ldots \ W_{k-1}))$ where W_j belongs to the vocabulary of the corpus and $p(x)$ is the conditional probability of the word W_j.

Fig. 5 Conv block

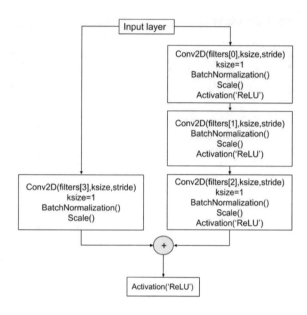

3.3.1 Long Short-Term Memory-Based Sentence Generator

In order to predict the next word in a sentence, LSTM layer takes all the tokens predicted until current point in time as input. These tokens are converted into word vectors and are passed to the LSTM cells of this layer. The LSTM cell as shown in Fig. 6 produces an output in time and an output in depth as a result obtained from the functions in the LSTM cell. The output in time is called as the cell state Ct that is passed to the next LSTM cell. The cell state plays a very crucial role in storing the context of the subject of the sentence being generated over long sequences which is generally lost in a RNN. The output in depth is used to compute the probability of the next word by using a categorical cross-entropy function, which gives a uniformly distributed representation of predicted probabilities of all the words in the dataset.

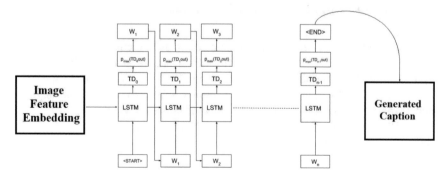

Fig. 6 Long short-term memory

The words with the top-m probabilities are chosen to be the most probable words where m is the beam size. The probabilities of top-m partial caption predicted till the last layer are evidential probabilities, considered in order to compute the conditional probabilities of top-m words from the current cell output and concatenate the words corresponding to these conditional probabilities with the partial captions corresponding to their evidential probabilities, in order to generate the new partial caption until the special '<END>' token is sampled.

3.3.2 LSTM Cell

The key to LSTMs is the cell state C_t, it runs straight down the entire chain, with only some minor linear interactions. The LSTM removes or adds information to the cell state, filtered down by sigmoid activation function followed by an element-wise multiplication operation called as gates.

The cell state is like a vector whose values refer to context from the previous cells. Due to the fact that the new information as well as new context has to be added by the newly predicted words the forget gate f_t has to make the cell state forget few of the holding contexts to various degrees defined by the sigmoid function (Fig. 7).

The tanh activation function is used to add new context from the current input word vector. The sigmoid used along with the tanh function ($i \odot g$) actually is supposed to ignore certain values in the vector obtained from tanh for reducing the noise in the

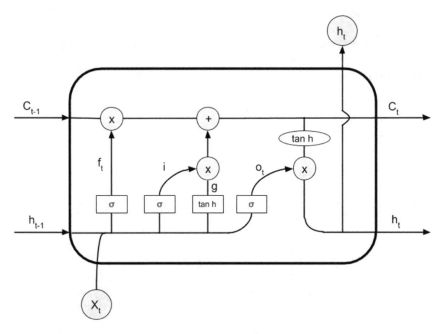

Fig. 7 LSTM cell [15]

new information. This selectively ignored information is excluded and the included candidate context from the new word vector is updated to the cell state ($f_t \odot c_{t-1} + i \odot g$). When it comes to generating h_t ($o_t \odot \tanh(c_t)$).

It has to contain only selective information in the cell state which is much useful for the next hidden cell or short term and avoid the cell state information that is relevant in the long term. The LSTM cell functions are defined as follows:

$$f_t = \sigma\left(W_{fx}x_t + W_{fh}h_{t-1}\right)$$
$$I = \sigma(W_{ix}x_t + W_{ih}h_{t-1})$$
$$g = \tanh\left(W_{gx}x_t + W_{gh}h_{t-1}\right)$$
$$o_t = \sigma(W_{ox}x_t + W_{oh}h_{t-1})$$
$$C_t = f_t \odot C_{t-1} + g \odot i$$
$$h_t = \tanh(C_t) \odot o_t \tag{4}$$

4 Experiments

4.1 Dataset Description

Our model is an end-to-end trained model and completely data driven. We used Flickr8k, i.e. a standard dataset used to train models for image captioning. It has 8000 images each of which is described by 5 reference captions, where every caption represents a different interpretation of the image. We split it such that the train and test datasets consist of 7000 and 1000 images, respectively.

4.2 Evaluation Metrics

We evaluated our model on test dataset of Flickr8k to measure the correctness of the results we obtained. In order to measure the correctness of the test dataset, we used several standard evaluation metrics that are BLEU, CIDEr, METEOR and ROUGE_L which are shown in Table 1.

Table 1 Performance evaluation metrics of our model for Flickr8k dataset

Model	BLEU-1	BLEU-2	BLEU-3	BLEU-4	CIDEr	METEOR	ROUGE_L
Our model	51.43	29.87	18.33	11.3	27.2	17.54	41.36

Table 2 Comparative analysis of VGG16 and ResNet-101

VGG16	ResNet-101
Top-1 error: 27.0%	Top-1 error: 19.87%
Time-taken: 64.31 ms (per forward pass)	Time-taken: 78.22 ms (per forward pass)
Vanishing gradient problem	No vanishing gradient problem
No weighted scaling after convolution operation	Weighted scaling after every convolution operation
More number of dense layers making it computationally expensive	Only one dense layer resulting in lesser space and time complexity

4.3 Comparative Analysis

We compared the two state-of-the-art feature extractors based on certain benchmarks. Both the models, i.e. VGG16 and ResNet-101 have their own unique features based on the parameters we selected to compare. Primarily, we considered the top-1 error and time complexity. It was noticed that VGG16 had 8% more top-1 error than ResNet-101 which indicates that the latter classifies more number of objects correctly than the former model on the same dataset, thus ResNet-101 has proved more efficient in feature extraction.

Vanishing gradient problem is quite prevalent in a model having a sequential stack of layers like VGG16, as you go deeper in this stack the gradient value that updates the weight parameters becomes infinitesimally small such that the update is equivalent to zero resulting in dead neurons which do not learn, whereas ResNet-101 overcomes it. We also observed that ResNet-101 performs weighted scaling of the values produced after each convolution operation, and this is responsible to transform the input values to make it suitable for the next operation which does a batch normalization on the scaled values. VGG16 has a disadvantage in computational complexity due to its final layers. It has two dense layers of higher dimensions as compared to ResNet-101 which has only one dense layer. Even though ResNet-101 has lower worst-case complexity than the shallower sequential models, it takes more time measured in milliseconds in terms of forward and backward pass of a single image than VGG16 as shown in Table 2.

4.4 Training the Model

We trained our model with a large number of epochs. After each epoch, we obtained the loss and accuracy of the model. To observe the nature of loss and accuracy with respect to epochs, we plotted graph of loss versus epochs and accuracy versus epochs over 500 epochs. By doing so we could analyze the range of epochs where the loss rapidly converged, the epoch from where the loss attained saturation, the minimal value of the loss.

Figure 8 depicts the trend of the loss with respect to epochs. At initial stages, the rate of improvement of loss is higher up to first 100–200 epochs. Then it is observed that the descent proceeds at a lower rate for next 100 epochs or so. Finally, the loss converges by an extremely low rate and thus the model saturates as it can be observed more clearly in Fig. 9 representing a graph of last 100 epochs where values attain sharp highs and lows which denote that the loss revolves around the optimum loss value and hence this graph makes it clear that the loss can not be further optimized.

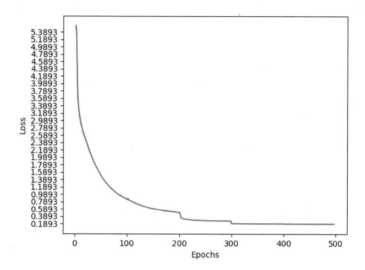

Fig. 8 Loss versus epochs

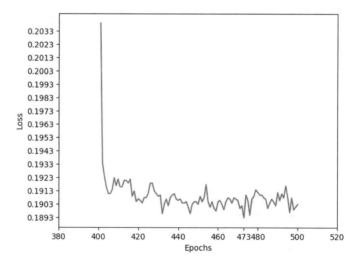

Fig. 9 Loss versus epochs (400–500)

Similarly, Figs. 10 and 11 depict the trend of the accuracy with respect to epochs. We can infer from these plots that in most of the cases the loss decreases as the accuracy increases and vice versa. Figure 11 is the plot of last 100 epochs where the accuracy values are positive and negative spikes which show that the accuracy has reached to a point after which it can not increase further.

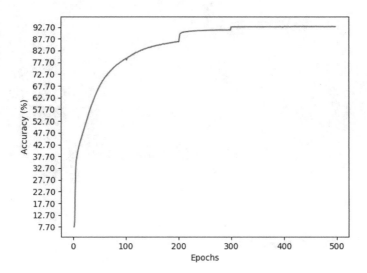

Fig. 10 Accuracy (%) versus epochs

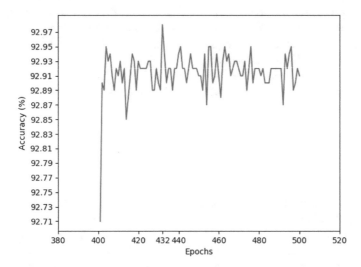

Fig. 11 Accuracy (%) versus epochs (400–500)

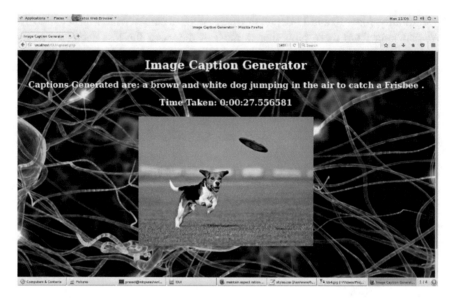

Fig. 12 Screenshot of actual output of our Web interface

4.5 Our Model in Action

Apart from the test images, we also evaluated the correctness of caption generation done by our model on random images given to it. Hence, we built an end-to-end Web interface where user can upload any image file and get the captions for the uploaded image as output as shown in Fig. 12. We found that the captions generated were quite satisfactory and some of the examples are shown in Fig. 13. Results in Figs. 12 and 13 are for ResNet-101 image model.

Our image captioning model was improved and optimized by well-designed quantitative and qualitative techniques which mainly consist of model evaluation, comparison, training and testing.

5 Conclusion

We have presented an image captioning system that uses convolutional neural network to generate image embeddings followed by recurrent neural network that finally produces description of the image in simple language understood by humans. We used two different standard CNN models VGG16 and ResNet-101 in our model. Both image captioning models were trained and tested on flickr8k dataset. It was found that they both generate almost same quality of results based on the computed BLEU score. The only difference found was the time complexity of both models where Resnet-101 takes more time compared to VGG16. Our system can be extended to be

Fig. 13 Qualitative results are shown on randomly chosen images on Flickr8k dataset. Our image captioning model was improved and optimized by well-designed quantitative and qualitative techniques

implemented in array of applications having a wide consumer base in domains like video surveillance, robotic system, education, etc.

References

1. Dai, J., He, K., Sun, J.: BoxSup: exploiting bounding boxes to supervise convolutional networks for semantic segmentation. In: IEEE International Conference on Computer Vision ICCV, pp 1635–1643 (2015). arXiv:1503.01640v2
2. Das, S.: CNN architectures: Lenet, alexnet, vgg, googlenet, resnet and more (2017). https://medium.com/@sidereal/cnns-architectures-lenetalexnet-vgg-googlenet-resnet-and-more666091488df5
3. Deng, J., Dong, W., Socher, R., Li, L.-J., Li, K., FeiFei, L.: Imagenet: a large-scale hierarchical image database. IEEE Comput. Vis. Patt Recogn. (CVPR) 248–255 (2009)

4. Fang, H., Gupta, S., Iandola, F., Srivastava, R., Deng, L., Dollar, P., Gao, J., He, X., Mitchell, M., Platt, J.C., Zitnick, C.L., Zweig, G.: From captions to visual concepts and back. In: IEEE Conference on Computer Vision and Pattern Recognition(CVPR), pp 1473–1482 (2015). arXiv: 1411.4952v3
5. Frossard, D.: VGG in tensorflow (2016). https://www.cs.toronto.edu/~frossard/post/vgg16/
6. Guillaumin, M., Verbeek, J., Schmid, C.: Multimodal semi-supervised learning for image classification. In: IEEE Computer Society Conference on Computer Vision and Pattern Recognition (2010)
7. He, K., Zhang, X., Ren, S., Sun, J.: Deep residual learning for image recognition. In: IEEE Conference on Computer Vision and Pattern Recognition (CVPR), pp 770–778, June 2016. arXiv:1512.03385v1
8. Hochreiter, S., Schmidhuber, J.: Long short-term memory. Neural Comput. 9(8), 1735–1780 (1997)
9. Hodosh, M., Young, P., Hockenmaier, J.: Framing image description as a ranking task: data, models and evaluation metrics. J. Artif. Intell. Res. 47, 853–899 (2013)
10. Jia, Y., Shelhamer, E., Donahue, J., Karayev, S., Long, J., Girshick, R., Guadarrama, S., Darrell, T.: Caffe: Convolutional architecture for fast feature embedding. ACM international conference on Multimedia, pp. 675–678, Nov 2014. arXiv:1408.5093v1
11. Karpathy, A., Fei-Fei, L.: Deep visual-semantic alignments for generating image descriptions. IEEE Trans. Patt. Anal. Mach. Intell. 14(8) (2015)
12. Kiros, R., Salakhutdinov, R., Zemel, R.: Multimodal neural language models. In International Conference on Machine Learning, pp 595–603 (2014)
13. Kiros, R., Salakhutdinov, R., Zemel, R.S.: Unifying visual-semantic embeddings with multimodal neural language models. In: Computing Research Repository (CoRR) in Machine Learning (2014). arXiv: 1411.2539v1
14. Lin, T.-Y., Maire, M., Belongie, S.. Bourdev, L., Girshick, R., Hays, J., Perona, P., Ramanan, D., Dollar, P., Zitnick, C.L.: Microsoft COCO: common objects in context. In: Springer European Conference on Computer Vision ECCV, pp. 740–755. Springer, Berlin (2014)
15. Olah, C.: Understanding LSTM networks (2015). http://colah.github.io/posts/2015-08-Understanding-LSTMs/
16. Redmon, J., Farhadi, A.: YOLO 9000: better, faster, stronger. In: IEEE Conference on Computer Vision and Pattern Recognition (CVPR), July 2007. arXiv:1612.08242v1
17. Simonyan, M., Zisserman, A.: Very deep convolutional networks for large scale image recognition. In: International Conference on Learning Representations (ICLR) (2015). arXiv:1409.1556v6
18. Soh, M.: Learning CNN-LSTM architectures for image caption generation. In: Stanford University, Proceeding (2016)
19. Szegedy, C., Toshev, A., Erhan, D.: Deep neural network for object detection. In: Proceeding of Neural Information Processing Systems (NIPS 2013)
20. Venugopalan, S., Xu, H., Donahue, J., Rohrbach, M., Mooney, R., Saenko, K.: Translating videos to natural language using deep recurrent neural networks. In: Annual Conference of the North American Chapter of the Association for Computational Linguistics (ACL) (2015). arXiv:1412.4729v3
21. Vinyals, O., Toshev, A., Bengio, S., Erhan, D.: Show and Tell: A Neural Image Caption Generator (2015). arXiv:1411.4555v2
22. Wang, X., Zhu, Z., Yao, C., Bai, X.: Relaxed multiple-instance SVM with application to object discovery. In: IEEE International Conference on Computer Vision (2015)
23. Xu, K., Ba, J.L., Kiros, R., Cho, K., Courville, A., Salakhutdinov, R., Zemel, R.S., Bengio, Y.: Show, Attend and Tell: Neural Image Caption Generation with Visual Attention (2016). arXiv: 1502.03044v3

Face Classification Across Pose by Using Nonlinear Regression and Discriminatory Face Information

Kumud Arora and Poonam Garg

Abstract Face recognition across pose cripples with the issue of non-availability of few important facial features. Some of the facial key features undergo occlusion during pose variations. The sole application of linear regression model in face recognition across pose is unable to predict the occluded features from the remaining visible features. With the approach like discriminative elastic-net regularization (DENR), the training sample's discriminatory information, is embedded into regularization term of the linear regression model. Classification is realized using least sqaure regression residuals. However, the existence of nonlinear mapping between frontal face and its counterpart pose limits the application of DENR. In this paper, discriminative elastic-net regularized nonlinear regression (DENRNLR) is proposed for face recognition across pose. DENRNLR learns discriminant analysis-based kernelized regression model constrained by elastic-net regularization. The effectiveness of the proposed approach is demonstrated on UMIST and AT&T face database.

Keywords Discriminant analysis · Discriminative elastic-net linear regression (DENLR) · Face recognition · Least square regression (LSR)

1 Introduction

Face recognition in the presence of pose variations remains one of the challenging issues in reliable face recognition. To maintain reliable recognition results in the presence of pose variations, a plethora of approaches has been proposed in the last two decades.

Among those proposed approaches, one native approach is to perform face synthesis so that the input face and enrolled face lie in same feature space, thereby allowing

K. Arora (✉)
Inderprastha Engineering College, Ghaziabad, India
e-mail: kumud.arora76@gmail.com

P. Garg
Institute of Management and Technology, Ghaziabad, India
e-mail: pgarg@imt.edu

© Springer Nature Singapore Pte Ltd. 2020
S. Agarwal et al. (eds.), *Machine Intelligence and Signal Processing*,
Advances in Intelligent Systems and Computing 1085,
https://doi.org/10.1007/978-981-15-1366-4_4

Fig. 1 Blurring and ghost
effects

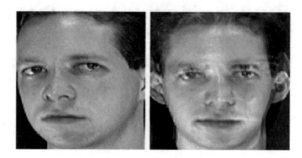

conventional face recognition methods to work. Successful 'Eigen-face approach'
[1] has led to the basis of framework where the probe face can be posed as a general regression problem. Face classification through general regression framework
attempts to utilize the regression residuals learned from the linear combination of
regression coefficients.

Linear combination of non-frontal faces to learn the frontal face, however, is
unable to yield reliable results. The learned regression coefficients used for feature
estimation cannot capture the inherent nonlinearity of face pose variations. This
results in artifacts like blurring and ghost effects (Fig. 1) in the predicted image of
the required pose. These artifacts degrade the discriminant boundaries between the
images of different classes and affect the classification accuracy.

Xiong et al. [2] proved that the geometrical structure of the data in the feature
space and kernel space is similar. However, due to limited number of samples, the estimated regression coefficients (β_i) are not completely reliable for predicting unseen
samples, and hence, overfitting takes place. The trade-off between the bias and variance is controlled by the strength of regularization. To avoid overfitting, penalty
terms/regularized terms are introduced in the least squares method. Though traditional regularization approaches like ridge regression are able to map the training
face images from each individual near their individual targets, however they fail to
embed the discriminatory features in the classification process. Zou and Hastie [3]
proposed the well-known elastic-net regularization model that utilizes both dense
and sparse regularization. The hybrid regularization helps in the selection of correlated features, but still, it is failed to utilize the discriminative features. Xue et al.
[4] proposed discriminatively regularized least squares classification (DRLSC) that
directly embeds the discriminative and local geometry information of train samples
into regularization terms. This discriminatory information is crucial for classification. With the direction of embedding the discriminative information in the output
space of classifier to be designed, this paper aims to group samples that have same
labels and maximize the margins between samples of different classes. In this paper,
we propose the use of elastic-net regularization model along with the discriminant
analysis of kernel subspace. Before the building of kernel space, feature space was
pre-processed with pose correction measure. With this kind of optimization of regression coefficients, we can simultaneously generate nonlinear regressors from feature
to parameter space along with optimal feature correlations between the train set and

test targets. Semi-definite positive kernels are used to generate nonlinear feature subspace. Optimized regression coefficients matrix is employed to transform both train set and test set. Face recognition (an inherent multi-category image classification) is obtained by simple label matching.

The main contribution of this paper is the proposal of utilizing the simple approach like Procrustes analysis for pose correction along with the learning of discriminative kernel (nonlinear) features in a framework capable of performing face recognition invariant of pose variations. Though our proposed approach is based on discriminative elastic-net regularized linear regression (DENLR) approach proposed by Zhang et al. [5], still the proposed approach differs from DENLR. The proposed approach formulates an optimization criterion and relaxed regression target matrix for kernel space rather than pixel space. Also, DENLR approach does not utilize the Procrustes analysis as a pre-processing step as used in the proposed paper for face pose correction. The proposed scheme is applied subject wise.

2   Related Work

In pose-invariant face recognition, one of the pioneer works that represented face synthesis as a linear regression problem was presented by Beymer and Poggio [1]. They represented face image holistically as the linear combination of a set of training faces under the same pose. Regression coefficients are calculated from linear combination and then employed to approximate a new unknown input pose. The residual error term represents the error in predicting the probe image as a linear combination of the regressors and linear prediction coefficients. The application of this approach remains limited due to the requirement of dense pixel-wise correspondence between face images.

Another most cited work in face pose synthesis using regression framework is by Chai et al. [6]. Instead of using holistic features, they proposed local linear regression (LLR), which utilized the local features from small input face patches. Their work was based on the assumption that local manifold structure remains the same for different poses. Each local patch was synthesized with simple linear regression giving ('β') coefficients for each patch. The final virtual view was generated by combining all the patches. This approach, however, failed to represent the manifold structure of the desired pose if the strict semantic correspondence between the patches is not fulfilled.

Naseem et al. [7] proposed linear regression classification (LRC) algorithm, where they represented a probe image as a linear combination of class-specific gallery images. In spite of the simplicity of the approach, it requires multiple pose images per subject in order to obtain a decent recognition rate. Also, this approach cannot be used in the case of a single gallery image for each subject. Lo and Chen [8] expanded the conventional LRC by localizing the manifold subspace. In their approach, they enforced locality regularization on LRC by only including the k closest images to the query image measured using Euclidean distance.

The above-mentioned regression-based approaches do not imbibe the discrimination of subjects in the face pose normalization/classification. Discriminant analysis forms a predictive model for set membership by using the discriminant function. To improve the robustness of the LRC algorithm, Huang and Yang [9] proposed a linear discriminant regression classification (LDRC) algorithm for face recognition. Fisher criterion was embedded into the LRC. To mitigate high variance in coefficients due to large pose variation, a regularization term was added (λ) to the regression function. Xue et al. [4] used ridge regression (RR)/L2 regression locally at different face components for classification to model the linear dependency between covariate variables and labels. When applied holistically, ridge regression was found to be not effective in recognizing the face with illumination variations/poses/partial occlusions.

Cross-validation algorithm was commonly adopted to select the regularization parameters in each local region. Ridge regression ignores meaningful ordering of the features (like specification of consecutive predictors). To compensate the ordering limitations of the ridge regression, L2 norm was replaced with an L1 norm. Tibshirani [10] has shown that least squares with an L1 penalty comes as close as subset selection techniques like Least Absolute Shrinkage and Selection Operator (LASSO). Maritato et al. [11] proved that L1 norm subspaces offer significant robustness to face image variations and disturbances. Zou and Hastie [3] proposed elastic-net regularization where they combined L2 and L1 norm. They also proved elastic net as a stabilized version of the LASSO. Zhang et al. [5] developed the discriminative and robust marginal linear regression models using elastic-net regularized linear regression. Classification accuracy is improved by increasing the margins between the classes. Feng et al. [12] put forward kernelized elastic-net regularization. They proved for some datasets; a fair amount of improvement is achieved in the classification accuracy than in the case of classical elastic-net regularization. The relaxed regression target matrix for pixel space proposed in [5] can be improved upon by the consideration of kernel space.

3   Problem Formulation and Proposed Approach

Table 1 describes the various notations used in the proposed approach.

The goal of regression is to minimize the residual error, and the goal of classification is to increase the separation plane as much as possible between the samples of different classes. In the proposed approach, an attempt is made to fuse the two separate aims by utilizing the discriminative least squares framework.

The proposed framework is broadly divided into four tasks:

(a)   Procrustes transformation for pose alignment.
(b)   Transformation to nonlinear space using 'Kernel Trick'.
(c)   Regression using discriminative elastic-net regularized nonlinear regression framework.
(d)   Calculation of classification accuracy (Fig. 2).

Table 1 Notations used in the paper

Notation used	Description
x_i	Training Set, $X = [x_1, x_2 \ldots, x_n] \in \mathfrak{R}^{D \, X \, N}$
N	Number of training instances
C	Number of classes in dataset
Y_i	Associated binary class labels, $Y = [Y_1, Y_2, Y_c] \in \mathfrak{R}^{D \, X \, C}$
k	Kernel function
Φ	Nonlinear mapping associated with kernel function k(.,.)
β	Linear projection matrix comprising regression coefficients
r	Rank of training set
Ω	Regularization term
f^*	Optimization function
M	Learned non-negative matrix
E	Constant matrix with $+1$ or -1 (depending upon dragging direction)
C_1	Lagrange's multiplier
μ	De-correlation penalty parameter
δ_c	Similarity function between Kernel and Class indicator function
S	Training instances available per class

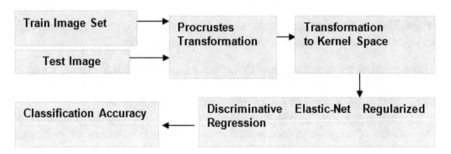

Fig. 2 Framework of discriminative elastic-net regularized nonlinear regression (DENRNLR)

With the proposed formulation, the problem is divided into the subproblems equal to the number of classes (C) and are grouped together to share a unique learning model. The proposed nonlinear model of the form:

$$f_\beta(x) = \langle \Phi(x), \beta \rangle \qquad (1)$$

The proposed nonlinear model is further constrained with the regularization terms $\Omega(f) = \lambda_1 \|\beta\|_* + \frac{\lambda_2}{2} \|\beta\|_F^2$, where λ_1 and λ_2 are the constant regularization parameters. Thus, the objective function with the regularization term takes the form of:

$$f^* = f_\beta(x) + \Omega(f) \tag{2}$$

Clearly, the objective function is dependent upon the sum of singular values of learned projection matrix. Based on convex optimization theory, objective function (2) can be considered as convex problem, which has a unique optimal solution. An optimization procedure corresponding to an iterative projection algorithm is derived. Due to variable decoupling in the proposed formulation, only a group of linear equation needs to be solved during every step of iteration. Optimization problem for the objective function can be expressed as:

$$\min_{\beta, M} \left\| \Phi(x)'\beta - (Y + E\circ M) \right\|_F^2 + \lambda_1 \|\beta\|_* + \frac{\lambda_2}{2}\|\beta\|_F^2 \tag{3}$$

Utlizing the Singular Value Decomposition (SVD) of $\beta = U\sum V'$, $\|\beta\|_*$ which is represented as $\mathrm{tr}(\Sigma)$ can be represented as

$$\frac{1}{2}tr(U\sqrt{\Sigma}\sqrt{\Sigma}U') + \frac{1}{2}tr(\sqrt{\Sigma}V'V\sqrt{\Sigma}) \text{ or } \frac{1}{2}(\|\beta_1\|_F^2 + \frac{1}{2}(\|\beta_2\|_F^2) \tag{4}.$$ The min-

imization of $\frac{1}{2}(\|\beta_1\|_F^2 + \frac{1}{2}(\|\beta_2\|_F^2)$ is equivalent to the minimization of $\|\beta\|_*$ under the constraint $\beta = \beta_1\beta_2$, where $\beta_1 \in R^{(D \times r)}$ and $\beta_2 \in R^{(r \times C)}$. Due to variable decoupling in proposed formulation, only a group of linear equation needs to be solved during every step of iteration. The class identity is established by the function $(\delta_i) \forall i = 1,\ldots,C$ on the basis of Euclidean distance value.

Proposition 1 *Proposed elastic-net regularization of learned projection matrix (β) highlights the components where the fundamental information lies.*

The regularization term, a penalty on the learned projection matrix, naturally analyzes the learned projection data via singular values. The combination of nuclear norm and Frobenius norm $\|\beta\|_*$ *and* $\|\beta\|_F^2$ can be represented as $\sum_i^r \|\sigma_i\|$ *and* $\sum_i^r \|\sigma_i^2\|$ where σ_i represents the ith singular values of β. Low values of singular terms points to the presence of noisy or redundant information whereas the large singular term values indicate the presence of fundamental information components. Hence, the elastic net regularization of singular values aid in incessant reduction of redundant components while enabling the collection of principal components.

Algorithm: DENRLNR	
Input:	*N training samples with label value*, i.e. $\{x_i, Y_i\}$ where $x_i \in R^d$ and Y_i as their class labels
Output:	Classification accuracy

(continued)

(continued)

Algorithm: DENRLNR	
1.	For $i = 1: C_i$ Perform Procrustes analysis to derive Procrustes transformation parameters for the alignment of e train images as per the the probe pose $(T_i, \theta_i, s_i) \leftarrow$ *Procrustes Analysis*(X_i, Yi) Align all the other train samples of the class with (T_i, θ_i, s_i) End For
2.	Normalization of data to zero mean and unit variance. Application of the Kernel function maps feature space to Hilbert space. $\Phi: X \rightarrow H$ (Hilbert space). The choice of kernel is left as free parameter but should satisfy *Mercer's Criterion*.
3.	Given the initial value of the relaxed target matrix (M) $M = 0^{N X C}$, the relaxed target matrix is given by using ε-dragging technique. The goal is to learn linear projection matrix from kernel space.
4.	By using the discriminative elastic-net regularized linear regression (DENRLR) approach, the objective function is formulated as the optimization problem with relaxed regression target matrix.
5.	The relaxed target matrix is given by using ε-dragging technique, thereby reformulating the binary target matrix (Y) as $\widetilde{Y} = Y + E^\circ M$ (5) where M is the learned non-negative matrix, E is the constant matrix with values ($+1$ or -1) corresponding to the dragging direction i.e. $E\_(i,j) = \{+1,$ if associated class label $Y_{ij} = 1$, and -1 if associated with class label $(Y_{ij}) = 0\}$.
6.	Call ALM_Solve to obtain optimal projection score matrix.
7.	The class identity of probe sample (Y) is established on the minimization of the residuals: $\delta_c = \left\| \phi(x)'\beta - \widetilde{Y}_c \right\|_F^2$, where δ_c is function that computes the Euclidean distances from probe target to all the train targets.

Sub Algorithm: ALM_Solve	
Input:	$\Phi(x)$, Relaxed target matrix (M) , λ_1, λ_2
Output:	Optimal projection score matrix

The following steps are used to derive an optimal projection score matrix. Augmented Lagrangian Method (ALM) is utilized to solve optimization function f* by alternatively maximizing the dual of original problem and minimizing the Augmented Lagrangian of the original problem

The Augmented Lagrangian function of the problem (3) is

$$\mathcal{L}(\beta, M, \beta_1, \beta_2, C_1) = \left\| \Phi(x)'\beta - (Y + E \circ M) \right\|_F^2 + \frac{\lambda_1}{2}\left(\|\beta_1\|_F^2 + \|\beta_1\|_F^2 \right) + \frac{\lambda_2}{2}\left(\|\beta\|_F^2 \right)$$

$$+ \langle C_1, \beta - \beta_1\beta_2 \rangle + \frac{\mu}{2}\|\beta - \beta_1\beta_2\| (6)$$

where C_1 is a Lagrange's multiplier, μ is a de-correlation penalty parameter and $\langle a, b \rangle = tr(a^T b)$

(continued)

(continued)

Sub Algorithm: ALM_Solve	
1.	Block coordinate descend (BCD) method is used for finding minimum points of $\mathcal{L}$ with respect to primal variables $(\beta,\ \beta_1, \beta_2, M, C_1)$
2.	Augmented Lagrangian is minimized by updating one coordinate in each direction. The algorithm during first phase solves β by fixing M and then in second phase solves M by fixing β.
3a.	Update β_1, by setting derivative of equation (6) w.r.t β_1 to zero. $$\beta_1^+ = \mathrm{argmin}_{\beta_1}\frac{\lambda_1}{2}\left(\|\beta_1\|_F^2 + \frac{\mu}{2}\left\|\beta - \beta_1\beta_2 + \frac{C_1}{\mu}\right\|_F^2\right) \quad (7)$$ The obtained equation being least square problem with regularization, its solution can be expressed as follows $$\beta_1^+ = (C_1 + \mu\beta)\beta_2^T \cdot \left(\lambda_1 I + \mu\beta_1\beta_2^T\right)^{-1} \quad (8)$$
3b	Update β_2, by setting derivative of equation (6) w.r.t β_2 to zero. $$\beta_2^+ = \mathrm{argmin}_{\beta_2}\frac{\lambda_1}{2}\left(\|\beta_2\|_F^2\right) + \frac{\mu}{2}\left\|\beta - \beta_1\beta_2 + \frac{C_1}{\mu}\right\|_F^2\right) \quad (9)$$ The obtained solution being least square problem with regularization, its solution can be expressed as follows: $$\beta_2^+ = \left(\lambda_1 I + \mu \cdot \beta_1\beta_1^T\right)^{-1} \cdot \beta_1^T \cdot (C_1 + \mu \cdot \beta) \quad (10)$$
3c	Update β, by setting the derivative of equation (6) w.r.t β to zero. $$\beta^+ = \mathrm{argmin}_\beta \left\|\Phi(x)'\beta - (Y + E \circ M)\right\|_F^2 + \frac{\lambda_2}{2}\left(\|\beta\|_F^2\right)$$ $$+ \frac{\mu}{2}\left\|\beta - \beta_1\beta_2 + \frac{C_1}{\mu}\right\|_F^2 \quad (11)$$ Being least square problem with regularization, the solution is $$\beta^+ = \left(2\Phi(x)'\Phi(x) + (\lambda_2 + \mu)I\right)^{-1} \cdot (2\Phi(x)(Y + E \circ M) + \mu\beta_1\beta_2 - C_1) \quad (12)$$
3d	Update M, $M^+ = \mathrm{argmin}_M\left\|\Phi(x)'\beta - (Y + E \circ M)\right\|_F^2$, s.t $M > 0$ (13) The optimal solution for (13) is $M^+ = \mathrm{Max}(E \circ M, 0)$
4.	Update Lagrange's multiplier C_1 by $C_1 = C_1 + \mu(\beta - \beta_1\beta_2)$ (14)
5.	Repeat steps 3a, 3b, 3c, 3d until $\left\|\beta^{k+1} - \beta^k\right\| \leq 10^{-4}$ (15)

4 Experiments and Results Analysis

All the below experiments were carried out in a computer with a core i5 2.3 GHz processor and 8 GB DDR4 RAM running MATLAB 2017b. The problem of learning is to predict effectively, given the training set and an estimator function 'f', the label of a query sample. This problem of classification translates in finding a coefficient vector β such that its correlation with actual class is maximized, and the residual correlation is minimized. Pre-processing step of Procrustes superimposition step aligns the database images class-wise according to the orientation of the probe sample. Estimation of Procrustes transformation parameters of one class is found to be sufficient to transform the images of different classes with same orientation. As an example, case in Fig. 3, scaling (S), rotation (θ) and translation (T) parameters extracted from one class are used to transform the images of the other class. Scaling and rotation parameter is a single value, whereas translation is a single dimension vector of 10,304 values, which are the same as the total number of pixels in an image under consideration.

We conducted experiments for the proposed approach on two standard data sets, namely: UMIST database [13] and AT&T face database [14]. The AT&T database contains ten different images of each of 40 distinct subjects. For some subjects, the images have varying lighting, facial expressions (open/closed eyes, smiling/not smiling) and yaw pose variations up to 30° facial details (glasses/no glasses). All the images have a dark homogeneous background. The UMIST, now known as Sheffield Face Database, consists of 564 images of 20 individuals (mixed race/gender/appearance). Each individual is shown in a range of poses from profile to frontal views—each in a separate directory labelled 1a, 1b,…1t. However, for the experiment part here, we selected images with yaw variation up to 65°. Both the database images are in PGM format of size of 112 × 92 pixels with 256-bit grey-scale. For each test, the mentioned percentage of images was randomly selected as the training set and the remainder as the test set. For every experiment, ten runs were taken into account due to which every accuracy value is associated with the corresponding standard deviation.

The best value is indicated along with the standard deviation value. A fixed value of $\mu = 0.25$ is taken throughout the experiments. For mapping to kernel space, fractional polynomial kernel [15] is used, which is defined as: $\Phi(x_i, x_j) = (\beta x_i \cdot x_j + c)^d$, where β is adjustable parameter, c is constant and d ($0 < d < 1$) is polynomial degree with $\beta = 1$, $c = 0.002$ and $d = 0.85$. The fractional polynomial kernel is

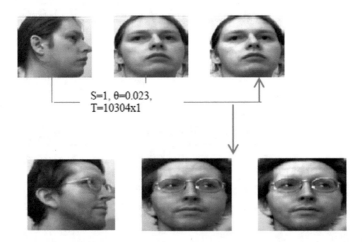

S=1, θ=0.023, T=10304x1

Fig. 3 Example of using Procrustes superimposition parameters of one class to transform another class

well suited where the training data has been normalized. Tables 2 and 3 indicate the classification accuracy with 50% train set and 50% test set selection of database when the values of regularization parameter (λ_2) is varied in the range of 0.00001, 0.0001, 0.001, 0.01, 0.1, 1 and 10.

It can be observed that (λ_1 and $\lambda_2 = 0.001$) steadily yields the best results for the two datasets. Also, the accuracy is affected more by the variation in λ_2 than λ_1, which is in accordance with the regularization parametric function used: $\Omega(f)$.

Figures 4 and 5 present the performance comparison of classification accuracies obtained by the proposed method and other state-of-the-art regression-based classification approaches for different train data and test data splits (40%—train data 60%—test data, 50%—train data 50%—test data, 60%—train data 40%—test data).

A good amount of gap is there in the accuracy between the principal component regression, which is a linear regression method for both the AT&T and UMIST face database. Accuracy for AT&T database is found slightly higher than for UMIST database. The underlying reason for the same can be accounted for Procrustes estimation step as there are large pose variations in UMIST database. In low-rank regression, the classification accuracy falls down if the rank value is near to the input database dimensionality. Also, there is a clear distinction in the superiority of results obtained in discriminant regression classification over linear regression-based classification.

Table 2 Classification accuracy obtained by varying λ_2, keeping λ_1 and μ fixed

λ_2	AT&T dataset	UMIST dataset
0.00001	86.67 ± 0.90	81 ± 0.597
0.0001	91.001 ± 1.50	86 ± 0.90
0.001	94.664 ± 1.343	90.04 ± 3.16
0.01	87.664 ± 1.51	87 ± 2.01
0.1	83.67 ± 0.97	80.034 ± 1.59
1	81.89 ± 2.18	81.67 ± 2..678
10	78 ± 2.35	75.45 ± 3.21

Table 3 Classification accuracy obtained by varying λ_1, keeping λ_2 and μ fixed

λ_1	AT&T dataset	UMIST dataset
0.00001	82.67 ± 2.02	78 ± 1.497
0.0001	89.91 ± 2.15	84.89 ± 0.90
0.001	94.664 ± 1.343	90.04 ± 3.16
0.01	90.76 ± 2.51	86.89 ± 0.976
0.1	83.67 ± 0.97	80.034 ± 1.77
1	80.966 ± 0.21	80.34 ± 0.445
10	78 ± 2.35	73.45 ± 4.25

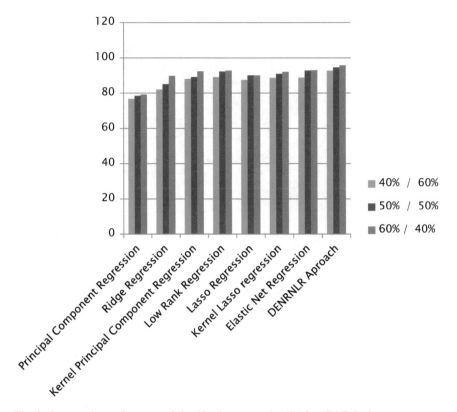

Fig. 4 Comparative performance of classification accuracies (%) for AT&T database

5 Conclusion and Future Work

Improved version of elastic-net linear regression (ENLR), i.e. discriminative elastic-net regularized nonlinear regression (DENRNLR) is proposed to capture the subtle details, in the form of singular values, present in different classes of the database. The regression targets are made relaxed regression targets by the ϵ-dragging approach. The proposed approach with relaxed regression targets enhances both the flexibility and discrimination ability, thereby leading to minimum residual error and better classification accuracy. The proposed formulation can deal effectively with the small sample size problem through the regularization as well as low dimensional statistical feature extraction simultaneously. The block coordinate descend method that loops over the kernel data in the main loop of ALM, is however time-consuming, but still, it is necessary to avoid convergence issue. Experimental results are done with preliminary information about the estimated rotation angles of training samples, which helped in smooth Procrustes transformation for aligning training images according to the probe sample alignment. However, in future, we will try to estimate alignment angles.

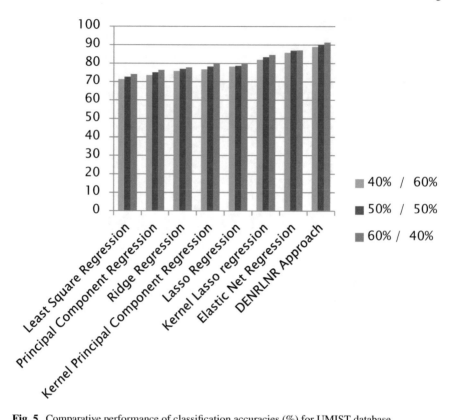

Fig. 5 Comparative performance of classification accuracies (%) for UMIST database

References

1. Beymer, D., Poggio, T.: Face recognition from one example view. In: Fifth IEEE International Conference on Computer Vision Proceedings, pp. 500–507 (1995)
2. Xiong, H., Swamy, MNS., Ahmad, MO.: Optimizing the kernel in the empirical feature space. IEEE Trans. Neural Networks **16**(2), 460–474 (2005)
3. Zou, V., Hastie, T.: Regularization and variable selection via the elastic net. J. R. Statistical Soc. Ser. B (Statistical Methodol.) **67**(2), 301–320 (2005)
4. Xue, H., Chen, S., Yang, Q.: Discriminatively regularized least-squares classification. Pattern Recogn. **42**(1), 93–104 (2009)
5. Zhang, Z., Lai, Z., Xu, Y., Shao, L., Wu, J., Xie, G.S.: Discriminative elastic-net regularized linear regression. IEEE Trans. Image Process. **26**(3), 1466–1481 (2017)
6. Chai, X., Shan, S., Chen, X., Gao, W.: Locally linear regression for pose-invariant face recognition. IEEE Trans. Image Process. **16**(7), 1716–1725 (2007)
7. Naseem, I., Togneri, R., Bennamoun, M.: Linear regression for face recognition. IEEE Trans. Pattern Anal. Mach. Intell. **32**, 2106–2112 (2010)
8. Lo, L.Y., Chen, J.C.: Manifold synthesis: predicting a manifold from one sample. In: 11th International Conference on Hybrid Intelligent Systems (HIS), pp. 505–510 (2011)
9. Huang, S.M., Yang, J.F.: Linear discriminant regression classification for face recognition. IEEE Signal Process. Lett. **20**, 91–94 (2013)

10. Tibshirani, R.: Regression shrinkage and selection via the lasso. J. Roy. Statist. Soc. Ser. B **58**, 267–288 (1996)
11. Maritato, F., Liu, Y., Colonnese, S., Pados, DA.: Face recognition with L1-norm subspaces. In: Compressive Sensing V: From Diverse Modalities to Big Data Analytics, p. 98570L. Intenational Society for Optics and Photonics (2016)
12. Feng, Y., Lv, S.G., Hang, H., Suykens, J.A.: Kernelized elastic net regularization: generalization bounds, and sparse recovery. Neural Comput. **28**, 525–562 (2016)
13. UMIST Face Database: Retrieved from: https://www.sheffield.ac.uk/eee/research/face
14. AT&T—The Database of Faces (formerly "The ORL Database of Faces"). Retrieved from https://www.cl.cam.uk/research/dtg/attarchive/facedatabase.html
15. Scholkopf, B., Smola, A.: Learning with Kernels: Support Vector Machines, Regularization, Optimization and Beyond. MIT, Cambridge (2002)

Emotion Recognition from Facial Images for Tamil Language Speakers

Cynthia Joseph and A. Rajeswari

Abstract In the field of image processing, it is very interesting and challenging to recognize the human emotions in Tamil language for general life applications. Facial expressions are the visual methods to convey emotions. Facial expressions of a person at different instances are not the same. Emotions can be understood by text, vocal, verbal and facial expressions. Automatic recognition of facial expressions is important for human–machine interfaces. If a machine can identify human emotions, then it can understand human behavior better, thus improving the task efficiency. Nowadays, one of the challenging parts is face detection and recognition of wide variety of faces, complexity of noises and image backgrounds. To solve this problem, this paper aims to deal with facial emotion recognition using effective features and classifiers. Databases are created with large set of images and feature extraction is implemented. SVM classifier is used for classifying the emotions. The facial expression recognition rate is obtained by using a database especially created in Tamil language. Emotion recognition is analyzed for different emotions using facial expressions.

Keywords Human–machine interaction · Principal component analysis · Canny edge detection · Support vector machine

1 Introduction

Facial expressions are the most important in facilitating human communications and interactions. The expression is the most visual method to convey emotions. In present-day technology, human–machine interaction (HMI) is growing in demand so the machine needs to understand human gestures and emotions. The emotions can be understood from a text, verbal and facial expressions. Nowadays, one of the

C. Joseph (✉) · A. Rajeswari
Coimbatore Institute of Technology, Coimbatore, India
e-mail: cynthuj.mails@gmail.com

A. Rajeswari
e-mail: rajeswari@cit.edu.in

© Springer Nature Singapore Pte Ltd. 2020
S. Agarwal et al. (eds.), *Machine Intelligence and Signal Processing*,
Advances in Intelligent Systems and Computing 1085,
https://doi.org/10.1007/978-981-15-1366-4_5

challenges in facial emotion recognition is to recognize the face from the complexity of noises. This is used to compress the dimension and shorten the training of images. The purpose of a study is to develop an intelligent system for facial images based on expression classification in Tamil language speaker by using a fisherface algorithm in a support vector machine classifier. Facial expressions play an important role in smooth communication among individual person. Emotions can be communicated by various modalities including speech, language, gesture, and posture as well as face expression. The recognition method begins by acquiring the image by using a web camera. The facial expression image is studied from the database created. In this proposed system, it focuses on human face for recognizing expressions for Tamil native speakers. From the expressions recognized, the following are the facial emotions such as a happy, surprise, neutral, sad and anger. In human–human interaction (HHI) spoken words only account for 7% of what a listener comprehends; the remaining 93% consist of the speaker's nonverbal communicative behavior. It is found that limited work is done in the field of real-time emotion recognition using facial images and there is no database in Tamil language. Therefore, the interaction between humans and computers will be more natural if computers are able to understand the nonverbal behavior of their human counterparts and recognize their effective state of a person's emotion.

2 Existing Methods

Makioka [1] proposes a model that the facial expressions on the left side of face appear stronger than the right side of face. This model uses the algorithm of an effective selection method that is based on supervised learning of multilayer neural networks for facial emotion recognition. The neural networks for facial emotion recognition are constructed to extract the perceptually important feature pixels for 224 kinds of window dimensions. The facial emotion recognition rate is improved from 78.8 to 83.1% using the emotion mask. It eliminates the irrelevant data and improves the classification performance using backward feature elimination.

Hussain [2] aims to identify two different methods of feature extraction for facial expression recognition from images. For feature extraction, Gabor filter and local binary pattern operator are used. Based on the minimum redundancy and maximum relevance algorithm, the optimum features were selected. The images were classified using naïve Bayesian classifier. Selecting a small subset of features out of the thousands of features is important for accurate classification. The percentage of accurate classification varied across different expressions from 62.8 to 90.5% for the log Gabor filter approach and from 71.8 to 94.0% for the LBP approach.

Suchitra et al. [3] proposed real-time emotion recognition system that recognizes the basic emotions such as happy, sad, surprise, neutral, anger and disgust. This model uses the CMU multi-pie database consisting of 2D images with different illumination. The software system used in this model is Raspberry pi II which can be used with robots. As the size of Raspberry pi II is very small, lightweight and very less power supply is needed for it. The proposed system design uses Raspberry pi II with external webcam, keyboard and display monitor. In real time, a person looks into the webcam, his/her image will be taken and given to Raspberry pi II. Emotion recognition software that is already deployed will recognize emotions and displays the recognized emotions into the display monitor. Recognition accuracy of 94% is achieved with average processing time of 120 ms using Raspberry pi II.

Ali [4] proposed a model that uses radon and wavelet transform for facial emotion recognition. The radon transform preserves the variations in pixel intensities. It is used to project the 2D image into radon space before subjected to discrete wavelet transform. The second-level decomposition is extracted and used as informative features to recognize the facial emotions. PCI is applied on the extracted features due to the large number of coefficients. The k-nearest neighbor classifier is used to classify seven facial emotions. In order to evaluate the effectiveness of this model, the JAFFE database has been used. The accuracy of facial emotion recognition rate is 91.3%.

Khan [5] proposed a model that the facial expressions are considered to be an effective and non-verbal by means of expressing the emotional states of humans in more natural and non-intrusive way. The features are extracted from the images using log Gabor filters. The co-clustering-based algorithm is used to select distinguished and interpretable features to deal with curse of dimensionality issues. SVM is used for classifying the emotions. The selected features are accurate and effective for the facial emotion recognition. It reduces not only the dimensionality but also identifies the face regions on images among all expressions.

3 Proposed Methodology

Emotion is one of the most important elements that express the state of mind of a person. The interaction between human beings and computers will be natural if computers are able to perceive and respond to human emotions. It is widely accepted from psychological theory that human emotions can be classified into the following basic emotions: surprise, anger, neutral, happiness and sadness. The purpose of a present-day study is to develop an intelligent system for facial image-based expression classification in Tamil language speaker by using a Fisherface algorithm in a support vector machine classifier. Facial expressions play an important role in smooth communication among individual person. Facial emotion and the tone of the speech play a major role in expressing emotions.

3.1 Proposed System

In real-time applications, the image is captured, trained and updated in the database (see Fig. 1). The method is to extract image from video sources. Capturing of image

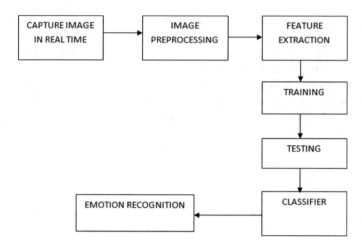

Fig. 1 Proposed system

is done in two different methods and then noise is removed from the real-time images by using preprocessing concept. The image is captured from cameras and also a web camera is used in laptop to provide corresponding results to extract features, which is connected as an input data or a format [6]. Modern frame grabbers are typically able to store multiple frames and compress the frames in real time using JPEG. The extraction process is to extract images which are in a real time form as a video frame. This extracted image is fed to an image preprocessing block; this process improves an image data which contains noisy images in real-time applications and preprocessed images is processed for training phase. It trains the images in the database for different classification of emotions.

500 images are being trained and then it is updated in the database. The training phase output is fed to the classifier. The classifier used is the combination of fisherface and SVM classifier. By using a classifier, different emotions are recognized as a happy, sad, surprise, anger and neutral.

3.2 Image Acquisition

An image deals with a projection of a 3D scene into a 2D projection plane. Image can be defined as a two-variable function $I(x, y)$, for each position (x, y) in the projection plane, $I(x, y)$ defines the light intensity at this point. Image acquisition is based on classification of the image characteristics and the image digitization. Image digitization contains two domains; they are the spatial domain and the intensity domain. Image acquisition is the creation of photographic images, such as a physical scene or the interior structure of an object [7]. The term is often assumed to imply or include the processing, compression, storage, printing and display of images.

Image sensing and acquisition are used for processing the analog images of physical scenes or the interior structure of an object and converting it into digital.

3.3 Database Creation

Face recognition is most popular in research areas. A facial expression database is a collection of images from video clips with facial expressions. Well-annotated (emotion tagged) media content of facial behavior is essential for training, testing, algorithms for the development of expression recognition systems [8]. The emotion annotation can be done in discrete emotions labels or in a continuous scale.

In this paper, around 500 images are collected in the database (see Fig. 2). The database consists of only the images of Tamil native speakers, since there is no separate database available for them. The existing databases are JAFFE, Cohn-Cande, DCAB, Yale and CASME. There is no work existing for recognizing emotions especially for Tamil native speakers. The emotions are recognized for many persons who speak Japanese, Chinese, English and Danish easily. The accuracy of the results of

Fig. 2 Database images of Tamil speakers

Table 1 Size of Database

Number of subjects	Videos	Images
Male	50	250
Female	50	250

facial recognition is highly depending upon the versatility of the database. A facial recognition system is a computer application, and it is capable of identifying or verifying a person from the captured image in a real-time applications or video frame from a video source. One of the ways to do this is by comparing the selected features from the image of the database to the trained phase.

In this paper, database is created in real-time emotions by using the mobile camera and also a digital camera as shown in Table 1. These images are extracted manually by using different software. This database provides the detection of each element such as eyes, nose, mouth and lips to recognize different emotions.

3.4 Face Detection

Face detection is also called as a face-priority autofocus, which is a process of the camera that detects human faces, so that the camera can set the focus and appropriate exposure for the shot automatically. In computer technology, face detection is being used in a variety of applications that identifies human faces in digital images. Face detection also refers to the psychological process by which human locate and attend to face a visual scene. In this, principal component analysis (PCA) is used for modeling a geometric structure of face and also Canny edge detection is used. Fusion of PCA-based geometric modeling provides higher face detection accuracy and it improves the time complexity. The equation for the formation of matrix is given in Eq. (1)

$$C = \begin{bmatrix} \sum I_x^2 & \sum I_x I_y \\ \sum I_x I_y & \sum I_y^2 \end{bmatrix} \tag{1}$$

where sum of the overall image gradients is performed, (I_x, I_y), in the window, and then find the eigenvalues of the matrix. The first eigenvalue gives the size of the first principal component, and the second eigenvalue gives the size of the second component.

The main concern of face detection is to locate all image regions which contain a face regardless of its orientation, background and lighting conditions. Such task is tricky since faces can have a vast assortment regarding shape, color, size or texture. At present time, a lot of automatic approaches involve detecting faces in an image and subsequently, detecting lips within each detected face. But most of the face detection algorithms are able to detect faces that are oriented in upright frontal view. These approaches cannot detect faces that are rotated in-plane or out-of-plane with respect to the image plane. Also, it cannot detect faces in case when only part of face is

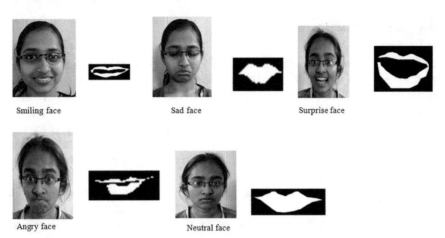

Smiling face   Sad face   Surprise face

Angry face   Neutral face

Fig. 3 Face detection along with Canny edge detection

visible. As a result by using threshold, the skin region is separated from an image for face detection. In face detection, template generation is also done in different images of emotion detection. In this paper, each element with different dimensional values is extracted, and it also extracts the elements automatically (see Fig. 3). The associated probability of face is calculated as given in Eq. (2)

$$P = \frac{a}{((a + b)^w)} \tag{2}$$

Here 'a' represents pixels in face region, 'b' represents pixels in background and 'w' is the weight.

3.5 Training and Testing Phase

The image preprocessing output is fed to the training and testing phase. In the training phase, 80% of images in database are trained, while the remaining 20% is used for testing. The main part of training helps to recognize the images captured in real time accurately by comparing with the trained images. The last part of SVM classifier is a fisherface classifier. This part is used to compress the image dimension and also takes the shortest time of training. Again this builds the features of the training sample face, which recognizes the face images [9]. The results show the accuracy rates are, respectively, 97.75% and the feature recognition time is 9.8 ms, which also demonstrates that the fisher algorithm is superior to eigenfaces on feature extraction.

3.6   Classification

In the general process of classification, ideas and objects are recognized, differentiated and also classified. Classification technique (or classifier) is a systematic approach of building classification models from an input data set. Classifier has many classifications of facial images such as gender, age, facial expressions and the identity [10]. For classification, support vector machine (SVM) is used to recognize the facial emotions. Since SVM is associated with supervised learning models, it analyzes data for classification [11]. SVM is used for real-time decisions. In this paper, the emotions are recognized in real time; therefore, SVM is preferred. This work provides a classification of emotions to be recognized in Tamil language speaker, since there is no work done in recognizing emotions for Tamil speakers. Fisherface classifier is used in this paper, to update all the emotions and SVM classifier is used for recognizing the images.

3.7   Emotion Recognition

Emotion often mediates and facilitates interactions among human beings. Thus, understanding emotion often brings context to seemingly bizarre or complex social communication. Emotion can be recognized from a variety of communication such as a voice intonation and body language [12]. Emotion recognition identifies the human emotions. The task of emotion recognition is difficult for two reasons, first reason is a large database that does not exist and second reason is a classification of the emotion is difficult depending on whether the input image is static or a dynamic frame for facial expression. The implemented system can automatically recognize the expressions in a real time. The real-time emotions include anger, neutral, happy, sad and surprise. This paper provides the result for facial emotion recognition in Tamil language speaker and also it shows an accuracy of 97.75%.

4   Experimental Results

Videos are taken at many instances and with different persons by using digital camera (see Fig. 4).

Images of different persons are extracted from videos by using different software. Various images are collected for database creation.

Capturing work is done by taking videos of different persons by using mobile phones and digital cameras (see Fig. 5). Capturing is done in two ways. They are

- Controlled environment (without environmental disturbances).
- Uncontrolled environment (with noisy environment).

Fig. 4 Videos of different persons

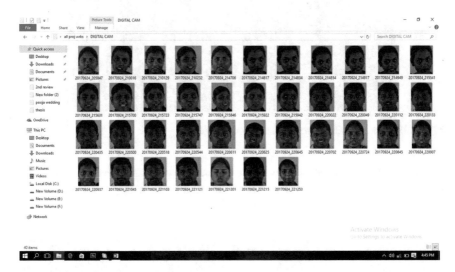

Fig. 5 Images extracted from videos

Comparison is made between mobile phones and digital camera (see Figs. 6 and 7). The resolution of the images is compared between mobile phones and digital cameras. The resolution of mobile phone is 640 × 1136 pixels and that of digital camera is 1024 × 768 pixels. In digital camera images even with noise, occlusion can be captured clearly. Images are extracted from the videos for database creation.

In real-time scenario, a person expresses his/her happiness, sadness, anger and neutral emotions in front of webcam, then the emotion of a particular person will be

Fig. 6 Different emotions of images using mobile phones

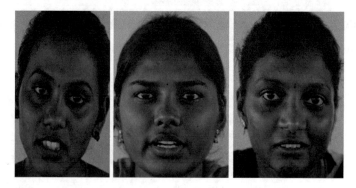

Fig. 7 Different emotions of images using digital camera

compared with the trained images of several emotions in the database. The emotion can be recognized by SVM classifier and the output will be displayed within 3 s in Tamil language (see Figs. 8, 9, 10, 11 and 12).

Fig. 8 Display of happy emotion in Tamil language

Fig. 9 Display of sad emotion in Tamil language

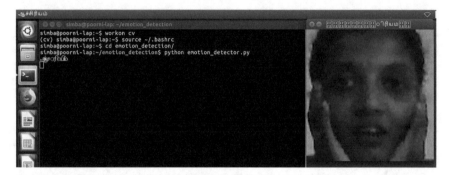

Fig. 10 Display of surprise emotion in Tamil language

Fig. 11 Display of anger emotion in Tamil language

5 Conclusion and Future Work

The emotion recognition in facial images for Tamil language speakers plays a vital role in image processing applications. The image which is to be used for training is extracted from videos of different persons. Database is created for separate emotions

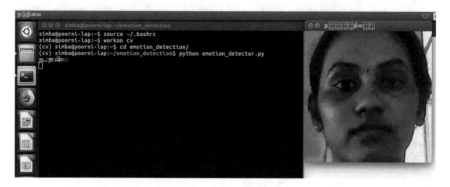

Fig. 12 Display of neutral emotion in Tamil language

like happy, angry, sad, neutral and surprise. The images are trained in the training phase. For recognizing emotions in real time, the webcam starts capturing person who stands in front. Support vector machine is used for classifying the emotions. The image which is captured is fed to the training phase and compared with images in the training phase. The compared image is given to recognition system. The proposed method is also used for training more images by updating the classifier accordingly. This method achieves accuracy of 97.75%. The image recognized by using this method is gray image instead of color image. Classifier used in the proposed method allows only minimal amount of images for training. In future work, different algorithm can be implemented to improve recognition the accuracy. Robots can also be made to recognize emotion by neurological inspiration. Other modalities like speech, biosignals can be combined along with image for accurate emotion recognition.

References

1. Makioka, T., Kuriyaki, Y., Uchimura, K.: Quantitative study of facial expression asymmetry using objective measure based on neural networks. In: International Symposium on Intelligent Signal Processing and Communication Systems, pp. 1–4 (2016)
2. Lajevardi, S.M., Hussain, Z.M.: A novel gabor filter selection based on spectral difference and minimum error rate for facial expression recognition. In: International Conference on Digital Image Computing: Techniques and Applications, pp. 137–140 (2010)
3. Suchitra, Suja, P., Tripathi, S.: Real time emotion recognition from facial images using Raspberry pi II. In: IEEE Transactions on International Conference on Signal Processing and Integrated Networks, pp. 666–670 (2016)
4. Ali, H., Sritharan, V., Hariharan, M., Zaaba, S.K., Elshaikh, M.: Feature extraction using radon transform and discrete wavelet transform for facial emotion recognition. In: IEEE Transactions on International Symposium on Robotics and Manufacturing Automation, pp. 1–5 (2016)
5. Khan, S., Chen, L., Zhe, X., Yan, H.: Feature selection based on Co-Clustering for effective facial expression recognition. In: IEEE Transactions on International Conference on Machine Learning and Cybernetics, vol. 1, pp. 48–53 (2016)
6. Rishin, C.K., Pookkudi, A., Ranjith Ram, A.: Face and expression recognition techniques: a review. Int. J. Eng. Innov. Technol. **4**(11), 204–208 (2015)

7. Fabian Benitez-Quiroz, C., Srinivasan, R., Martinez, A.M.: EmotioNet: an accurate, real-time algorithm for the automatic annotation of a million facial expressions in the wild. In: IEEE Conference on Computer Vision and Pattern Recognition (CVPR), pp. 5562–5570(2016)

8. Bhattacharya, A., Choudhury, D., Dey, D.: Emotion recognition from facial image analysis using composite similarity measure aided bidimensional empirical mode decomposition. In: IEEE First International Conference on Control, Measurement and Instrumentation (CMI), pp. 336–340 (2016)

9. Zhang, Y.-D., Yang, Z.-J., H-M, L., Zhou, X.-X., Phillips, P., Liu, Q.-M., Wang, S.-H.: Facial emotion recognition based on biorthogonal wavelet entropy, fuzzy support vector machine, and stratified cross validation. Spec. Sect. Emotion-Aware Mob. Comput. IEEE Access **4**, 8375–8385 (2016)

10. Joseph, A., Ramamoort, P.: Facial expression recognition using SVC classification & INGI method. Int. J. Sci. Res. **2**(4), 296–300 (2013)

11. Rathod, P., Gagnani, L., Patel, K.: Facial expression recognition: issues and challenges. Int. J. Enhanced Res. Sci. Technol. Eng. **3**(2), 108–111 (2014)

12. Nayak, J.S., Preeti, G., Vasta, M., Kadiri, M.R., Samiksha, S.: Facial expression recognition: a literature survey. Int. J. Comput. Trends Technol. **48**(1), 1–4 (2017)

A Phase Noise Correction Scheme for Multi-channel Multi-echo SWI Processing

Sreekanth Madhusoodhanan and Joseph Suresh Paul

Abstract In a GRE acquisition, effect of noise in each voxel can be minimized through the knowledge of prior information about the phase error, which can result in more robust estimation of susceptibility-related features. However, estimating the error in the given phase image of GRE acquisition is difficult. Moreover, the errors can significantly affect the resolution of the resulting phase mask at regions where the magnitude intensities are significantly low. Even though there are methods like non-linear least square fitting, smoothness introduced by these methods in the filtered phase reduces resolution and contrast-to-noise ratio (CNR). In order to maintain the resolution at voxel level, we propose a new filtering scheme where a series of local weights estimated from the probability density function of the phase measurements conditioned on the magnitude intensity and the noise variance. With an estimate of the local noise variance made available through a noise scan, the CNR improvement attained with the proposed noise correction scheme is compared using magnitude SWI images reconstructed at different acceleration factors.

Keywords Phase correction · Noise error · CNR · Spatial high-pass filter

1 Introduction

Susceptibility weighted imaging (SWI) [1, 2] in MRI uses the phase information in gradient echo (GRE) acquisition to improve the susceptibility-related contrast between different tissues. Phase images generated from GRE acquisitions consist of local field variation from tissues with different susceptibility and a background field due to inhomogeneities in the main field [3, 4]. In SWI, increasing the contrast between tissues with different susceptibility requires local field information after suppression of the background field [5]. This necessitates the requirement of

S. Madhusoodhanan (✉) · J. S. Paul
Indian Institute of Information Technology and Management-Kerala, Trivandrum, Kerala, India
e-mail: sreekanth.m@iiitmk.ac.in

J. S. Paul
e-mail: j.paul@iiitmk.ac.in

© Springer Nature Singapore Pte Ltd. 2020
S. Agarwal et al. (eds.), *Machine Intelligence and Signal Processing*,
Advances in Intelligent Systems and Computing 1085,
https://doi.org/10.1007/978-981-15-1366-4_6

a high resolution acquisition. However, this high resolution acquisition leads to low signal-to-noise ratio (SNR). Therefore, a filtering method capable of producing phase images with higher SNR and contrast between structures is highly desirable. In SWI processing, the noise error in the phase reduces the structural contrast in the magnitude SWI because of the application of high-pass (HP) filtering.

Most of the suggested SWI denoising algorithms [6–8] have the risk of losing the fine details due to over-smoothing. An alternate approach for GRE signal is that the reconstructed phase images have to be corrected with an estimate of the phase noise [9–11] using an error bound. However, this approach requires the removal of phase wraps which occur due to phase excursions when the signal component and phase contribution of the additive noise exceed $[-\pi, \pi)$ range at large echo times [12]. With the spatial filtering, the effect of this noise error can be corrected voxel-wise using a weight function estimated from the probability density function that describes the error in the local phase conditioned on the signal intensity and variance. In order to exploit the advantage of the new filtering HP filtering scheme, it will be required to correct the local noise errors introduced in the filtering process due to the neighborhood phase differencing operation. In this work, a weight is calculated for every voxel to reduce the effect of noise error propagation into the filtered phase. A region with large phase differences may have low intensities on the magnitude images and may not actually represent the true structural information in the phase. Consequently, the weights are dependent not only on the phase but also on the probability density function describing the dependence of signal magnitude and variance on the phase. Here, these weights associate venous structural edges and neighboring phase differences. These weights are applied on the neighboring phase difference during spatial HP filtering to suppress the noise components in the phase mask without deteriorating the structural information. With accelerated scans, the noise correction weights are shown to improve the CNR and overcome the effects of noise amplification due to decreased data sampling and k-space filling.

2   Theory

2.1   SWI Post-processing

With s_j^k denoting the GRE signal received in the j'th coil at echo time TE_k,

$$s_j^k(x, y) = \rho_j^k(x, y)\exp\left(i2\pi \Delta B(x, y)\text{TE}_k + v_j\right) + N_j^k(x, y). \tag{1}$$

Here, $\Delta B(x, y)$ is the local field variation, $\rho_j^k(x, y)$ is the spin density measured from the k'th echo, v_j is the phase angle contributed by the coil sensitivity, and $N_j(x, y)$ is the complex zero-mean Gaussian noise with variance σ_j^2. From the above GRE signal model, the computed phase angle can be written as

$$\psi_j^k = 2\pi \Delta B (\text{TE}_0 + k \Delta \text{TE}) + \nu_j + \Omega_j^k + 2\pi r_j^k, \tag{2}$$

where Ω^k is the contribution of additive noise term to the phase image, which is the function of echo time, and r_j^k is the phase wrapping term which folds the resultant phase back in the range $[-\pi, \pi)$ when the sum of the first two term in Eq. (2) falls outside the range $[-\pi, \pi)$.

To obtain the local field information from the raw phase image, phase wraps and background field are to be removed. This is usually performed using a homodyne high-pass (HP) filter [2, 13] that imposes a rigid constraint on the cut-off frequency without considering the shape and orientation of the object that determines the magnetization [14]. A mapping function is applied on the HP-filtered phase to generate a phase mask W in the range $[0, 1]$. For example in a right-handed system, a linear mask can be designed as

$$W(x, y) = \begin{cases} [\pi + \varphi_H(x, y)]/\pi & \text{for } -\pi < \varphi_H(x, y) < 0 \\ 1 & \text{otherwise}, \end{cases} \tag{3}$$

where $\varphi_H(x, y)$ is the HP-filtered phase at (x, y) [15]. On repeatedly multiplying the phase mask with the original magnitude image, the contrast-enhanced magnitude image M' is obtained.

$$M'(x, y) = W^m(x, y) M(x, y), \tag{4}$$

where M is the magnitude image, M' is the magnitude SWI, and m is the contrast parameter. Although the number of phase mask multiplications increases the contrast, the effect of noise limits the number of multiplication.

2.2 Phase Correction and Spatial Filtering

Depending on the orientation of venous structures relative to the main field, dipolar modulation of the main field may cause phase changes within and outside the venous structure. The phase change at vessel boundaries is corrected with the help of a predetermined weighting that can be applied to the phase difference of neighboring pixels. These weights are estimated using a non-linear function, such that the magnitude of the weights approaches to unity for large phase differences and tends to zero as the phase differences approach to zero. Although in principle, this helps to enhance faint structures by lowering the weights and the likelihood of noisy edges being enhanced is also high.

More specifically, a region with large phase differences may have low intensities on the magnitude images and may not be actually represent the true structural information in the phase. Therefore, the weights for HP filter are modulated with the change in phase and the probability density function describing the dependence

of signal magnitude and variance on the phase [16]. In an MR acquisition, based on
the prior information about the noise variance and signal intensity, probability of the
phase to take a value in the interval $[-\pi, \pi)$ can be estimated as [17]

$$
P(\varphi_j|M_j, \sigma_j) = \frac{\exp(-\alpha^2)}{2\pi} \times \left\{1 + \left[\sqrt{\pi}\alpha \cos(\varphi_j - \bar{\varphi}_j) \exp(\alpha^2 \cos^2(\varphi_j - \bar{\varphi}_j))\right]\right.
$$
$$
\left. \left[1 + \mathrm{erf}(\alpha \cos(\varphi_j - \bar{\varphi}_j))\right]\right\},
\tag{5}
$$

where $\alpha = M_j/\sqrt{2}\sigma$, M_j is the sinal intensity, and $\bar{\varphi}$ is the mean of the phase
image. Let φ_p be the phase at the periphery of the structures like veins and φ_c be the
phase at the center of the vein. By estimating $P(\varphi_p|M, \sigma)$ and $P(\varphi_c|M, \sigma)$ for a pair
of observations, the reliability of relating a phase difference which is comparatively
larger in magnitude with an edge of a venous object is determined. On the assumption
that the signal measurements at each voxel are statistically independent, then $w_c w_p$
be the influence of noise on the joint observation on a pair of neighboring phase
values.

The above probabilistic measures for each channel images have to be computed
separately since the noise model is satisfied only in the channel level. Taking into
consideration of the noise variance across channels, the weights for channel combined
measure at each voxel can be represented as

$$
w = \left(\left[w_1, w_2, \ldots w_{nc}\right]\begin{bmatrix} \sigma_1^2 & \cdots & 0 \\ \vdots & \ddots & \vdots \\ 0 & \cdots & \sigma_{nc}^2 \end{bmatrix}^{-1} \left[w_1, w_2, \ldots w_{nc}\right]^{\mathrm{T}}\right)^{\frac{1}{2}},
\tag{6}
$$

where n_C is the number of channel used for signal acquisition. At location (x, y),
the reliability index across the i' th neighborhood direction can be defined as [18]

$$
W_{\mathrm{Ri}}(x, y) = 1 - \frac{w(x+l, y+q)w(x, y)}{w_{\max}},
\tag{7}
$$

where Ri is the index for the i' th neighborhood direction at location pair
$((x+l, y+q), (x, y))$ and $W_{\max} = \{w(x+l, y+q)w(x, y): 1 \le x \le M, 1 \le y \le N - (N_r - 1)/2 \le (l, q) \le (N_r - 1)/2\}$. Application of these reliability indices for
in the phase filtering is given by

$$
\varphi_{\mathrm{HP}} = \sum_i W_{\mathrm{Ri}} W_{\mathrm{Fi}} \Delta \varphi_i,
\tag{8}
$$

where W_{Fi} denotes filter weights derived using the non-linear functions.

3 Method and Materials

3.1 Data Acquisition and Processing

SWI data were acquired on a GE Discovery MR 750W 3.0 T scanner equipped with 12 channel head array coil using multi-echo 3D GRE sequence. The other scan parameters were as: echo time (TE) = 24 ms; repetition time (TR) = 35 ms; flip angle (FA) = 15; readout bandwidth of 325.52 Hz/pixel; field of view (FOV) = 168 × 240 × 135 mm^3, with a matrix size of 384 × 288 × 56 voxels and slice thickness of 2.4 mm. All the subjects were scanned after obtaining informed written consent.

In the case of parallel imaging, GRAPPA [19] and SPIRiT [20] were used for uniform and non-uniform image reconstructions, respectively.

3.2 SWI Processing Pipeline

In the proposed SWI processing, the phase mask used to modify the contrast of the magnitude image is estimated from an HP-filtered phase, obtained through a spatial operation applied to the phase after background removal. In the spatial HP filtering operation, neighboring phase difference across all directions is multiplied with the filter weights and the weights for correcting the phase, prior to summation across all directions. The filter weights are determined using a non-linear mapping function of the neighborhood phase differences $\Delta\varphi_i$.

In order to obtain the phase correction weights, an estimate of noise level σ_j in each channel is required. In its original procedure, a noise scan is to be performed separately to account the non-stationarities caused due to pMRI reconstruction. Otherwise, an estimate of σ_j can be obtained by measuring the signal outside the brain region in the real and imaginary components of the complex SWI signal and takes the standard deviation of these intensities. Using these measurements, $P(\varphi_j|M_j, \sigma_j)$ is computed using Eq. (5) for each image location. To maintain the conditioning on σ_j, the weights w_j for each channel are computed by multiplying $P(\varphi_j|M_j, \sigma_j)$ by $P(M_j)$. The $P(M_j)$ can be computed from the normalized histogram M_j, where M_j is the channel magnitude. After combining these channel-wise weights w_j, the phase correction weights for weighting the neighboring phase difference can be estimated from normalized measure of product of pair of neighboring voxels in w. The proposed SWI processing pipeline is shown in Fig. 1.

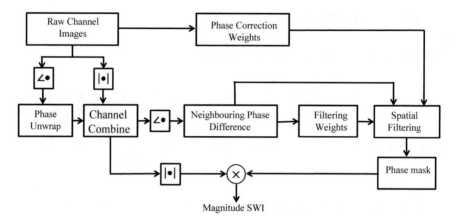

Fig. 1 Schematic block diagram representing the processing flow of incorporating phase correction scheme in multi-channel SWI processing using the spatial HP filter

4 Results

4.1 Application of Phase Correction Weights

Figure 2 shows sample mIP images for volunteer data with uniform sampling. Column-wise panels show mIP images obtained from fully sampled, uniformly under sampled with acceleration factor $R = 2$ and $R = 3$, respectively. Phase images of adjacent slices are HP-filtered with and without phase correction to obtain phase mask. mIP images obtained without phase correction (top row) appear more noisy compared to the images obtained with phase correction (bottom two rows)while filtering. The performance of noise correction weights improves if the weights are estimated from noise scan data than from the noise level estimated from the region outside the brain region of complex signal. This is clearly depicted in bottom two panels. With higher acceleration, the effects of noise become more critical in mIP images without phase correction. Application of phase correction weights in the SWI filtering process reduces the noise amplification and improves the visibility of structural information.

Figure 3 illustrates the noise reduction capability of phase correction weights in SWI processing applied to non-uniformly sampled data. Row-wise panels show the reconstructed mIP images from input data, sampled with random (top) and radial (bottom) sampling masks. In both cases, the input data is retrospectively under sampled and reconstruction is performed with only 33% samples. It can be visually inferred that the effect of noise in mIP images without phase correction is higher (first column). The visibility of structural information is improved in mIP images generated with phase correction. Similar to the observation in uniform sampling, the use of noise scan data estimates phase correction weights better than using noise

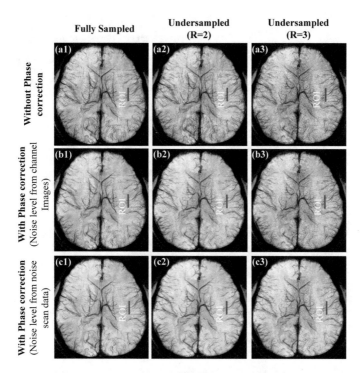

Fig. 2 Row-wise panels show mIP SWI images obtained (1) without phase error correction, (2) phase error correction with noise level estimated from complex channel images, and (3) phase error correction with noise scan data. All mIP images are generated from SWI data over eight slices each of thickness 2.4 mm. Column-wise panels show mIP images obtained from fully sampled data, uniformly undersampled data with $R = 2$ and $R = 3$, respectively. Red lines are used to indicate the ROI used for *VB_CNR* analysis

level estimated from the complex image and reduces the effect of noise in the mIP images.

4.2 Influence on CNR

As a quantitative measure of reduction in noise in the mIP images, vein-based CNR (VB_CNR) analysis is performed as in Eq. 9 [8].

$$VB\_CNR = \frac{S_{\text{OUT}} - S_{\text{IN}}}{\delta},\qquad(9)$$

where S_{IN} and S_{OUT} represent the mean value of pixel representing the venous structures (black square) and its surrounding, as shown in Fig. 4. δ be the standard deviation of a homogeneous region which does not contain any venous information. ROI used

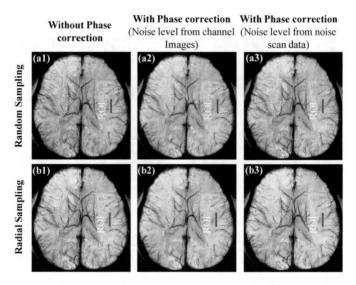

Fig. 3 Left-to-right panels show mIP SWI images obtained (1) without phase error correction, (2) with phase error correction and noise level estimated from complex channel images, and (3) with phase error correction and noise level estimated from noise scan data. Top and bottom rows show mIP images generated after sampling with random and radial sampling masks

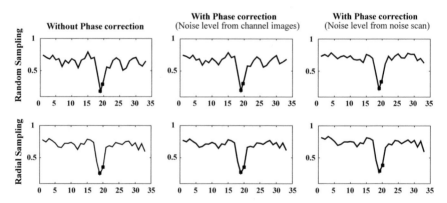

Fig. 4 *VB_CNR* analysis of mIP SWI reconstructed from undersampled data. Top (random sampling) and bottom (radial sampling) rows show voxel intensities in the ROI (indicated by red lines in Fig. 3)

for *VB_CNR* analysis is indicated by red lines in the mIP images shown in Figs. 2 and 3. The in-plane profiles of voxel intensities across the venous structures in ROI are plotted in Fig. 4. Panel arrangements are in the same order as in Fig. 3. The in-plane profiles of voxel intensities and CNR values clearly indicate the reduction in noise level achieved with phase correction scheme in SWI processing without disturbing the structural contrast.

Table 1 CNR values of mIP SWI images with and without phase correction for different acceleration factors and sampling masks

Methods		Without phase correction	With phase correction	Phase correction using noise scan
Fully sampled		6.151	7.021	8.754
Cartesian under sampled (uniform)	$R = 2$	4.872	5.918	8.638
	$R = 3$	4.301	5.183	7.284
Random sample [33% sampling]		5.500	6.588	9.472
Radial sample [33% sampling]		6.580	7.603	8.153

Table 1 shows the CNR value computed from mIP images with and without phase correction using different sampling patterns. In all the cases, CNR of mIP images computed from phase correction using noise scan data shows high value. This clearly indicates that the performance of phase correction approach in SWI processing can be improved with accurate estimation of noise level. The CNR values obtained from the reconstructed mIP images generated using noise scan data show an average CNR improvement of 57.01% over the mIP images without phase correction and 18.90% over the phase correction scheme with noise estimated from complex images.

4.3 Application to Clinical Data

Top and bottom rows in Fig. 5 illustrate the application of the phase error correction scheme in tumor and stroke cases. In both cases, the proposed method shows superiority in reducing noise and enhances the structural information. Red arrows are used to indicate the new structural information seen in the mIP obtained using the phase noise correction scheme.

5 Discussion

The macroscopic field present in the raw phase image masks the structural information present in the GRE phase images. Removal of the low spatial frequency will give the local susceptibility field which can be used to create a phase mask for improving the contrast of the magnitude image. In particular, SWI processing produces high-pass-filtered phase images in which the paramagnetic and diamagnetic regions exhibit phase values with opposite sign. But the contrast enhancement in SWI magnitude image is determined by the number of phase mask multiplication or the contrast parameter m. Usually a value of m is limited to four to optimize the CNR. In our proposed SWI processing, a choice of a higher value of the contrast

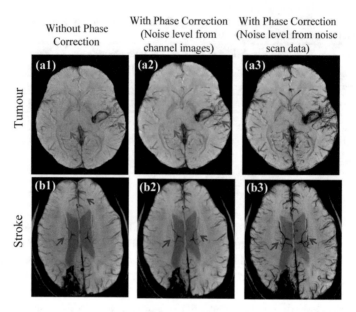

Fig. 5 Left-to-right panels show mIP SWI images obtained (1) without phase error correction, (2) with phase error correction and noise level estimated from complex channel images, and (3) with phase error correction and noise level estimated from noise scan data. Top and bottom rows show mIP images generated using clinical dataset indicating tumor and stroke

parameter is enabled by introduction of noise correction weights into the filtering process.

In the SWI processing discussed in this article, before performing any spatial operation, a weight is calculated for each voxel based on its magnitude and noise variance. This weight is used to discount the effects of low-signal pixels and increase the input of high-signal pixels for every neighborhood in the filtering process. Since the noise correction method requires weights that are localized to small neighborhoods of pixels, we observed that when small high-signal features were surrounded by large low-signal areas, the enhancement produced by the correction procedure often did not represent the venous objects effectively. However, by weighting each voxel's contribution individually, we were able to produce better results in these cases. As a drawback, incorporation of weights to correct the noise in the phase in the spatial filter necessitates the requirement for unwrapping the phase in each channel separately. Despite the computational complexity, the channel-by-channel phase processing is already reported to be advantageous for SWI in terms of eliminating the phase singularities, known as cusp artefacts. Moreover, the computational complexity of this processing increases as the number of channel increases. The second major disadvantage is the errors introduced in the unwrapping process at higher echo times. Nevertheless in pMRI with accelerated acquisition, these demerits are counterbalanced by the increased CNR brought forth by application of the noise correction scheme.

6 Conclusion

In conclusion, a novel algorithm for phase error correction in GRE images which leads to CNR improvement in the magnitude SWI was proposed. Phase error correction scheme using the magnitude and noise variance improves the image quality by removing the noise error without over-smoothing the fine details in the structural information. In the case of undersampled reconstruction, phase correction scheme controls the noise error amplification while generating the mIP images using magnitude SWI. More importantly, it is shown that the use of a separate noise scan can further improve the CNR performance of the phase error correction scheme. This is evaluated with different acceleration factors and sampling patterns

Acknowledgements The authors are thankful to the Council of Scientific and Industrial Research-Senior Research Fellowship (CSIR-SRF, File No: 09/1208(0001)/2018.EMR-I) and planning board of Government of Kerala (GO(Rt)No.101/2017/ITD.GOK(02/05/2017)), for financial assistance.

References

1. Reichenbach, J.R., Venkatesan, R., Schillinger, D.J., Kido, D.K., Haacke, E.M.: Small vessels in the human brain: MR venography with deoxyhemoglobin as an intrinsic contrast agent. Radiology **204**(1), 272–277 (1997)
2. Haacke, E.M., Xu, Y., Cheng, Y.C.N., Reichenbach, J.R.: Susceptibility weighted imaging (SWI). Magn. Reson. Med.: Off. J. Int. Soc. Magn. Reson. Med. **52**(3), 612–618 (2004)
3. Wang, Y., Yu, Y., Li, D., Bae, K.T., Brown, J.J., Lin, W., Haacke, E.M.: Artery and vein separation using susceptibility-dependent phase in contrast-enhanced MRA. J. Magn. Reson. Imaging **12**(5), 661–670 (2000)
4. Haacke, E.M., Mittal, S., Wu, Z., Neelavalli, J., Cheng, Y.C.: Susceptibility-weighted imaging: technical aspects and clinical applications, part 1. Am. J. Neuroradiol. **30**(1), 19–30 (2009)
5. Rauscher, A., Sedlacik, J., Barth, M., Mentzel, H.J., Reichenbach, J.R.: Magnetic susceptibility-weighted MR phase imaging of the human brain. Am. J. Neuroradiol. **26**(4), 736–742 (2005)
6. Borrelli, P., Tedeschi, E., Cocozza, S., Russo, C., Salvatore, M., Palma, G., Comerci, M., Alfano, B., Haacke, E.M.: Improving SNR in susceptibility weighted imaging by a NLM-based denoising scheme. In: 2014 IEEE International Conference on Imaging Systems and Techniques (IST), pp. 346–350 (2014)
7. Borrelli, P., Palma, G., Tedeschi, E., Cocozza, S., Comerci, M., Alfano, B., Haacke, E.M., Salvatore, M.: Improving signal-to-noise ratio in susceptibility weighted imaging: a novel multicomponent non-local approach. PLoS ONE **10**(6), e0126835 (2015)
8. Borrelli, P.: Denoising in magnetic resonance imaging: theory, algorithms and applications. Ph.D. thesis, UniversitàdegliStudi di Napoli Federico II
9. Ahn, C.B., Cho, Z.H.: A new phase correction method in NMR imaging based on autocorrelation and histogram analysis. IEEE Trans. Med. Imaging **6**(1), 32–36 (1987)
10. Bernstein, M.A., Thomasson, D.M., Perman, W.H.: Improved detectability in low signal-to-noise ratio magnetic resonance images by means of a phase-corrected real reconstruction. Med. Phys. **16**(5), 813–817 (1989)
11. Noll, D.C., Nishimura, D.G., Macovski, A.: Homodyne detection in magnetic resonance imaging. IEEE Trans. Med. Imaging **10**(2), 154–163 (1991)
12. Dagher, J., Reese, T., Bilgin, A.: High-resolution, large dynamic range field map estimation. Magn. Reson. Med. **71**(1), 105–117 (2014)

13. Rauscher, A., Barth, M., Herrmann, K.H., Witoszynskyj, S., Deistung, A., Reichenbach, J.R.:
    Improved elimination of phase effects from background field inhomogeneities for susceptibility
    weighted imaging at high magnetic field strengths. Magn. Reson. Med. **26**(8), 1145–1151
    (2008)
14. Ogg, R.J., Langston, J.W., Haacke, E.M., Steen, R.G., Taylor, J.S.: The correlation between
    phase shifts in gradient-echo MR images and regional brain iron concentration. Magn. Reson.
    Imaging **17**(8), 1141–1148 (1999)
15. Robinson, S.D., Bredies, K., Khabipova, D., Dymerska, B., Marques, J.P., Schweser, F.: An
    illustrated comparison of processing methods for MR phase imaging and QSM: combining
    array coil signals and phase unwrapping. NMR Biomed. **30**(4), e3601 (2017)
16. Chung, A.C., Noble, J.A., Summers, P.: Fusing speed and phase information for vascular
    segmentation of phase contrast MR angiograms. Med. Image Anal. **6**(2), 109–128 (2002)
17. Aja-Fernández, S., Tristán-Vega, A.: A review on statistical noise models for magnetic reso-
    nance imaging. Technical Report. LPI, ETSI Telecommunication, Universidad de Valladolid,
    Spain (2013)
18. Madhusoodhanan, S., Kesavadas, C., Paul, J.S.: SWI processing using a local phase difference
    modulated venous enhancement filter with noise compensation. Mag. Reson. Imaging **59**,
    17–30 (2019)
19. Griswold, M.A., Jakob, P.M., Heidemann, R.M., Nittka, M., Jellus, V., Wang, J., Haase, A.:
    Generalized autocalibrating partially parallel acquisitions (GRAPPA). Mag. Res. Med.: Off. J.
    Int. Soc. Magn. Res. Med. **47**(6), 1202–1210 (2002)
20. Lustig, M., Pauly, J.M.: SPIRiT: iterative self-consistent parallel imaging reconstruction from
    arbitrary k-space. Magn. Reson. Med. **64**(2), 457–471 (2010)

A Scanning Technique Based on Selective Neighbour Channels in 802.11 Wi-Fi Networks

Abhishek Majumder, Samir Nath and Sudipta Roy

Abstract In a network which is IEEE 802.11 based, handoff is a challenging issue. When mobility rate of the mobile station is very high, it becomes more challenging. If it takes too much time, Quality of Service (QoS) provided to the users will decrease. Handoff consists of three parts, namely scanning, authentication and association. Out of these, the scanning process consumes maximum time. It is very necessary to reduce the time required for scanning. In this paper, a selective scanning-based technique has been proposed to increase the efficiency of the network. The host Access Point (AP) assists the mobile station to select the channel which is to be scanned. The proposed scheme has been simulated in NS-3. From the performance comparison, it has been observed that the proposed scheme performs better than that of IEEE 802.11g with respect to mean delay, packet loss ratio, jitter and throughput.

Keywords Wireless LAN · Access point · Quality of service · Handoff · Scanning

1 Introduction

With increasing popularity of mobile services, mobility management in 802.11 wireless network is a critical issue. In present days, user of the network wants uninterrupted network services. When a mobile node is moving at high speed in the network, Quality of Service (QoS) also gets affected. Due to high mobility of mobile client more handoffs are faced. Quality of Service of the network can be improved if handoff time of the mobile node can be reduced. Handoff consists of three parts: scanning, authentication and re-association. Out of these three parts, scanning process takes the maximum time [1, 2]. Scanning is of two types, namely active scanning and passive scanning. In passive scanning, the mobile station (STA) waits for beacon message

A. Majumder (✉) · S. Nath
Department of Computer Science & Engineering, Tripura University,
Suryamaninagar, Agartala, India
e-mail: abhi2012@gmail.com

S. Roy
Department of Computer Science & Engineering, Assam University, Silchar, India

© Springer Nature Singapore Pte Ltd. 2020
S. Agarwal et al. (eds.), *Machine Intelligence and Signal Processing*,
Advances in Intelligent Systems and Computing 1085,
https://doi.org/10.1007/978-981-15-1366-4_7

from the AP. In active scanning, probe request message is sent by mobile station to AP. In response, the AP sends back probe response message. This paper mainly focuses on active scanning process.

In IEEE 802.11g [3], the mobile station needs to scan all the channels. This causes higher delay in scanning process. In this paper, a new scanning technique based on selective neighbour channels has been proposed to minimize the scanning delay and enhance the performance. Mobile node does not need to scan all the available channels specified in 802.11g network. It needs to scan the channels which are being used by the neighbour APs. The scheme has been designed for mobile nodes having single radio.

Paper is organized as follows. Second section discusses some of existing schemes. In Sect. 3, proposed scheme is presented. Proposed scheme is numerically analysed in Sect. 4. In Sect. 5, simulation and performance comparison has been presented. In sixth section, paper has been concluded and future work has been discussed.

2   Related Work

In this section, some existing techniques for performing handoff have been presented.

Handoff techniques in IEEE 802.11 can be classified in two ways, namely based on the phase in which handoff process works and based on the radios' number [4]. Based on radios' number there are two classes of techniques, namely single radio based and multi-radio based. On the other hand, depending on the phase where it operates, the techniques are categorized into three types, namely re-association-based handoff, authentication-based handoff and scanning-based handoff. The paper mainly focuses on single radio-based scheme that works in scanning process.

In handoff, scanning takes maximum time. IEEE 802.11g has 14 channels [5]. When IEEE 802.11g is used, during scanning the mobile node switches from one channel to another and extracts signal strength of the Access Points (APs) working on that channel. Thus, it switches to all the 14 channels. Therefore, significant amount of time is incurred when IEEE 802.11g is used.

Practical Schemes for Smooth MAC Layer Handoff (PSMH) [6] uses two mechanisms, namely greedy smooth handoff mechanism and smooth MAC layer handoff mechanism. In this scheme, groups of channels are formed. In smooth MAC layer handoff mechanism, the mobile station is switched to one of the channel groups and every channel of that group is scanned. Then, it switches back to its original channel and performs normal operation. After some time mobile station again switches to another group of channels and this process continues. In greedy smooth handoff, every channel of nearby group is scanned by the mobile. If an AP having better signal strength is found, the mobile station joins the AP.

Novel MAC Layer Handoff (NMLH) [7] uses two techniques, namely fast handoff avoiding probe wait and adaptive preemptive fast handoff. In fast handoff avoiding probe wait mechanism, the mobile station switches to all the channels one after another and sends probe request message. All the APs in the range of mobile STA's previous AP send probe response to previous AP. Mobile station then switches back to previous AP's channel for receiving all probe response messages in its range. In adaptive preemptive fast handoff mechanism, preemptive AP discovery is done. After that, it enters into re-authentication phase.

In Multimedia Ready Handoff Scheme (MRH) [8], there are two parts in handoff process, namely scanning process and handoff process. In handoff process, the mobile station sends power save mode (PS) frame to the AP it is connected with. Then, it switches to neighbouring channel and performs scanning process. After that, the mobile station switches back to the channel of current AP, the AP then sends the buffered packets the mobile station. This process continues periodically. When signal strength of current AP goes below threshold, mobile station gets associated with the AP having the best signal strength.

In Fast MAC Layer Handoff Protocol (FMH) [9], the neighbouring nodes help the mobile station to find the AP. The neighbouring nodes scan the available APs working in different channels. After scanning, the list of APs is forwarded to the mobile station. Mobile nodes cache the list and during handoff, and it gets associated with the best AP.

In behaviour-based Mobility Prediction for seamless handoffs [10], mobility behaviour of a mobile station is considered for selecting the next AP by the Characteristics Information Server (CIS). The CIS stores duration, time-of-day, group and location to predict the mobility behaviour of the mobile station.

3 Proposed Scheme

The section presents proposed technique.

3.1 Tables and Frame Formats

3.1.1 K-Neigh$_{Channls}$ Table

Each AP maintains a K-Neigh$_{Channls}$ table for recording the information of the channels used by neighbour APs. K-Neigh$_{Channls}$ table consists of two attributes, namely N-Chnnl_List and count. N-Chnnl_List contains channel numbers of neighbour APs. Count field stores the number of times STA or number of STAs that has switched to particular channel during handoff. An example of K-Neigh$_{Channls}$ table is shown in Table 1.

Table 1 K-Neigh$_{Channls}$ Table

Neighbour channel number (N-Chnnl-List)	Count (C)
Ch-1	8
Ch-6	5
Ch-11	4

Table 2 K-Neigh$_{APs}$ Table

Channel number	Id of AP	Signal-to-noise ratio (dB)
Ch-6	AP_a	250
Ch-6	AP_b	220
Ch-11	AP_c	200
Ch-1	AP_d	190

3.1.2 K-Neigh$_{APs}$ Table

K-Neigh$_{APs}$ table is maintained by mobile stations. It contains information of every active AP functioning in various channels. This table consists of three fields, namely channel number, Id of an AP and signal-to-noise ratio (SNR) value. Channel number field contains the channel number of the corresponding neighbouring AP. Id of an AP field stores identity of the neighbouring AP. SNR field stores SNR value of the corresponding neighbouring AP. An example of K-Neigh$_{APs}$ table is shown in Table 2.

3.1.3 Association Response and Re-association Response Frame

In IEEE 802.11 standard association response and re-association response frame consists of many fields like status code, capability info, seq-ctl, Basis Service Set ID (BSSID), Source Address (SA), Destination Address (DA), duration, frame control, association ID and supported rates. In addition to that in this scheme, one new field namely channel_list has been added. This field contains the entire K-Neigh$_{channls}$ table of the host AP. Figure 1 shows the association response and re-association response frame format.

3.1.4 Disassociation Frame

In IEEE 802.11 standard, disassociation frame consists of many fields, namely seq-ctl, BSSID, SA, DA, duration, frame control, reason code and Frame Check Sequence (FCS). In addition to that in the proposed scheme, one new field namely channel number has been added. This field contains the channel number of the new AP in which the mobile station has been associated. Figure 2 shows the disassociation frame format.

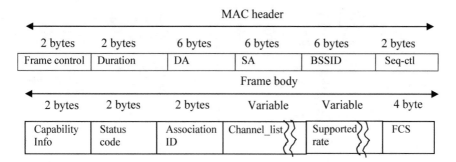

Fig. 1 Association response and re-association response frame format

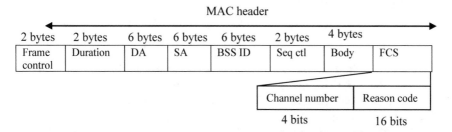

Fig. 2 Disassociation frame

3.2 Selective Neighbour Channel Scanning Procedure

In the proposed scheme, the STA can scan the available neighbour channels periodically one after another. This periodic scanning process continues until and unless current Access Point's signal goes below some predefined handoff threshold. The steps of the proposed scanning process are shown in Fig. 3. When a mobile station or STA registers into new Access Point, it gathers the neighbour channel list (Chnnl_list) from association or re-association response message and stores the list. Chnnl_list contains K-Neigh$_{Channls}$ table of the AP which is originator of that association or re-association response message. (1) After getting the Chnnl_list, STA sends a null frame with power management bit = 1 to the current AP. (2) After receiving null frame, the AP begins buffering of data or messages destined towards STA. STA switches to the neighbour channel which has highest count value. (3) & (4) After switching to the chosen channel, STA broadcasts probe request message towards APs on chosen channel and gathers information of each AP by probe response messages. (5) On getting the probe responses of working APs in visiting channel, the STA switches back to the home channel and (6) updates its K-Neigh$_{APs}$ Table (7). Then, the STA transmits null frame having power management bit set to 0 towards its current AP to forward buffered data or message. (8) Current AP forwards the buffered data and STA performs the normal data communication function for T time units.

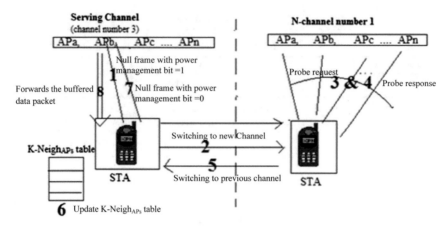

Fig. 3 Periodic scanning procedure

After T time unit, the STA switches to the next channel present in N-Chnnl_List which has the second highest counter value and performs step 1–8 and so on.

Using this procedure STA periodically scans all the neighbour channels stored in Chnnl_List one by one and updates its K-Neigh$_{APs}$ table.

When STA registers into the new AP, it executes this scanning algorithm and discovers the available APs in different neighbour channels. After every T time unit, STA switches to the nearby channel and collects APs' information. Before starting the scanning operation, STA collects the nearby channels' information, i.e. 'Chnnl_list' from association response or re-association response message transmitted by current AP. STA sends a null frame with power management bit = 1 to current Access Point. On receiving null frame with power management bit = 1, the current AP temporarily stores the data or message whose destination is STA. Now STA finds the channel from Chnnl_list whose value of count is the highest. After switching to the channel, STA broadcasts probe request (P_{Req}) message to APs. On receiving P_{Req} message all the APs currently active on the channel reply with a probe response (P_{Res}) message to the STA. STA gathers the information, i.e. which APs are currently operating on that channel and selects only APs whose SNR value greater than the predefined threshold value. Then, the selected APs' information is stored in K-Neigh$_{APs}$ table. STA switches back to home channel. Then, it transmits a null frame having power save mode set to 0 to currently serving AP. AP then forwards all the buffered data packets to the STA. Now STA performs normal data communication for T time unit. After T time unit, STA chooses the channel having next highest count value in Chnnl_list and does the same procedure. This process continues till current AP's SNR value goes below a predefined threshold. If content of Chnnl_list is null, STA performs full scanning. Most of the time, STA gets suitable AP from Chnnl_list. The algorithm for scanning process is shown in Fig. 4.

In this scanning algorithm, STA does not need to scan all the channels of IEEE 802.11g. Only selected nearest neighbour APs' channels are scanned. This scanning

---

N = Number of channels in Channel_list.
Ch$_i$ = ith channel in the Chnnl_list.
C$_i$ = count value for number of STAs moving to AP in channel Ch$_i$.
Thrs$_H$ = Received signal strength threshold for inclusion of an AP in K-Neigh$_{APs}$ table.
Power management bit = 1; set by STA, for leaving the current channel and informing the current AP to buffer the data whose destination is STA.
Power management bit = 0; set by STA, while sending null frame to current AP to forward the buffered data packets after STA switches back to current APs channel.

1.   Sort the entries of Chnnl_list in descending order of count value.
2.   For i = 1 to N
        2.1.   STA sends a null frame with Power management bit = 1 to the current AP;
        2.2.   STA switches to the Ch$_i$.
        2.3.   Scan available APs in Ch$_i$
                2.3.1.   If (SNR(APs)> Thrs$_H$)
                        2.3.1.1.   Store AP's information in K-Neigh$_{APs}$ table
                2.3.2.   End if
        2.3.   Switch back to the current AP's channel
        2.4.   Send null frame having Power management bit = 0 towards current AP;
        2.5.   Do normal data Communication for T time unit.
3.   End for.

Fig. 4 Scanning algorithm

algorithm reduces the scanning time. Most of the cases neighbour APs serve good quality signal.

3.3 Handoff Process

Handoff process is pictorially presented in Fig. 5. When current AP's signal strength reaches to handoff threshold value, handoff process is initiated. (1) STA selects the best AP considering SNR value from its K-Neigh$_{APs}$ Table (2 & 3) Then, STA performs authentication and re-association operation with newly selected AP. (4) STA sends a disassociation frame which contains the new AP's channel number to the old AP. (5) Upon receiving disassociation frame, old AP updates its K-Neigh$_{Channls}$ table by incrementing the count value corresponding to the new AP's channel number.

Handoff process is performed from old Access Point (AP$_o$) to new Access Point (AP$_n$) when the SNR of AP$_o$ is less than or equal to the predefined handoff threshold value γ. β_{SNR} = Threshold SNR value for selecting new AP. The algorithm for handoff process is shown in Fig. 6.

When handoff is initiated, STA selects the best AP (AP$_n$) from K-Neigh$_{APs}$ table on the basis of SNR value. Through periodic scanning, STA stores nearby APs' information in K-Neigh$_{APs}$ table before current AP's SNR value goes below the handoff threshold limit. STA directly sends authentication request (authentication-req) frame

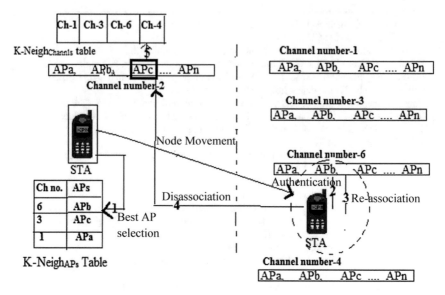

Fig. 5 Handoff process

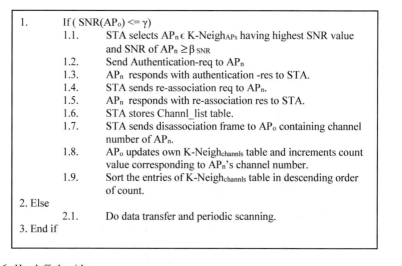

Fig. 6 Handoff algorithm

to AP_n. On receiving authentication-req, AP_n replies with authentication response to the STA. On completion of authentication process, STA sends re-association request towards AP_n. Like Inter-Access Point Protocol (IAPP) AP_n then sends security block to AP_o which sends back security block ack in response. Then, AP_n sends move notify message to AP_o. In response, AP_o sends back move response message to AP_n. Thus, the process is performed using IAPP messages. After that, the AP_n responds with positive re-association response message. With the re-association response message, the STA gets the K-Neigh$_{channls}$ table of the AP$_n$ through Channl_list and stores it. Now STA sends disassociation frame containing channel number of AP$_n$ to the AP$_o$. On receiving the disassociation frame from STA, AP$_o$ updates its K-Neigh$_{Channls}$ table.

4 Numerical Analysis

It is assumed that Access Point (AP) residence time of the STA follows exponential distribution with rate λ_s. Let, total number of APs in the WMN is M. Let, α be the average hop count between arbitrary AP and the mesh portal. Per hop communication latency is considered as γ. Let P_r is the probability that the mesh station broadcasts topology control message in its neighbourhood. Average number of packets sent and received by the STA per time unit is considered as λ_p.

When STA moves out of the coverage of its current AP and enters into another, new AP broadcasts routing information of the STA (through topology control message) using Optimizer Link State Routing (OLSR) protocol. In OLSR, only multipoint relay nodes diffuse control message. Therefore, per handoff cost is,

$$C_{\text{perhand off}} = P_r \times M \tag{1}$$

Handoff cost per time unit can be calculated as,

$$C_{\text{handoff}} = P_r \times M \times \lambda_s \tag{2}$$

The portal sends the packets to the destination STA using its routing table entry. Since the distance between the portal and an arbitrary AP is α, per-packet delivery cost can be calculated as,

$$C_{\text{per packet}} = \alpha \times \gamma \tag{3}$$

The per time unit packet delivery cost is,

$$C_{\text{packet}} = \alpha \times \gamma \times \lambda_p \tag{4}$$

5 Simulation Results

Using Network Simulator 3 (NS-3.25) [11], the simulation was performed. Topology used in simulation is shown in Fig. 7. The portal and root node is node 0. UDPEchoServer is installed in node 0. On the other hand, UDPEchoClient is installed in

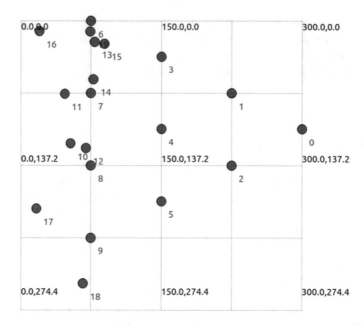

Fig. 7 Topology of the nodes in the scenario used

mobile clients, i.e. the STAs. For communication between the mesh stations, propagation model used is Range Propagation Loss model. For communication between the AP and the STAs, propagation model used is Long Distance Propagation Loss model with exponent 2.65. Optimized Link State Routing (OLSR) has been used to route the packets from STAs to the portal and from portal to STAs through the intermediate mesh STAs. STAs' movement follows Random Waypoint Mobility Model (RWMM). Table 3 shows the parameters used in simulation. The proposed technique has been compared with IEEE 802.11g standard with active scanning.

Figure 8 shows impact of nodes' number on mean delay. End to end delay of the proposed scheme is less than that of standard IEEE 802.11g. This is because; in the proposed scheme, the STA performs probing to selective channels. Thus, the STA

Table 3 Parameters used

Parameters	Value	Parameters	Value
Node speed	50 m/s	Pause time	1 s
Packet size	1024 bytes	Time to wait between packets	1 s
Number of mesh stations having Access Point capability	10	Simulation time	100 s
Distance between two neighbouring mesh stations	75 m	Radio range of mesh STA	100 m
Maximum number of packets the application will send			25

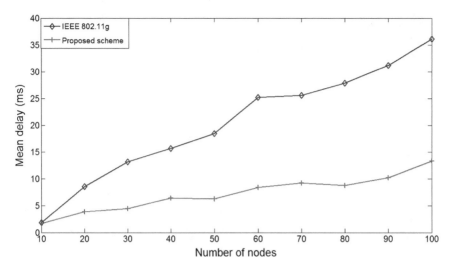

Fig. 8 Impact of nodes' number on mean delay

gets more time for data transmission and reception to and from the serving AP. On the other hand, in IEEE 802.11g probe requests are sent in all the channels periodically pausing normal data transmission and reception. So, a large amount of time is spent for probing of APs. As number of node increases mean delay of both the schemes increase. This is because of congestion caused by transmission of increased number of data and control packets by the STA.

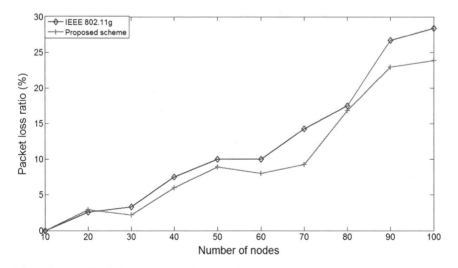

Fig. 9 Impact of nodes' number on packet loss ratio

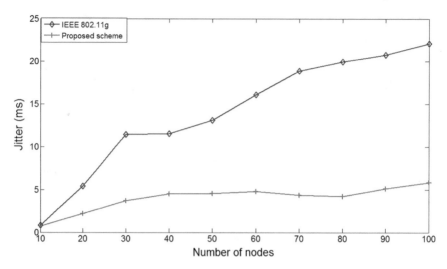

Fig. 10 Impact of nodes' number on jitter

Figure 9 shows impact of nodes' number on packet loss ratio. It can be observed that the packet loss ratio of proposed scheme is less than that of IEEE 802.11g. In the proposed scheme, the packets to and from the STA are queued for shorter time compared to IEEE 802.11g because in the proposed scheme the STA switches to one of the neighbouring AP's channel, performs probing and switches back to its serving AP's channel for normal data transfer. Whereas in IEEE 802.11g, the STA switches to all the channels one after another and performs probing. Then it switches back to serving AP's channel for data transfer. So, the data packets are queued for longer duration and queue gets overflowed. If number of node increases, packet loss ratio of both the schemes also increases. Congestion caused by the transmission of packets by increased number of nodes is the main reason behind this.

Figures 10 and 11 show the effect of number of nodes on jitter and throughput, respectively. It can be observed that the jitter of the proposed scheme is less than that of IEEE 802.11g. Higher throughput of the proposed scheme compared to IEEE 802.11g can also be observed. Due to selective scanning, the proposed scheme outperforms IEEE 802.11g with respect to jitter and throughput. Jitter of both the schemes increases with increase in number of nodes. This is because of increased amount of congestion caused by transmission of packets by increased number of STAs. On the other hand, throughput of both the schemes increases with increase in STAs' number. It happens because of increase in transmission of packets by increased number of STAs.

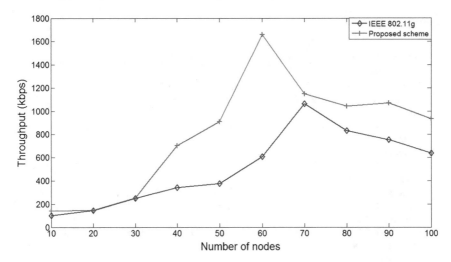

Fig. 11 Impact of nodes' number on throughput

6 Conclusion and Future Work

In this paper, a mobility management scheme for highly mobile clients or STA has been developed. The technique focuses on improvement of active scanning process in IEEE 802.11g. The proposed scheme has been simulated and compared with standard IEEE 802.11g. During simulation to reflect highly mobile nature of the STA, the STA speed has been considered as 50 m/s. Simulation results demonstrate that proposed scheme outperforms IEEE 802.11g standard with respect to mean delay, packet loss ratio, jitter and throughput. More extensive study of the proposed scheme in different network scenario remains as future work. Though the paper only focuses on scanning phase, work can be done in future to improve authentication and association phase. Enhancing the proposed scheme for multi-radio mobile STA also remains as future work.

Acknowledgements The work has been supported by Ministry of Electronics and Information Technology (MeitY), Government of India, Vide no. 14(8)/2014-CC&BT, Dated: 03.09.2015. Authors are also thankful to Mobile Computing Laboratory, Department of Computer Science & Engineering, Tripura University for providing necessary infrastructure.

References

1. Pei, C., et al.: Why it takes so long to connect to a WiFi access point. In: IEEE INFOCOM, pp. 1–9. IEEE (2017)
2. Murray, D., Dixon, M., Koziniec, T.: Scanning delays in 802.11 networks. In: International Conference on Next Generation Mobile Applications, Services and Technologies (NGMAST), pp. 255–260 (2007)
3. Singh, G., Atwal, A.P.S., Sohi, B.S.: Effect of background scan on performance of neighbouring channels in 802.11 networks. Int. J. Commun. Netw. Distrib. Syst. **1**(1), 19–32 (2008)
4. Majumder, A., Nath, S.: Classification of handoff schemes in a Wi-Fi-based network. In: Enabling Technologies and Architectures for Next-Generation Networking Capabilities, pp. 300–332. IGI Global (2019)
5. Enterprise Mobility 4.1 Design Guide, pp. 3-1–3-36. https://www.cisco.com/c/en/us/td/docs/solutions/Enterprise/Mobility/emob41dg/emob41dg-wrapper/ch3_WLAN.pdf. Last accessed 23 Mar 2019
6. Liao, Y., Gao, L.: Practical schemes for smooth MAC layer handoff in 802.11 wireless networks. In: International Symposium on World of Wireless, Mobile and Multimedia Networks, pp. 181–190 (2006)
7. Chintala, V.M., Zeng, Q.A.: Novel MAC layer handoff schemes for IEEE 802.11 wireless LANs. In: IEEE Wireless Communications and Networking Conference, pp. 4435–4440 (2007)
8. Singh, G., Singh, A.P., Sohi, B.S.: Multimedia ready handoff scheme for 802.11 networks. In: 4th International Conference on Mobile Technology, Applications, and Systems and the 1st International Symposium on Computer Human Interaction in Mobile Technology, pp. 230–237 (2007)
9. Zhang, Z., Richard, W.P., Boukerche, A.: A fast MAC layer handoff protocol for WiFi-based wireless networks. In: IEEE Local Computer Network Conference, pp. 684–690 (2010)
10. Wanalertlak, W., Lee, B., Yu, C., Kim, M., Park, S.M., Kim, W.T.: Behavior-based mobility prediction for seamless handoffs in mobile wireless networks. Wirel. Netw. **17**(3), 645–658 (2011)
11. Carneiro, G.: NS-3: Network Simulator 3. http://www.nsnam.org/tutorials/NS-3-LABMEETING-1.pdf. Last accessed 15 Mar 2019

Persistent Homology Techniques for Big Data and Machine Intelligence: A Survey

Milan Joshi, Dhanajay Joshi and Vivek Sharma

Abstract *Persistent homology* (PH) is a powerful burgeoning technique from Topological data analysis (TDA) that leverages machinery drawn from algebraic topology. PH records the appearance and disappearance of essential topological features of an object that persist across various scales or resolutions, and it is immune to noise. PH is independent of parameters, dimension and coordinates. In recent years, PH is catching an eye of the machine learning community due to the challenges offered by nature of data available today. But due to unapproachable introductory literature on PH, beginners who have the mathematical aptitude but no background in algebraic topology find it difficult to understand the concepts and techniques involved. On the other side, researchers are working effortlessly to bring together machine learning and TDA, as both combined can do wonders. This paper is an attempt to introduce the theory of PH step by step for the beginner and illustrate the concepts involved by using toy data sets. The main purpose of this research work is to introduce extensions of PH that helps researchers to apply tools from statistics and machine learning.

Keywords Machine learning · Big data · TDA · Algebraic topology · Homology · Betti numbers · Persistent homology · Point cloud · Simplicial complex

1 Introduction

Data is getting generated at an exponential rate. Size of data is not the problem. The real challenge is in the form of data. Data today is full of surprises. It is noisy, dirty,

M. Joshi (✉) · V. Sharma
Department of Applied Science and Humanities, Mukesh Patel School of Technology
Management and Engineering Shirpur, SVKM's NMIMS (Deemed to be University), Shirpur,
India
e-mail: mlnjsh@gmail.com

D. Joshi
Department of Computer Engineering, Mukesh Patel School of Technology Management and
Engineering Shirpur, SVKM's NMIMS (Deemed to be University), Shirpur, India

© Springer Nature Singapore Pte Ltd. 2020
S. Agarwal et al. (eds.), *Machine Intelligence and Signal Processing*,
Advances in Intelligent Systems and Computing 1085,
https://doi.org/10.1007/978-981-15-1366-4_8

heterogeneous, imponderable, unstructured and high-dimensional which endangers the modelling of artefacts. From the past several years, we have been analysing data that comes in the form of point clouds which seats inside higher-dimensional space and which carries some geometric structure. Various techniques from statistics and machine learning help to reveal underlying geometric structures of given data. Most of these techniques assume that the given data has a simple underlying geometry. The only topological information these techniques study is connectivity information which tells how many disjoint parts are there in objects. Moreover, these techniques require parameter estimation or sometimes coordinates, and more importantly they require a right question to ask to get an answer. But natural data offers very hard challenges, so we need strong mathematical, statistical and algorithmic methods which aim to infer and analyse relevant qualitative and quantitative topological structures directly from data. One way to approach these problems is to use methods from Topological data analysis (TDA). TDA is a relatively new area of research which revolves around the principle that data has shape and shape has meaning [1]. TDA has its root in topology. Topology is a very old branch of pure mathematics that studies shapes. See [1, 2] more detailed introduction on topology. PH is a technique from TDA that transforms a complex, high-dimensional data into collection of intervals (over the real line) called barcodes that characterizes connectivity information of data points or equivalently topological properties which includes things like the number of connected components, cycles or two-dimensional holes, three-dimensional holes and their counterparts in a higher dimension [3, 4]. These topological properties or features of a data are to be computed at various scales or resolutions. By construction, those features are more likely to represent true characteristics of the underlying shape. In other words, PH helps to infer higher-dimensional structure from low-dimensional representations and gathers discrete points into global structures. The features that persist for short period of time are called noise. We can compute PH using linear algebra [5]. There is a very nice and simple explanation of PH on YouTube [6]. PH has extensive applications in computer vision, image analysis, machine learning, astrophysics, signal processing, sensor, and social network analysis and AI. Some applications can be found in [7–11]. Various types of data sets including finite metric spaces or also called point cloud data sets, digital images, level sets of real-valued functions, matrix and networks can be studied with PH. There are lot of R and Python packages which are available for doing PH, few of them are R-TDA, R-TDA stats, Python-Scikit TDA, Javaplex GUDHI and online tool called Ripser. Refer [12] for more information on s/w and packages.

This paper is organized in to four sections. In Sect. 1, we introduce the concept of homology with simple example. Section 2 is about some mathematical background that is needed to build the concept behind PH. Section 3 introduces PH and its applications in machine learning. Finally, in Sect. 4, we will discuss various extensions of PH that helps to apply machine learning techniques using toy data sets.

1.1 Homology

The notion of homology is used to count number of connected components (clusters), holes and voids present in the general topological space. Homology groups are used to find holes in a topological space [13]. It is very hard, for a given arbitrary topological space to compute the homology [12]. But, it can be computed for any point cloud data with suitably chosen metric. In practical, we are not given the topological space X, but we are always given finite number of points sampled from space X, on the top of which we build an algebraic structure called simplicial complex (see Sect. 2). These algebraic structures are robust and do not change even if we transform, deform or stretch the underlying space. In technical terms, they are *homotopy* invariant (Fig. 1).

2 Mathematical Background

In this section, we will discuss some basic definitions and theorems that will help us in building the concept of homology group of simplicial complex and persistent homology.

Definition 2.1 A set $\{x_0, \ldots, x_n\}$ of points in $\mathbb{R}^N$ is called affinely linearly (geometrically) independent if for any scalars λ_i, the equations

$$\sum_{i=0}^{n} \lambda_i = 0 \text{ And } \sum_{i=0}^{n} \lambda_i x_i = 0 \text{ imply that } \lambda_0 = \lambda_1 = \ldots = \lambda_n = 0$$

Therefore, $\{x_0, \ldots, x_n\}$ are affinely linearly (geometrically) independent if and only if $\{x_1 - x_0, \ldots, x_n - x_0\}$ are linearly independent.

Definition 2.2 Let $\{x_0, \ldots, x_n\}$ be affinely (geometrically) independent set in $\mathbb{R}^N$. Then n-simplex σ spanned by $x_0, \ldots, x_n$ denoted by $\langle x_0, \ldots, x_n \rangle$ and is defined as

$$\sigma = \left\{ x = \sum_{i=0}^{n} \lambda_i x_i \,\middle|\, x \in \mathbb{R}^n, \sum_{i=0}^{n} \lambda_i = 1 \right\} \text{ and } \lambda_i \geq 0 \text{ for all } i$$

Fig. 1 Both A and deformed A have the same topological features or the same homology group, namely single connected component and one cycle or one-dimensional hole. Figure source [14]

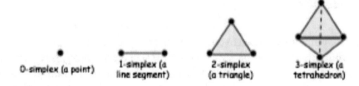

Fig. 2 Simplices for $n = 0, 1, 2, 3$. Figure Ref. [15]

For $n = 1, 2$ and 3, in $\{x_0, \ldots, x_n\}$ the corresponding n-simplex is just a regular vertex, edge, triangle and tetrahedron, respectively, shown in Fig. 2.

The points $x_0, \ldots, x_n$ that span σ are called the vertices of σ; n is the dimension of σ. A face of σ is the convex hull of a non-empty subset of the $\{x_0, \ldots, x_n\}$, i.e. for any $\{x_{i_1}, x_{i_2}, \ldots, x_{i_k}\} \subseteq \{x_0, \ldots, x_n\}$ then $\langle x_{i_1}, x_{i_2}, \ldots, x_{i_k} \rangle$ is called a face of σ, and it is denoted by τ.

Definition 2.3 A simplicial complex (also called complex) K in $\mathbb{R}^N$ is a collection of simplices such that

(1) $\sigma \in K$ and τ is face of σ ($\tau \leq \sigma$) then $\tau \in K$
(2) $\sigma_1, \sigma_2 \in K \Rightarrow \sigma_1 \cap \sigma_2 \neq \emptyset$ or $\sigma_1 \cap \sigma_2$ is a face of both σ_1 and σ_2 (Fig. 3).

Definition 2.4 An abstract simplicial complex is a finite collection of sets A such that $\alpha \in A$ and $B \subseteq \alpha$ imply $\beta \in A$. The sets in A are its simplices.

Definition 2.5 Let X be a topological space. A *cover* of X is a collection of sets. $U = \{B_i\}_{i \in I}$, such that, $X \subset \bigcup_{i \in I} B_i$, where I is any indexing set.

Definition 2.6 Given any cover of topological space $X, \mathcal{U} = (U_i)_{i \in I}$, its nerve is an abstract simplicial complex $\mathcal{N}(U)$ whose vertex set is I and such $\{U_{i_0}, \cdots, U_{i_k}\}$ $\epsilon \mathcal{N}(U)$ if and only if. $U_{i_0} \cap U_{i_1} \cap \cdots \cap U_{ik} \neq \emptyset$ (Fig. 4).

A natural question to ask is how well is the simplicial complex representation of my data set capturing the true topological features of my data? Luckily, we have the following results.

Theorem 2.1 *For the finite collection of closed, convex sets U in Euclidean space. the nerve of U and the union of the sets in U have the same homotopy type. This theorem is called a **Nerve Theorem**. Loosely speaking both of them have many topological invariants in common.*

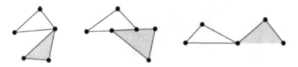

Fig. 3 Simplicial complex (left) and a collection of simplices (middle and right) which do not form a simplicial complex. Figure Ref. [6]

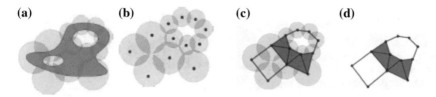

Fig. 4 **a** Cover of a finite metric space, **b** vertex set generated by cover, **c** edges and faces generated by non-empty intersections of the cover **d** the nerve of the cover). Figure source [16]

Definition 2.7 Let $X \subseteq \mathbb{R}^N$ be a finite subset. For each $x \in X$, define $B_\epsilon(x) = \{y \in \mathbb{R}^N : d(x, y) \leq \epsilon\}$ be a closed ball centred around x with $\epsilon \geq 0$. The Čech complex of X and ϵ is the nerve of $\{B_\epsilon(x)\}_{x \in X}$, namely $\text{Cech}(X, \epsilon) = \{\sigma \subset X : \bigcap_{x \in \sigma} B_\epsilon(x) \neq \emptyset\}$ (Fig. 5).

Definition 2.8 Let $X \subseteq \mathbb{R}^N$ be a finite subset. For each $x \in X$, define $B_\epsilon(x) = \{y \in \mathbb{R}^N : d(x, y) \leq \epsilon\}$ be a closed ball centred around x with $\epsilon \geq 0$. We define Vietoris–Rips complex as follows $V(X, \epsilon) = \{\sigma \subset X : B_\epsilon(x_i) \cap B_\epsilon(x_j) \neq \emptyset \text{ for all } x_i, x_j \in \sigma\}$ (Fig. 6).

$VR(X, r)$ Consists of all subsets of X whose diameter is no greater than $2r$. From the definitions, it is not hard to see that $\text{Cech}(X; \epsilon) \subseteq VR(X; \epsilon)$, one of the main drawbacks of the Cech and Vietoris–Rips constructions is that the dimension gets very large, even if $X \subseteq \mathbb{R}^d$ for d small.

There are several other complexes like *Alpha complex, witness complex, Delaunay complex, graph induced complex, cubical complex*. That can be used for different purposes. The details can be found here [2].

Definition 2.9 Let $\mathbb{F}_2 = \{0, 1\}$ be a finite field with two element (binary field). We define p-chain $C_p(K)$ as a vector space over field $\mathbb{F}_2$. Thus, $C_p(K)$ is the set generated by all p-simplices of K over binary field $\mathbb{F}_2$ and

$$C_p(K) = \left\{ c = \sum_j \gamma_j \sigma_j \mid \gamma_j \in \mathbb{F}_2 \text{ and } \sigma_j \text{ are } p\text{-simplices in } K \right\}$$

Fig. 5 Čech complex, if any two balls intersect, we draw an edge between them. If three balls intersect in all, then we draw a triangle and fill in that triangle

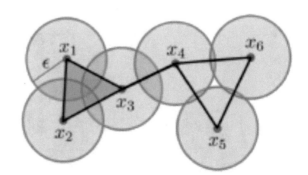

Fig. 6 Vietoris–Rips
complex unlike Čech
complex we fill in the
triangle here even if we do
not have common
intersection of three balls

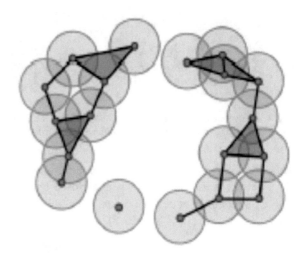

Let $c, \grave{c} \in C_p(K)$ with $c\grave{c} = \sum_j \delta_j \sigma_j$, then we can define $c + c\grave{c} = \sum_j (\gamma_j + \delta_j)\sigma_j$, where $\gamma_j, \delta_j \in \mathbb{F}_2$. Thus C_p with above-defined addition form a group is called group of p-chain denoted as $(C_p, +)$.

Consider an example of a simplicial complex given in the following figure with 5 0-simplices $\{a, b, c, d, e\}$, 6 1-simplices $\{A, B, C, D, E, F\}$ and one 2-simplex $\{\tau\}$.

Example 2.9 For K in Fig. 7, $C_0(K) = \langle a, b, c, d, e \rangle, C_1(K) = \langle A, B, C, D, E, F \rangle, C_2(K) = \langle \tau \rangle$.

An example of 0-chain is $a + b + e$, 0-chain is $A + C + D + E$ and so on.

Definition 2.10 Let $\sigma = \langle x_0, \ldots, x_n \rangle$ be a p-simplex. We define its boundary as

$$\partial_p \sigma = \sum_{i=0}^{n} (-1)^i \langle x_0, \ldots \hat{x}_i, \ldots, x_n \rangle = \sum_{i=0}^{n} \langle x_0, \ldots \hat{x}_i, \ldots, x_n \rangle,$$

Fig. 7 Simplicial complex
K: figure reference [17]

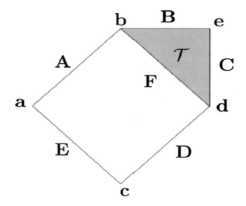

where $\hat{x}_i$ means that x_i is dropped. ∂_p is called boundary operator.

Let $c = \sum_j \gamma_j \sigma_j$ be a p-chain . Therefore, the boundary is defined as linear combination of boundaries of its—simplices $\partial_p c = \sum_j \gamma_j \partial_p \sigma_j$.

The boundary operator maps a p-chain group C_p to a $(p-1)$-chain group C_{p-1}. The collection of chains and boundary maps together forms a chain complex shown below

$$\dots \xrightarrow{\partial_{p+2}} C_{p+1} \xrightarrow{\partial_{p+1}} C_p \xrightarrow{\partial_p} C_{p-1} \xrightarrow{\partial_{p-1}} \dots \to C_2 \xrightarrow{\partial_2} C_1 \xrightarrow{\partial_1} C_0$$

Moreover, $\partial_p : C_p \to C_{p-1}$ is homomorphism, since $\partial_p(c + \grave{c}) = \partial_p(c) + \partial_p(\grave{c})$.

For defining homology, we define two special types of chains boundary and cycle in the following.

Example 2.10 For K in Fig. 7 $(\tau) = B + F + C$, $\delta(A) = a + b$, also the chain complex of K is given by

$$\langle \tau \rangle \xrightarrow{\partial_2} \langle A, B, C, D, E, F \rangle \xrightarrow{\partial_1} \langle a, b, c, d, e \rangle \xrightarrow{\partial_0} 0.$$

Definition 2.11 A p-cycle is a p-chain with empty boundary, i.e. $\partial_p c = 0$. The set of all p-cycles from a subgroup Z_p of C_p in other words

$$Z_p = \ker \partial_p = \{c \mid \partial_p(c) = 0\}$$

Example 2.11 For K in Fig. 7 $\partial(C + B + F) = \partial(C) + \partial(B) + \partial(F) = d + e + b + e + e + d = 2(d + e + b) \mod 2 = 0$ implies $C + B + F$ is 1-chain, i.e. $C + B + F \in Z_p$.

Definition 2.12 A p-boundary c is a—chain, which is the boundary of a $(p+1)$—chain d, i.e. $c = \partial d$. The set of all p-boundaries forms a subgroup B_p of the chain group C_p. More specifically $B_p = $ Image of map ∂_{p+1}

The fundamental property of homology procedure is the following.

Theorem 2.2 *The boundary of boundary is Null i.e* $\partial_p \partial_{p+1} d = 0$.

Definition 2.13 Using the above notations, we define the p-homology group as $H_p := Z_p / B_p$. $Z_p / B_p = \{c + B_p, c \in Z_p\}$ is called quotient group, and each element in H_p is an equivalence class of H_p is a collection of p-chains by adding p-boundaries from B_p to a given cycle, $c + B_p$ with $c \in Z_p$.

Definition 2.14 $\text{rank}(H_p) = \log_2(|H_P|)$ where, $|H_p|$ is cardinality of H_p.

Definition 2.15 The rank of p-homology group $rank(H_p)$ is called p-Betti number β_p.

Betti numbers are used to describe homology of space or shape. For $p = 0, 1,$ 2, the β_p can be interpreted as. β_0 Represents a number of clusters or connected components, β_1 the number of holes or loops and β_2 the number of two-dimensional holes or voids. For example, a circle has $\beta_0 = 1$, $\beta_1 = 1$, $\beta_2 = 0$, as it has one loop or one one-dimensional hole, and no void or two-dimensional hole.

3 Persistent Homology

When simplicial complexes are formed for given scale parameter, for small value of scale parameter the simplicial complex is all points in point cloud data, i.e. all vertices unconnected. For sufficiently large value, the simplicial complex becomes one big cluster. What value of reveals the "correct" structure?

PH addresses the above problem. It tries to create a filtration on point cloud data with increasing sequence of nested simplicial complexes at the same time we keep track of which topological feature is born and died. Using filtration, we study the connectivity information and the robustness. Topological features born (appear) and died (disappear) as resolution parameter value increases. The features which persist over a longer range of the resolution parameter are considered as signal and short-lived features as noise.

Figure 8 shows the pipeline for persistent homology.

Definition 3.1 Given simplicial complex, filtration of K is a nested sequence of subcomplexes, $\emptyset \subseteq K^0 \subseteq K^1 \subseteq, \ldots \subseteq K^n = K$. A simplicial complex K with filtration is called filtered complex (Fig. 9).

Definition 3.2 Let K^l be a filtration of simplicial complex, and $Z_k^l = Z_k(K^l)$ and $B_k^l = B_k(K^l)$ be the k-th cycle and boundary group of K^l, respectively. The k-th homology group of K^l is $H_k^l = Z_k^l / B_k^l$. The k-th Betti number β_k^l of K^l is the rank of H_k^l.

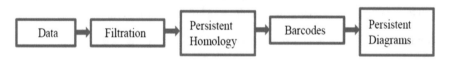

Fig. 8 Pipeline persistent homology

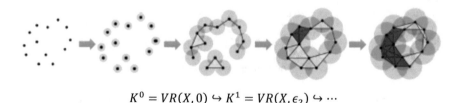

$$K^0 = VR(X, 0) \hookrightarrow K^1 = VR(X, \epsilon_2) \hookrightarrow \cdots$$

Fig. 9 Filtration of finite point cloud process with VR-complex. Figure Ref. [12]

Definition 3.3 The p-persistent k-homology group at filtration time l is represented by $H_k^{l,p} = Z_k^i / \left(B_k^{i+p} \cap Z_k^l \right)$. The p-persistent k-th Betti number $\beta_k^{l,p}$ of K^l is the rank of $H_k^{l,p}$.

3.1 Persistent Barcodes

In [17], the author introduces the concept of persistent barcodes and persistent diagrams. Persistent homology tracks the connectivity information of spaces or objects or shapes, such as number of connected components or clusters, holes in different dimensions across multiple scales. We diagrammatically represent it using barcodes and persistent diagrams. In persistent barcodes, each bar in the barcode represents a cycle or loop in some H_i. The starting point of the bar is the birth of loop, and the end point of the bar is death of loop. We represent this concept using a data set roughly sampled form three annuli. As you can see, there are three long orange bars indicating three one-dimensional cycles or loops, so $\beta_1 = 3$. The blue bar represents the element in H_0 (i.e. 0-dimensional cycles = vertices). There is one long red bar indicating one component, i.e $\beta_0 = 1$ (Figs. 10 and 11).

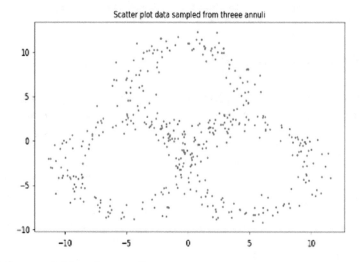

Fig. 10 Data sampled from three annuli

Persistence intervals in dimension 0:

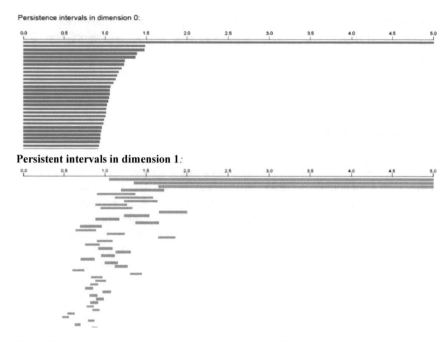

Fig. 11 Persistent barcode for data sampled from three annuli computed using Ripser live online tools shows three orange bar for three loops and one long blue bar for one connected cluster or component. The shorter bars here are noise

3.2   Persistent Diagram (PD)

In [3], authors have introduced persistence diagrams that summarize the birth and death of each topological feature. A point (x, y) in persistent diagram plane with multiplicity m represents m features that all appear for the first time at scale x and disappear at scale y. The points which are close to the diagonal $(x = y)$ indicate that, the feature is born and dead in very short span of time. The difference $y - x$ is called the persistence of a feature. For each cycle in H_i = bar in barcode, we can plot the point (birth, death) where birth equals birth time of this cycle and death equals death time of this cycle red point = cycle in H_0. Green triangle = cycle in H_1. All the points near the diagonal are either noise or artefacts of sampling. From the diagram, it is clear that we have one connected component and three loops or one-dimensional hole indicated by green triangle (Fig. 12).

The homology of a three circle data is as follows: Rank of $H_0 = \beta_0 = 1$ since a three circle data has only one component Thus, we expect one persistent (long) bar in the 0-dim barcode plus some shorter bars that we can "ignore". Rank of $H_1 = \beta_1 = 3$ since a three circle has a three one-dimensional component or holes. Thus, we expect one persistent (long) bar in the 1-dim barcode plus possible some shorter bars that we can "ignore". Rank of $H_2 = \beta_2 = 0$ since we do not have any two-dimensional holes.

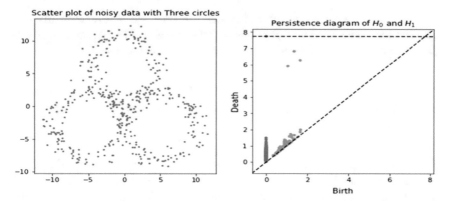

Fig. 12 Persistent homology of three circle PD showing three dots for three holes

3.3 Distance Between Persistent Diagrams

Given two persistent diagram X and Y, there Bottleneck distance is defined as $W_\infty(X, Y) = \inf_{\eta:X \to Y}(\sup_{x \in X}||x - \eta(x)||_\infty)$, where η is the set of all possible bijections, and $||x, y||_\infty = \max\{|x|, |y|\}$ is the usual L_∞ norm; similarly, we define the qth Wasserstein distance as follows $W_q(X, Y) = \left(\inf_{\eta:X \to Y}||x - \eta(x)||_\infty^q\right)^{\frac{1}{q}}$ (Fig. 13).

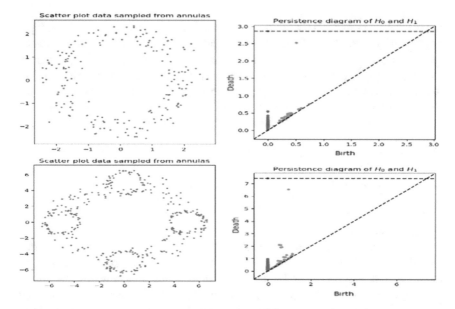

Fig. 13 Persistent diagrams for two different point clouds

Then, we have computed the **Bottleneck distance** and **Wasserstein distance** between these two persistent diagrams as **2.012968 and 5.76258,** respectively, as these two data sets do differ in most of the topological properties.

3.4 Challenges with PH Persistent Barcodes and Persistent Diagrams

Standard *persistent homology* is limited to a single scale parameter. PH can be extended for multidimensional parameters. More detailed can be found in [18]. Recently, researchers are combining PH with statistical and machine learning methods. But the space of barcodes and persistent diagrams is not amiable to statistics. Data only being metric space is insufficient for many ML techniques. We can define mean called Frechet means on space of barcodes, but they are not unique and difficult to compute [19]. An alternative idea is to embed the space of barcodes into a larger and better behaved space, in which means are well-defined and simple to compute. So, we extend PH to persistent landscape that helps to apply statistical and machine learning methods with PH.

3.5 Persistent Landscape (PL)

In [19], author introduces the concept of PL a new topological summary for data that developed by Peter Bubenik, a mathematics professor at Cleveland State University. PL converts the points from persistent diagram to points in vector space, so that we can easily apply statistics and machine learning techniques, in addition to the standard topological summaries. PL is random variable with values in a Hilbert space, this summary obeys a strong law of large numbers and a central limit theorem, and this summary is stable. We formally define persistent landscape as follows

Definition 3.5.1 Let (b, d) be a persistent interval, so $b \leq d$. Consider the function $f_{(b,d)} : \mathbb{R} \to \mathbb{R}$ such that $f_{(b,d)}(t) = \min(t - b, d - t)_+$ that is $c_+ = \max(c, 0)$, then the persistent landscape of set of intervals $\{(b_i, d_i)\}_{i=1}^n$ represents a map $\lambda : \mathbb{N} \times \mathbb{R} \to \mathbb{R}$ whose profile $\lambda_k : \mathbb{R} \to \mathbb{R}$ at a fixed $k \in \mathbb{N}$ is defined as $\lambda_k(t) = k$th largest value $of \left\{ f_{(b_i,d_i)} \right\}_{i=1}^n$ for $t \in \mathbb{R}$. Furthermore, $\lambda_k(t) = 0$, for $k > n$ (Figs. 14 and 15).

Persistent landscape captures two most important pieces of information. PL carries the information of length of barcode intervals, which is important since long bar represents features. Secondly, PL carries the information of how many intervals are overlapped at a fixed filtration time, which is important because regions with high number of overlap indicate short-lived bars which are usually considered as noise. The map from persistent diagram to persistent images is bijective.

Fig. 14 Contours of persistent landscape constructed on barcodes (0, 16), (4, 14) and (2, 6)

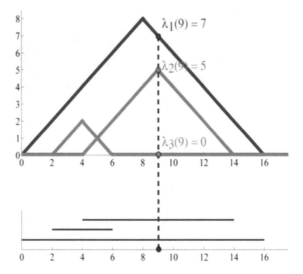

Fig. 15 Three-dimensional plot of persistent landscape constructed on barcodes (0, 16), (4, 14) and (2, 6)

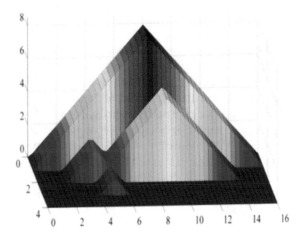

3.6   Persistent Images (PI)

In [20], author introduces persistent images (PI). PI is a representation of persistence as a vector in $\mathbb{R}^n$, this representation is stable with respect to input noise, and computationally efficient, the representation maintains an interpretable connection to the original PD, and the representation allows one to adjust the relative importance of points in different regions of the PD. PIs live in Euclidean space, and hence, are amenable to a broader range of ML techniques.

For constructing PI, we associate each component of the **persistence diagram** to the centre of a two-dimensional Gaussian distribution of a specific variance. By

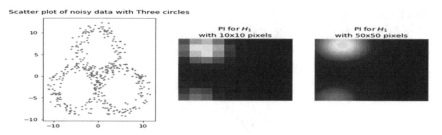

Fig. 16 Persistent image of data sampled from three annuli using Python Persim library

creating the corresponding matrix of distribution superposition, you obtain the image (Fig. 16).

Once PI is obtained, and we can extract features to apply machine learning [19, 20].

3.7　PH and Machine Learning (PHML)

The connection between persistent homology and machine learning is studied in recent years. We extract topological features from data then apply machine learning that includes both supervised and unsupervised learning, as illustrated in Fig. 17, given the data set, once we apply persistent homology to extract topological features of data. Then, we look for barcodes and persistent diagrams, we convert persistent diagram into persistent landscape or persistence images to have stable vector representation of PH to do further statistical analysis and machine learning on the data, and there are lot of other extensions like multidimensional persistent, zigzag persistence, persistence terrace that we have not discussed here. The pipeline for doing machine learning on PH analysis is shown below. The detailed introduction on this topic can be found in [19].

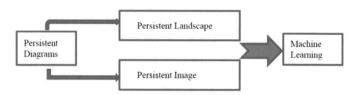

Fig. 17 Machine learning after persistent homology

4   Discussion

Persistent homology is a new technology from mathematics that has received attention from inside and outside mathematics particularly in the field of data science and machine learning and computer vision. We introduced PH and all extensions of its step by using toy data sets as an example. This will definitely help beginner to start with.

Acknowledgements The author highly acknowledges the co-authors for their support and suggestions.

References

1. Carlsson, Gunnar: Topology and data. Bull. Am. Math. Soc. **46**(2), 255–308 (2009)
2. Zomorodian, A.: Topological data analysis. Adv. Appl. Comput. Topol. **70**, 1–39 (2012)
3. Carlsson, G., Collins, A., Guibas, L., Zomorodian, A.: Persistence barcodes for shapes. Int. J. Shape Modeling **11**(2), 149–187 (2005)
4. Ghrist, R.: Barcodes: the persistent topology of data. Bull. Am. Math. Soc. **45**(1), 61–75 (2008)
5. Zomorodian, A., Carlsson, G.: Computing persistent homology. Discrete Comput. Geometry. **33**(2), 249–274 (2004). https://doi.org/10.1007/s00454-004-1146-y. ISSN 0179-5376
6. http://iopscience.iop.org/article/10.1088/1742-5468/2009/03/P03034/fulltext/
7. Adams, H., Chepushtanova, S., Emerson, T., Hanson, E., Kirby, M., Motta, F., Neville, R., Peterson, C., Shipman, P., Ziegelmeier, L.: Persistence images: a stable vector representation of persistent homology. J. Mach. Learn. Res. **18**, 1–35 (2017)
8. Edelsbrunner, H., Harer, J.: Persistent homology—a survey. Contemp. Math. **453**, 257–282 (2008)
9. De Silva, V., Ghrist, R.: Coverage in sensor networks via persistent homology. Algebr. Geom. Topol. **7**, 339–358 (2007)
10. Lee, H., Chung, M.K., Kang, H., Kim, B., Lee, D.S.: Discriminative persistent homology of brain networks. In: 2011 IEEE International Symposium on Biomedical Imaging: From Nano to Macro, Chicago, IL, pp. 841–844 (2011)
11. Kovacev-Nikolic, V., Bubenik, P., Nikolic, D., Heo, G.: Using persistent homology and dynamical distances to ´analyze protein binding (2014). arXiv:1412.1394
12. Otter, N., Porter, M.A., Tillmann, U., et al.: A roadmap for the computation of persistent homology. EPJ Data Sci. **6**, 17 (2017)
13. Edelsbrunner, H., & Harer, J.: Computational Topology: An introduction. American Mathematical Society (2010)
14. https://www.ayasdi.com/blog/topology/three-properties-of-topological-analysis/
15. http://outlace.com/TDApart1.html
16. http://www.dyinglovegrape.com/math/topology_data_1.php
17. Course Notes from Sara kalisnik. http://brown.edu/Research/kalisnik/appliedtop.html
18. Carlsson, G., Zomorodian, A.: The Theory of Multidimensional Persistence. Discrete Comput. Geom. **42**(1), 71–93 (2009)
19. Pun, C.S., Xia, K., Lee, S.X.: Persistent-homology based machine learning and its applications—a survey. arXiv:1811.00252
20. Bubenik, P.: Statistical topological data analysis using persistence landscapes. Mach. Learn. Res. **16**, 77–102 (2015)

Concave Point Extraction: A Novel Method for WBC Segmentation in ALL Images

Vandana Khobragade, Jagannath Nirmal and Suprava Patnaik

Abstract Analysis of hematological disorder of leukocyte (WBC) is common in medical diagnoses. The counting and classification of blood cells enables the estimation and identification of large number of diseases. Morphological analysis of blood cells requires skill operators, is time-consuming, and less accurate. This paper discusses the Computer-Aided Acute Lymphoblastic Leukemia (ALL) diagnosis technique based on image examination. Here, methods for the segmentation and categorization of white blood cell (WBC) in the interior of acute lymphoblastic leukemia (ALL) smear images are presented. The segmentation method consists of contour evidence extrication to detect the apparent part of each entity and contour estimation to evaluate the final object contours. Then, some attributes were extracted from each segmented cell. In this paper, detection of hematological disorders like leukemia (blood cancer) is carried out which is based on geometric, texture, and mixed features. Classification is done to categories each segmented cell to be normal (benign) or abnormal (malignant).

Keywords Acute leukemia · WBC segmentation · Classification · Multilayer perceptron · Backpropagation algorithm

1 Introduction

The cancer of the blood or the bone marrow (the tissue in the interior of bones, where blood cells are produced) is termed as leukemia. In this disease, WBC are undeveloped. These immature cells are named "lymphoblasts," which blocks robust

V. Khobragade (✉)
Department of Electronics and Telecommunication Engineering, LTCE, Mumbai, India
e-mail: vandanamotghare@gmail.com

J. Nirmal
Department of Electronics Engineering, KJCE, Mumbai, India

S. Patnaik
Department of Electronics and Telecommunication Engineering, XIE, Mumbai, India

© Springer Nature Singapore Pte Ltd. 2020
S. Agarwal et al. (eds.), *Machine Intelligence and Signal Processing*,
Advances in Intelligent Systems and Computing 1085,
https://doi.org/10.1007/978-981-15-1366-4_9

red cells, platelets, and fully developed white cells (leukocytes) from being generated. Four recognized leukemia types are acute lymphoblastic leukemia (ALL), acute myeloid leukemia (AML), chronic lymphocytic leukemia (CLL), and chronic myeloid leukemia (CML) [1].

The automated recognition and segmentation of cell nuclei is one of the most interesting fields in cytological image analysis. In this, nucleus is the very important structure within the cell and its structure represents remarkable reforms when the cell is affected by a disorder. Therefore, accurate definition of the nucleus boundary is a crucial task. The WBCs are classified as normal and abnormal based on the recognition and quantification of the reforms in the nucleus morphology and mass [2].

An arrangement which is set to conduct an automated counting of blood cells include segmentation, feature extraction, and classification of cells. According to [3], segmentation allows artifacts to be separated without the inclusion of unnecessary material by describing the borders of the blood cells. Pixel-based image segmentation and region-based segmentation are the two types which are widely discussed in literature. In [4], it is specified that the shape segmentation methods are categorized into threshold-based, edge-based, and region-based techniques.

Thresholding and clustering are the straightforward segmentation approaches [5, 6]. Thresholding is appropriate only for the items that do not touch one another. Clustering has been extensively utilized to segment gray-level images [6, 7]. Hari et al. [7] have used the auto-segmentation of blood cells for counting. Other techniques that have been used for the segmentation of blood cells are chessboard distance measure and watershed algorithm. Kernel-induced rough c–means (KIRCM) clustering algorithm is developed in [8] for the segmentation of human lymphocyte images. In higher-dimensional feature space, rough c–means clustering (RCM) is executed to acquire enhanced segmentation accuracy and to enable automated acute lymphoblastic leukemia (ALL) detection.

The study in [9] intends to use a new stained WBC image segmentation technique utilizing stepwise merging rules. These rules are based on mean-shift clustering and boundary elimination rules with a gradient vector flow (GVF) snake. The system suggested in [10] incorporates a preprocessing step, nucleus segmentation, cell segmentation, feature extraction, feature selection, and classification. Here, the segmentation technique comprises of thresholding, morphological operation, and ellipse curve fitting. Various features were extricated from the segmented nucleus and cytoplasm patch. Distinguished features were selected by a greedy search algorithm also known as sequential forward selection. Finally, with a set of selected important features, both linear and Naïve Bayes classifiers were applied for performance comparability.

Nucleus of the WBC along with the cytoplasm is known as whole leukocyte. In [11], a method is recommended which separates the whole leukocyte from a microscopic image. Color, shape, and texture features are extracted to train different models of classification. In this paper, a new methodical approach is used for the segmentation of WBC from blood smear images. The contours form roughly elliptical shapes. The method which has shown promising results in segmenting nanoparticles, i.e., the work presented in [12] is implemented in our work for WBC segmentation. In sections ahead, object segmentation which includes contour evidence extraction and contour estimation are discussed in detail in Sects. 2 and 3. In Sect. 4, assessment and outcomes are examined. Lastly, the conclusion and future work are discussed in Sect. 5 followed by references.

2 Methodology

In order to accomplish the categorization of acute leukemia cells, three main steps involved are image segmentation, feature extraction, and classification using artificial neural network. In this work, acute lymphoblastic leukemia (ALL) images used for the first step of segmentation are taken from ALL-IDB. It is a free of cost widely available dataset provided by Department of Information Technology, precisely conceived for the assessment and comparison of algorithms for segmentation and image classification.

The ALL-IDB database has two definite types (ALL-IDB1 and ALL-IDB2). The ALL-IDB1 dataset is formed of 108 images containing about 39,000 blood components captured with different magnifications of the microscope varying from 300 to 500. The latter is a collection of cropped area of interest of normal or blast cells from ALL-IDB1. ALL-IDB2 which comprises of single WBC in each image is used for testing quantitative implementation of our method. It has 260 images. The first half is from ALL patients (lymphoblasts) and the last half is from non-ALL patients (normal WBCs). The comprehensive planning of the leukemia detection system is summarized in Fig. 1. Further details of proposed image segmentation for ALL are discussed in the following section.

3 Proposed Leukocyte (WBC) Segmentation Technique

Usually, three main regions of blood sample are white blood cells (WBC), red blood cells (RBC), and background areas. Nucleus and cytoplasm regions form entire WBC contain salient information to be detected by hematologists. However, the RBC and background regions hold no information and can be excluded from the image.

This segmentation of convex-shaped cell images is accomplished using concave points and ellipse characteristics. The technique comprises of two main steps: contour evidence extraction to detect the perceptible element of individual object and contour estimation to evaluate the objects contours. The mechanism used bank only on edge statistics and can be implemented on objects that have an approximately elliptical formation. As shown in Fig. 2, a usual configuration for segmenting the WBC composed of two steps: (1) contour evidence extraction and (2) contour estimation.

The segmentation process starts with preprocessing a grayscale input image in order to build an edge map. To achieve this, first the binarization of the image is

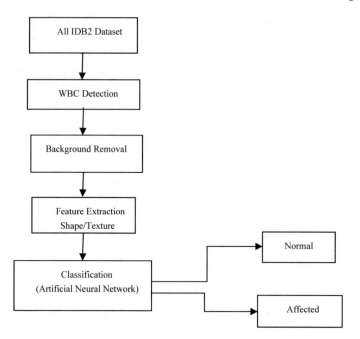

Fig. 1 Proposed lymphoblastic leukemia diagnosis structure

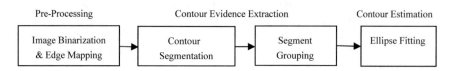

Fig. 2 Method for WBC segmentation [12]

acquired by background elimination which has its origin in Otsus method [13]. The Canny edge detector [14] is employed to create an edge map.

The initial measure is to obtain the contour evidence comprising the noticeable parts of the object's boundaries. The contour evidence extraction demands two separate functions: contour segmentation and segment grouping.

Contour Segmentation

After extracting the image edge by Canny edge detector [14], the detected concave points are acquired employing the curvature scale space (CSS) technique (based on curvature analysis [15]). Concavity test is applied on the corner points. This ensures the retention of the detected corner points in the concave region.

Segment Grouping

Let us represent an identified corner point by x_i and its two kth adjoining contour points by x_{i-k} and x_{i+k}. The corner point x_i is said to be concave if the path joining

x_{i-k} to x_{i+k} does not lie in the interior of the object. The acquired concave points are utilized to divide the contours into contour segments. Due to the overlap between the objects, a single object may produce multiple contour segments. To unite all the contour sections belonging to the similar object, grouping of segments is required. In segment grouping, the target is to detect a class of contour segments that jointly construct an object with elliptical form. To optimize the categorization, the contour segments under grouping procedure are only inspected with the connecting segments. Two segments are neighbor if the Euclidean distance connecting their center points is less than the predefined threshold value. The contour segment grouping is conducted through the technique of ellipse fitting.

The contour segments $S = \{s_1, s_2, \ldots, s_{ns}\}$ are separated into the target and the candidate segments. A segment with length more than (12 pixels length) is defined as target segment ($s_{t, i} \in S_t$). A candidate segment ($s_{c, i} \in S_c$) is the one which is still not assigned to any other segments group. Two segments can be grouped if one of them is target segment. The contour segment assembling is conducted through the exercise of ellipse fitting implemented on the target, candidate, and joint target-candidate contour segments and thereby investigating the goodness of fit (GoF) of ellipse with respect to the original segments. The purpose is to locate the group of contour segments that altogether form a specific ellipse shape object. Shown in Fig. 3 are some of the WBC images from ALL-IDB2 database on which concave points are detected and segment grouping is being performed to separate the WBC from the blood smear images.

WBC segmentation is followed by the step of feature extraction. The extracted features will yield beneficial data for classification of WBC into abnormal and normal cell. The extrication of distinguished features plays key role in the performance augmentation of the classifier and for decreasing the computational difficulty as compared to a high-dimensional features space [16]. In this work, geometric features of WBC like area, major axis, minor axis, perimeter, and texture features with GLCM like contrast, correlation, energy, homogeneity, and entropy statistics are obtained from the GLCM matrix of the nucleus region.

4   Results and Discussion

The estimate of similarity for the segmentation performance of the suggested technique with established methods such as thresholding (T) and K-means method (K) is shown in Table 1. It can be seen that the concavity-based method (C) performs better than other two methods.

The classification of segmented WBC as lymphoblast and normal is done by employing classifiers specifically multilayer perceptron neural networks. The categorization of WBC is carried out for geometric and texture features using the MLP. Here, in totality 4, 7 input features have been taken into consideration which represents the geometric and texture features of the WBC, respectively. In constructing

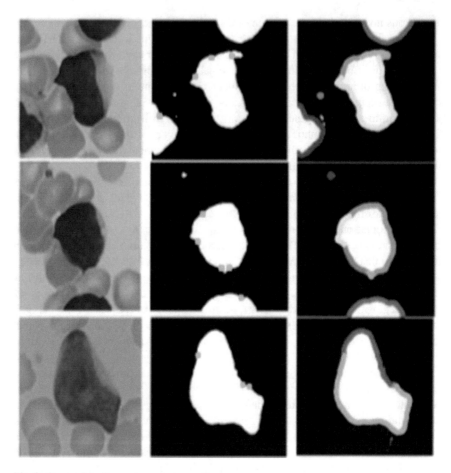

Fig. 3 Corner detection and segment grouping

Table 1 Segmentation performance comparison

Methods	Accuracy	Sensitivity	Specificity
Thresholding-based	92.92	98.65	91.63
K-means-based	89.93	97.41	89.21
Concavity-based	95.4	99.29	94.34

ANN-based leukemia detection arrangement, major assignment is to detect best design for ANN. We have carried out test on 3-layer, 4-layer, and 5-layer networks. The architectures are supplied with number of nodes in every layer and the output function employed for that layer. For the classification with MLP using backpropagation neural networks, the configuration of the network is set to I:H:2 (input node: hidden node: output node). For example, 4:10:2 indicates that it is a 3-layer network with 4 input and 2 output nodes with 10 nodes in the hidden layer.

Various investigations that will supervise the classification of WBC are:

- For optimum numbers of hidden layers and hidden node for MLP network.
- For optimum numbers of training epoch and learning rate.

A number of input features to be positioned directly decide the number of input nodes. The inputs will be normalized within the range of 0 and 1 to avoid features dominating in the training stage. For MLP network, the output nodes for the two groups will be set as lymphoblast (1) and normal cell (0). Here, there are 260 WBC in totality that have been segmented from acute leukemia images. From this, 4 geometric features, 7 texture features, and total of 11 mixed input features have been taken from the completely segmented WBC and will be given as inputs to MLP network.

Figure 4 represents the comparison of classification performance using geometric (Fig. 4a), texture (Fig. 4b) and mixed features (Fig. 4c) obtained from C, K and T, WBC segmentation for 10 nodes in hidden layer, no. of Epocs = 100, and learning rate = 0.001. The analysis is carried out by increasing the number of hidden layers. It is noticed that the classification accuracy for mixed features is better and is giving comparable results for our proposed method of segmentation as compared to existing thresholding method.

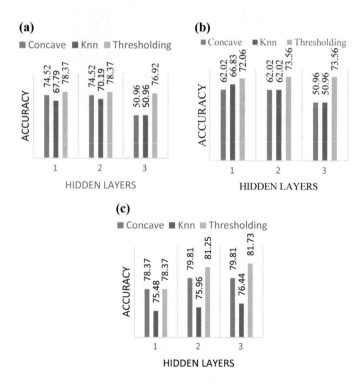

Fig. 4 Comparison of classification performance by changing the hidden layers

Table 2 Comparison of classification accuracy (%) using geometric, texture, and mixed features obtained from concavity-based (*C*), Knn (*K*), and thresholding (*T*) WBC segmentation (no. of hidden layers: 3, no. of Epocs: 100, learning rate: 0.001)

Nodes	10			25			40		
Features	Segmentation type								
	C	*K*	*T*	*C*	*K*	*T*	*C*	*K*	*T*
Geometric	50.96	50.96	76.92	75.48	70.67	78.37	76.44	71.63	78.37
Texture	50.96	50.96	73.56	50.96	65.87	73.56	61.06	66.35	73.56
Mixed	**79.81**	**76.44**	**81.73**	**79.81**	**78.37**	**82.69**	**81.25**	**75.48**	**81.73**

Table 2 exhibits the comparison of classification performance using MLP. Here, the number of nodes in hidden layer are changed and the classification accuracy is calculated for no. of Epocs = 100, learning rate = 0.001. Here, we have done the analysis by increasing the number of hidden nodes (10-25-40). It is observed that the classification accuracy for mixed features extracted from the segmented results using concavity method (C), Knn (K), and thresholding (T) is better as compared to accuracy results obtained for geometric and texture features.

Further, as the number of nodes is increased (to 40), it is discovered that the classification accuracy acquired by suggested concavity method of segmentation is 81.25% for mixed features extracted. As can be seen from Table 2, this value is very much close to 81.73% obtained using mixed features extracted for thresholding method of segmentation.

Figure 5 exhibits the analysis of classification performance of MLP by changing the number of Epocs. For this analysis, we have kept number of hidden layers = 3, no. of hidden nodes = 40, and learning rate = 0.001. A number of Epocs are changed from 100–200–300. It is observed that for all the three segmentation methods, namely concavity-based method (C), Knn (K), and thresholding (T), only geometric and mixed features extracted are giving more than 70% classification accuracy. Also, mixed features extracted from our proposed method of segmentation is giving comparable (with thresholding method) results, i.e., 81.25 and 81.73% for Epocs = 100, and 80.77% for Epocs = 200.

The analysis of classification performance by changing the learning rate is shown in Table 3. This is done by using geometric, texture, and mixed features obtained from concavity-based, Knn, and thresholding method of WBC segmentation. For this analysis, we have kept No of hidden layers = 3, No. of hidden nodes = 40, and number of Epocs = 300. The learning rate is changed for two values 0.05 and 0.09 and then the analysis is done. It is observed that only for mixed features extracted from WBC segmented from proposed concavity-based method and thresholding methods, the classification accuracy is highest, i.e., 79.33 and 83.65%, respectively, for learning rate of 0.09.

5   Conclusion and Future Work

The proposed method segments WBC in single-cell image (sub-images from ALLIDB2 database). This segmentation of convex-shaped cell images is performed utilizing the concave ends and ellipse properties. The mechanism consists of contour evidence extraction to detect the noticeable part of every object and contour estimation to approximate the final objects contours. Investigations performed on ALL-IDB database displayed that the recommended method reached high detection and segmentation accuracies. Analyses performed during the classification of WBC are:

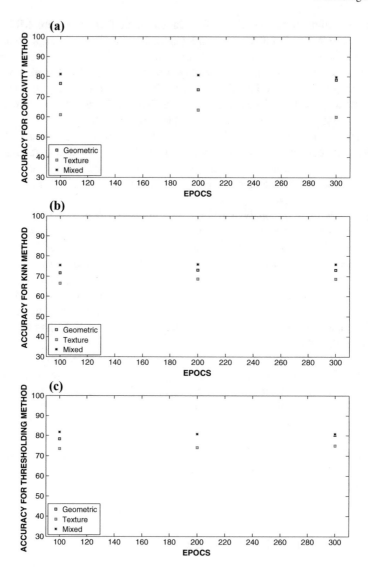

Fig. 5 Comparison of classification performance using geometric, texture, and mixed features obtained from concavity-based (**a**), Knn (**b**), and thresholding (**c**) WBC segmentation (no of hidden layers: 3, no of hidden nodes: 40, learning rate: 0.001)

- For optimum numbers of hidden layers and hidden node for MLP network.
- For optimum numbers of training epoch and learning rate.

From the experimental results, it is observed that the classification accuracy is high only for mixed features extracted from WBC segmented using proposed concavity method and thresholding method.

Table 3 Comparison of classification accuracy (%) using geometric, GLCM, and mixed features obtained from concavity-based (*C*), Knn (*K*), and thresholding (*T*) WBC segmentation (no. of hidden layers: 3, no. of hidden nodes: 40, no. of Epocs: 300)

Features	Classification accuracy (%) for *C*		Classification accuracy (%) for *K*		Classification accuracy (%) for *T*	
Learning rate	0.05	0.09	0.05	0.09	0.05	0.09
Geometric	78.37	78.37	74.04	71.15	77.40	80.29
GLCM	60.10	60.10	64.90	64.90	75.48	75
Mixed	79.33	79.33	72.60	76.44	80.77	83.65

Classification results can be improved by using other feature extraction techniques and machine learning methodology. For validation, proposed method can be applied on different database. Future work will be to improve segmentation and classification accuracy in case of overlapping WBC in blood smear images.

References

1. Mohapatra, S.: Hematological image analysis for acute lymphoblastic leukemia detection and classification. Dissertation (2013)
2. Kothari, S., Chaudry, Q., Wang, M.: Automated cell counting and cluster segmentation using concavity detection and ellipse fitting techniques. In: IEEE International Symposium on Biomedical Imaging, pp. 795–798 (2009)
3. Sabino, D.M.U., Costa, L.F., Rizzatti, E.G., Zago, M.A.: A texture approach to leukocyte recognition. In: Real-Time Imaging—Special Issue on Imaging in Bioinformatics, vol. 10, pp. 205–216 (2004)
4. Zuva, T., Olugbara, O.O., Ojo, S.O., et al.: Image segmentation, available techniques, developments and open issues. Can. J. Image Process. Comput. Vision, 20–29 (2011)
5. Kim, K., Jeon, J., Choi, W., Ho, W.: Automatic cell classification in human's peripheral blood images based on morphological image processing. In: Australian Joint Conference on Artificial Intelligence, pp. 225–236 (2001)
6. Dorini, L.B., Minetto, R., Leite, N.J.: White blood cell segmentation using morphological operators and scale-space analysis. In: Proceedings of the XX Brazilian Symposium on Computer Graphics and Image Processing (2007)
7. Hari, J., Prasad, A.S., Rao, S.K.: Separation and counting of blood cells using geometrical features and distance transformed watershed. In: 2nd International Conference on Devices, Circuits, and Systems (ICDCS), pp. 1–5 (2014)
8. Mohapatra, S., Patra, D., Kumar, K.: Kernel induced rough c-means clustering for lymphocyte image segmentation. In: IEEE Proceeding on Intelligent Human Computer Interaction (2012)
9. Byoung, C., et al.: Automatic white blood segmentation using stepwise merging rules and gradient vector flow snake. Micron, 695–705 (2011)
10. Prinyakupt, J., Pluempitiwiriyawej, C.: Segmentation of white blood cells and comparison of cell morphology by linear and naïve Bayes classifiers. In: Biomedical Engineering Online, pp. 63 (2015)
11. Putzu, L., Caocci, G., Di Ruberto, C.: Leucocyte classification for leukaemia detection using image processing techniques. Artif. Intell. Med., 179–191 (2014)
12. Zafari, S., Eerola, T., Sampo, J., Kälviäinen, H., Haario, H.: Segmentation of Partially Overlapping Nanoparticles Using Concave Points, vol. 9474, pp. 187–197. Springer, Heidelberg (2015)

13. Otsu, N.: A threshold selection method from gray-level histograms. Automatica **11**(285–296), 23–27 (1975)
14. Canny, J.: A computational approach to edge detection. IEEE Trans. Pattern Anal. Mach. Intell., 679–698 (1986)
15. He, X., Yung, N.: Curvature scale space corner detector with adaptive threshold and dynamic region of support, vol. 2. In Proceedings of the 17th International Conference on Pattern Recognition. ICPR 2004, pp. 791–794 (2004)
16. Markiewicz, T., Osowski, S., Marianska, B., Moszczynski, L.: Automatic recognition of the blood cells of myelogenous leukemia using SVM, vol. 4. In: Proceedings of IEEE International Joint Conference on Neural Networks, pp. 2496–2501 (2005)

Speech Emotion Recognition for Tamil Language Speakers

V. Sowmya and A. Rajeswari

Abstract Emotional information in speech signal is an important information resource that expresses the state of mind of a person. For robots to plan their actions autonomously and interact with people, recognizing human emotions is a critical parameter required for emotion recognition. In this paper, the emotion is recognized and identified from Tamil speech signal of the speaker so as to make conversation between human and computer more native and natural. The most human non-verbal cues such as pitch, loudness, spectrum and speech rate are efficient carriers of emotion. In this proposed work, there are three important aspects of designing a speech emotion recognition system. For developing a speech emotion recognition system, three aspects are taken into consideration. They are database, feature extraction and emotion classifiers. A database is created for Tamil language, and the Mel-Frequency Cepstral Coefficient (MFCC) and energy are calculated from both recorded audio signal and real-time audio signal. From these extracted prosodic and spectral features, emotions are classified using support vector machine (SVM) with overall 85.4% accuracy.

Keywords Emotion recognition · MFCC · SVM

1 Introduction

Speech is signal of communicating and expressing the thoughts and feelings to articulate spoken words including formal verbal and non-verbal vocal sounds. Similarly, emotions are the non-verbal cues to find the mental state of a person while delivering the speech signal [1]. Specifically, in India, Tamil language speakers express more emotions than other speakers. This work focusses on the recognition of emotional state of Tamil language Speakers.

V. Sowmya (✉) · A. Rajeswari
Coimbatore Institute of Technology, Coimbatore, Tamil Nadu, India
e-mail: Sowmya.v@cit.edu.in

A. Rajeswari
e-mail: rajeswari@cit.edu.in

© Springer Nature Singapore Pte Ltd. 2020
S. Agarwal et al. (eds.), *Machine Intelligence and Signal Processing*,
Advances in Intelligent Systems and Computing 1085,
https://doi.org/10.1007/978-981-15-1366-4_10

Initially, people interested in creating machines to assist their works like playing music and taking small things for them, and later on, they are interested in developing robots where people find difficult to work like nuclear power plants, chemical factories, militaries during wars, etc. As the technology improvised their growth, more number of robots are installed in industries to increase production. As the technology keeps on advancing, people are interested in involving them in every field to assist them like in hotel for serving customer, in home for cleaning windows, in hospital automatic wheelchair controller for physically challenged patients and in education to assist student to learn concept better and easy way, etc [2].

Major emotional information from the speech signal is retrieved through different features like paralinguistic and prosodic features such as pitch, utterances, intensities, tone, continuity, loudness, tempo, frequency, intonation and stresses. From these features, the mental state of a person like sad, anger, surprise, disgust, fear, anticipation, trust, joy, neutral and some more emotions intermediate to major emotions like joyful crying, painful silence, etc. have been observed.

Python coding platform is utilized for the work. Python Anaconda (conda) version has the capability to process the signal, extract the feature and can classify based on the signal capabilities.

2 Literature Survey

In this section, a systematic literature review is carried out in the area of "Speech Emotion Recognition" that utilizes database, feature extraction and classifiers.

Mustafa et al. [3] presented a survey result in the emotion recognition from the speech signal and revealed that most of the work is carried out in spontaneous database with 44% of database used is German language, 17% for English language and less for other languages. Among the classifier, 27% of the work used SVM, while neural network constitutes 19% and Hidden Markov Model contributes to 9%. The emotions recorded were 17% anger, 14% sadness, 13% neutral, 12% happiness and 10% fear.

In a study conducted by Ke et al. [4], emotions were classified using SVM and ANN classifiers. After extracting the features, feature reduction is done by principal component analysis (PCA), and forward selection is done to remove the irrelevant or redundant features from the original dataset. The accuracy using ANN is 75% and SVM is 83%, which shows that SVM outperforms neural network.

Abdollahpour et al. [5] observed that in Malay language, the emotions are recognized for different aged people namely children, young adult and middle-aged people. The accuracy of emotion recognition of young adult is identified to be very low around 50%, and for children, it is 73%; for middle-aged, it is found to be 90% [6].

From Albornoz and Milone [7], it is observed that the emotions are extracted and classified using Emotion Profile Ensemble Classifier. Here, cross-corpus is used, and for this, RML database with 6 different languages is used as dataset. The system produces a performance of 70.58% for language dependent and 62.47% for real application.

Joe [8] extracted the emotions from Tamil sounds of both male and female acted voices to analysis the psychological state for business development. The sounds are mainly modulated with anger, fear, neutral, sad and happy emotions. Features of emotional sound deliver MFCC, energy that has been extracted and classified through support vector machine (SVM). Around 68.73% accuracy has been achieved for classifying Tamil speech sound emotions [9].

Identified the emotional state of a person through utterance of sound in Malayalam language containing four emotions alike neutral, angry, sad and happy [10]. The database consists of 250 uttered words from 7 speakers and extracted features like averages, standard deviation of the intensity, MFCC and fundamental frequencies of sound signal. The extracted features were classified using probability functions system which contains standard deviation and averages. As the result, system shows accuracy of classifying neutral, angry, happy and sad emotions of Malayalam sounds were 36, 72, 44 and 70%, respectively.

Chernykh and Prikhodko [11] in a study inferred two approaches: Connectionist Temporal Classification (CTC) and Deep Recurrent Neural Network approaches. By these two approaches, four emotions are classified as neutral, anger, sad and excitation and compared with emotional Judging Experts and found that they were almost similar in most of the cases and concluded that the proposed system would produce significant knowledge for emotional speech recognition.

El Ayadi et al. [12] surveyed the emotion classification through feature extraction and classification algorithm of spoken utterances. The different analysis for database creation, feature extraction and classification algorithm was conducted for better recognition. The conclusion was made such that some classifier works well for particular emotional features and would produce much better accuracy while combining the features.

Garg and Sehgal [13] explained various classifiers for speech emotion recognition. Lexical and paralinguistic features were considered for feature extraction and classifiers such as ANN, SVM, K-nearest neighbors (KNN), hidden Markov model (HMM), Bayes and decision tree were taken and analyzed. Among them, SVM outperforms other classifiers, but still, the accuracy range of anger was poor while considering paralinguistic feature.

Joseph and Sridhar [14] extracted features namely discrete wavelet transform (DWT), linear prediction coefficient (LPC) and formant feature extraction and were classified through random forest classifier and neural networks. Four emotions namely Neutral, Sad, Happy, and Angry, are used for classifying and generating data set for speech recognition system. The classifier random forest performs better than neural network-based classifier.

Javier et al. [15] published their research paper, and in that, they classified the already extracted features like MFCC and energy through decision tree, classification using SVM and BayesNet. Developed model, is capable of adapting to the emotions that are classified and inferred that, the efficiency of the system has been improved for the emotions anger and happy.

From the survey, it is identified that there is no standard database for Tamil language, and the emotions are also dependent on speakers which may be either due to age factor or due to the gender of the person. And more work has been carried out only for few emotions.

3 Proposed Methodology

A systematic methodological approach is utilized in this paper which consists of pre-processing, feature extraction and classification as shown in Fig. 1.

3.1 Pre-processing

The pre-processing stage in speech recognition systems is used in order to increase the efficiency of subsequent feature extraction and classification stages and also improve the overall recognition performance. Commonly, the pre-processing includes sampling, windowing and denoising process. At the end of the pre-processing, the compressed and filtered speech frames are forwarded to the feature extraction stage.

The speech signal is sampled at a rate of 8000 kHz. According to the Nyquist–Shannon sampling theorem, a continuous-time signal $x(t)$ that is bandlimited to a certain finite frequency f_{max} needs to be sampled with a sampling frequency of at least $2f_{max}$, and it can be reconstructed by its time-discrete signal. The frequency of a human speech signal is around 4 kHz [16].

Speech is a non-stationary time variant signal. To achieve stationarity, the speech signal is divided into number of overlapping frames to reduce the discontinuities in frames. When the signal is framed, it is necessary to consider the edges of the frame. Therefore, to tone down the edges windowing is used. In speech processing, however,

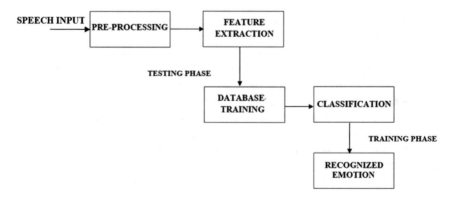

Fig. 1 Block diagram of proposed model

the shape of the window function is not that crucial but usually some soft window, i.e., hamming window is used in order to reduce discontinuities of the speech signal at the edges of each frame. The hamming window is defined in Eq. (1),

$$w(n) = 0.56 - 0.46 \cos\left(\frac{2\pi(n-1)}{k-1}\right), \quad \text{with } n = 0 \ldots k - 1. \tag{1}$$

The stage of denoising or noise reduction, also referred to as enhancing of speech degraded by noise, aims to improve the quality of speech signals. Speech signals may be corrupted due to microphone-related noise, electrical noise and environmental noise. Conclusively, the general measure of quality of a noise reduction system is its improvement in signal-to-noise ratio (SNR), but with respect to speech recognition, the best measure is the improvement in recognition performance.

3.2 Feature Extraction

3.2.1 Mel-Frequency Cepstral Coefficients (MFCCs)

Among the different feature extraction methods, Mel-Frequency Cepstral Coefficient (MFCC) is utilized as it is widely accepted and used model. The main point to understand about speech is that the sounds generated by a human are filtered by the shape of the vocal tract including tongue, teeth, etc [17]. This shape determines what sound comes out. If the shape can be determined accurately, this should give us an accurate representation of the phoneme being produced. The shape of the vocal tract manifests itself in the envelope of the short time power spectrum, and the job of MFCCs is to accurately represent this envelope.

3.2.2 Steps to Calculate the MFCCs

The various steps for calculating MFCC is depicted in Fig. 2. An audio signal is constantly changing, so to simplify things, by assuming that on short time scales the audio signal does not change much. This is the reason for framing the signal into 20–40 ms frames. If the frame is much shorter, then there will not be enough samples

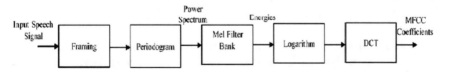

Fig. 2 Steps to calculate MFCC

to get a reliable spectral estimate, and if it is longer, the signal changes too much throughout the frame.

The next step is to calculate the power spectrum of each frame. Periodogram identifies which frequencies are present in the frame. The Mel scale tells exactly how to space the filter banks and how wide to make them. Once the filter bank energies are there, logarithm of them can be taken. The final step is to compute the DCT of the log filter bank energies. Filter banks are all overlapping, and the filter bank energies are quite correlated with each other. The DCT de-correlates the energies.

3.2.3   Energy

Energy is another feature extracted from a speech signal which is a measure of how much signal exists at a given time. Signal energy is seen as the strength or power or voice volume. Voice energy is associated with the respiratory system. The energy of audio signal is influenced by conditions in which the record was created. The distance between the mouth and the microphone or lower sensitivity of hardware has a significant impact on the quality of digitalized sound. The energy parameter is calculated using the short-term signal energy which is defined in Eq. (2),

$$E = \left[\frac{1}{N}\right] \sum_{n}^{N-1} [x(n)]^2 \tag{2}$$

The logarithm is often used to demonstrate minor changes of energy. With energy, the speech activity and the division of voiced and unvoiced speech sounds are determined. On the other hand, signal energy sometimes cannot separate voiced and unvoiced sounds perfectly, and thus, low-energy voiced sounds are lost.

3.3   Classification—SVM Classifier

Classifier gives a decision based on the parameters from trained database with test speech parameters. Different types of classifiers are used for classifying emotions from speech signal. From the survey, the accuracy of SVM is found to be better than other classifiers, and also, SVM provides minimum empirical classification error with high geometric margin. So the work is carried out with SVM classifier in Python platform.

Support vector machine (SVM) is a supervised machine learning algorithm which can be used for both classification and regression challenges. However, it is mostly used in classification problems. In this algorithm, the plot of each data item as points in n-dimensional space with the value of each feature being the value of a particular coordinate. Then, it performs classification by finding the hyper-plane that differentiates the two classes very well.

SVM is a supervised learning algorithm-based classifier. Training and testing section comes under classifier section. First, train the SVM classifier using the features obtaining from the MFCC and energy values. The goal of SVM vector machine is to seek out the optimal separating hyper-plane, which maximizes the margin of the training data.

SVM has higher accuracy. In multi-class SVM, divide the class problem into several binary sub-problems and builds a standard SVM for each. For testing, take speech sample of a speaker. Then apply it to the SVM as input. Then extract features of this test input and compare these features with trained feature set and recognize the speaker.

Even though SVM provides better performance, its training cost is very high. So to decrease the training cost, twin support vector machine (TWSVM) can be used. SVM and TWSVM work better for balanced datasets, and it results in a biased classifier toward the majority class in imbalanced dataset. So to give importance to minority dataset, an entropy-based fuzzy TWSVM for class imbalanced dataset is proposed in [18].

But TWSVM has two drawbacks namely overfitting and convex quadratic programming problems. To solve these problems, two smoothing approaches by formulating a pair of unconstrained minimization problems (UMP) in dual variables [19] is incorporated. The solution will be obtained by solving two systems of linear equations rather than solving two quadratic programming problems.

4   Results and Discussions

4.1   Database Creation

First step taken for this work is database creations for training the classifier phase which includes nearly 25 uttered sound signals from 6 different people (4 females and 2 males) of four different emotions.

4.2   Feature Extraction

Second step of this work is extracting features from both recorded and real-time signal. There are two major features which are taken into account for recognizing the emotional state of a person; one is prosodic feature and other is spectral feature. In prosodic feature, energies of the sound signal are taken into consideration and similarly, for spectral feature MFCCs are considered.

(a) **(b)**

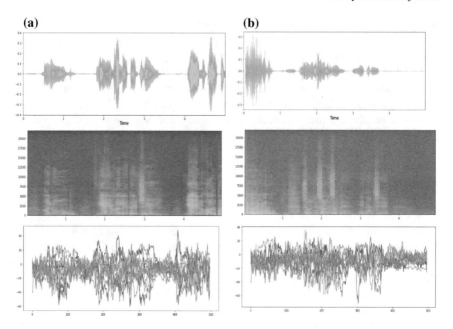

Fig. 3 Spectrogram and MFCC plot of a speech signal **a** Plot signal for Anger **b** Plot signal for Happy

Figure 3a shows the representation of sound signal which has the duration of 5 s with the emotion as anger and its corresponding spectrogram and MFCC plot. The energy of such signal is calculated and found as 61.4847026 using Python coding platform. Also, the MFCC of the signal is calculated, and it is found to be 0.010745288969. Figure 3b shows the representation of sound signal which has the duration of 5 s with the emotion as happy and its corresponding spectrogram and MFCC plot. The energy and MFCC of happy signal are calculated, and the values are 49.0393990469 and 0.0225676628858, respectively.

Figure 4a shows the representation of sound signal which has the duration of 5 s with the emotion as neutral and its corresponding spectrogram and MFCC plot. The energy of such signal is calculated and found that 28.442599228 using Python coding platform. The MFCC of the signal is found to be 0.0184275263467. Figure 4b shows the representation of sound signal which has the duration of 5 s with the emotion as sad and its corresponding spectrogram and MFCC plot. The energy and MFCC of sad signal are calculated, and the values are 17.4770927124 and 0.0434828660627, respectively. The spectrogram is also used to differentiate the voiced signal from unvoiced signal. The MFCC coefficients are plotted with respect to the amplitude for each frame.

From each speaker, four different emotion signals are recorded, and the energies of each signal and Mel filter bank are calculated. The average values of energies and Mel filter bank are calculated for each emotion which is tabulated in Tables 1 and 2. From Table 1, the average value of energies for anger is the highest and sad is the

(a)   **(b)**

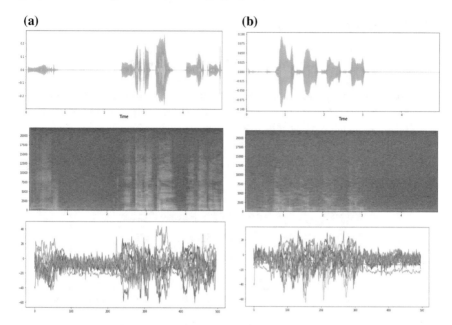

Fig. 4 Spectrogram and MFCC plot of a speech signal **a** Plot signal for Neutral **b** Plot signal for Sad

Table 1 Energy of 6 speakers with 4 emotions

Emotion	Speaker_1	Speaker_2	Speaker_3	Speaker_4	Speaker_5	Speaker_6	Average
Anger	61.484	60.593	62.890	60.239	61.756	59.998	61.160
Sad	17.477	16.903	18.020	16.930	17.735	16.920	17.330
Neutral	28.442	27.542	29.456	27.567	28.234	26.945	28.031
Happiness	49.039	48.567	49.456	48.754	49.903	47.976	48.949

Table 2 MFCC of 6 speakers with 4 emotions

Emotion	Speaker_1	Speaker_2	Speaker_3	Speaker_4	Speaker_5	Speaker_6	Average
Anger	0.0107	0.00972	0.0107	0.0117	0.01072	0.0103	0.0106
Sad	0.04342	0.04223	0.04348	0.04438	0.04523	0.04148	0.0433
Neutral	0.018427	0.01798	0.01832	0.01824	0.00998	0.01892	0.0169
Happiness	0.022567	0.02567	0.02156	0.02626	0.02367	0.02265	0.0213

lowest among these four emotions. Table 2 shows the Mel filter bank values, where there is a large variation for the values of anger and sad.

4.3 Emotion Classification

In this paper, support vector machine (SVM) machine learning algorithm is used for classifying the emotions for both recorded and real-time audio signal from sample of six speakers and classified using conda platform. Using SVM, KNN and artificial neural network (ANN), the performance measures such as accuracy, recall and precision are calculated using Eqs. (3)–(6).

$$Recall = \frac{True\ Positive}{True\ Positive + False\ Negative} \tag{3}$$

$$Precision = \frac{True\ Positive}{True\ Positive + False\ Postive} \tag{4}$$

$$F1\_Score = 2 * \frac{Precision \times Recall}{Precision + Recall} \tag{5}$$

$$Accuracy = \frac{True\ Positive + True\ Negative}{True\ Positive + False\ Positive + True\ Negative + False\ Negative} \tag{6}$$

True positive is the number of emotions predicted positive and recognized as negative. False positive is the number of emotions predicted positive but are actually negatives. True negative is the number of emotions predicted negative and recognized as negative. False negative is the number of emotions predicted negative but are actually positives.

$F1$ score is the weighted average of precision and recall. This score takes both false positives and false negatives into account. For uneven class distribution, $F1$ is more useful than accuracy. Accuracy works best if false positives and false negatives have similar cost. If the cost of false positives and false negatives are very different, it is better to look at both precision and recall. Accuracy is calculated from confusion matrix using equation.

Table 3 shows the class precision, class recall, $F1$ score and accuracy are calculated by using Eqs. (3)–(6) for KNN, ANN and SVM classifier. Among these three, KNN has its lowest values but ANN and SVM provide the same result. Even though they provide the same result, SVM is found to be better than ANN. The reason is that ANN often converges on local minima rather than global minima and also ANN overfit if training is very large which causes some noise.

Table 3 Performance measures of different emotions from different classifiers

Classifier	KNN	SVM	ANN
Precision	0.90	0.94	0.94
Recall	0.85	0.92	0.92
$F1$-score	0.85	0.92	0.92
Accuracy (%)	66.67	85.7	85.7

5 Conclusion

Speech emotion recognition for Tamil language speaker has been done using SVM classifier. The overall accuracy of detecting the emotional signal using KNN, SVM and ANN classification with prosodic energy feature and spectral MFCC feature of this system are 66.67, 85.7 and 85.7%, respectively. The accuracy of the system falls because the cross-validation of two emotions is quite similar which makes the model to confuse in predicting category of the signal. This can be further improved by including more parameters in both prosodic and spectral features of the signal which makes the model to train much better to recognize the emotion more accurately when compared to training model with less features. In future, including database of large sets by extracting various features from emotional signal will provide the model to predict the signal more accurate.

References

1. Agrawal, A., Mishra, N.K.: Fusion based emotion recognition system. In: Proceedings—2016 International Conference on Computational Science and Computational Intelligence, CSCI 2016, pp. 727–732. Institute of Electrical and Electronics Engineers Inc. (2017). https://doi.org/10.1109/CSCI.2016.0142
2. Desai, D.: Emotion recognition using speech signal: a review. Int. Res. J. Eng. Technol. (IRJET) **05**(04), 1599–1605 (2018)
3. Mustafa, M.B., Yusoof, M.A.M., Don, Z.M., Malekzadeh, M.: Speech emotion recognition research: an analysis of research focus. Int. J. Speech Technol. **21**(1), 137–156 (2018). https://doi.org/10.1007/s10772-018-9493-x
4. Ke, X., Zhu, Y., Wen, L., Zhang, W.: Speech emotion recognition based on SVM and ANN. Int. J. Mach. Learn. Comput. **8**(3), 198–202 (2018). https://doi.org/10.18178/ijmlc.2018.8.3.687
5. Abdollahpour, M., Zamani, J., Rad, H.S.: Feature representation for speech emotion recognition. In: 2017 25th Iranian Conference on Electrical Engineering, ICEE 2017, pp. 1465–1468. Institute of Electrical and Electronics Engineers Inc. (2017). https://doi.org/10.1109/IranianCEE.2017.7985273
6. Jamil, N., Apandi, F., Hamzah, R.: Influences of age in emotion recognition of spontaneous speech. In: 2017 International Conference on Speech Technology and Human-Computer Dialogue (SpeD) (2017)
7. Albornoz, E.M., Milone, D.H.: Emotion recognition in never-seen languages a novel ensemble method with emotion profiles. IEEE Trans. Affect. Comput. **8**(1), 43–53 (2017)
8. Joe, C.V.: Developing Tamil emotional speech corpus and evaluating using SVM. In: 2014 International Conference on Science Engineering and Management Research, ICSEMR 2014. Institute of Electrical and Electronics Engineers Inc. (2014). https://doi.org/10.1109/ICSEMR.2014.7043627
9. Fragopanagos, N., Taylor, J.G.: Emotion recognition in human-computer interaction. IEEE Trans. Speech Audio Process **18**(9), 389–405 (2017)
10. Chandran, A., Pravena, D., Govind, D.: Development of speech emotion recognition system using deep belief networks in Malayalam language. In: 2017 International Conference on Advances in Computing, Communications and Informatics, ICACCI 2017, Jan 2017, vol. 2017, pp. 676–680. Institute of Electrical and Electronics Engineers Inc. (2017). https://doi.org/10.1109/ICACCI.2017.8125919

11. Chernykh, V., Prikhodko, P.: Emotion recognition from speech with recurrent neural networks (2017). Retrieved from http://arxiv.org/abs/1701.08071
12. El Ayadi, M., Kamel, M.S., Karray, F.: Survey on speech emotion recognition: features, classification schemes, and databases. Pattern Recognit. **44**(3), 572–587 (2011). https://doi.org/10.1016/j.patcog.2010.09.020
13. Garg, P., Sehgal, S.: Comparison of emotion recognition models in spoken dialogs. Int. J. Softw. Hardw. Res. Eng. **3** (2015)
14. Joseph, A., Sridhar, R.: Performance evaluation of various classifiers in emotion recognition using discrete wavelet transform, linear predictor coefficients and formant features (2017). https://doi.org/10.1007/978-981-10-2525-9
15. Javier, G., Sundgren, D., Rahmani, R., Larsson, A., Moran, A., Bonet, I.: Speech emotion recognition in emotional feedback for human-robot interaction. Int. J. Adv. Res. Artif. Intell. **4**(2) (2015). https://doi.org/10.14569/ijarai.2015.040204
16. Srikanth, M., Pravena, D., Govind, D.: Tamil speech emotion recognition using deep belief network (DBN). In: International Symposium on Signal Processing and Intelligent Recognition Systems (SIRS), pp. 328–336 (2017)
17. Dave, N.: Feature extraction methods LPC, PLP and MFCC in speech recognition. Int. J. Adv. Res. Eng. Technol. **1**(VI), 1–5 (2013)
18. Gupta, D., Richhariya, B., Borah, P.: A fuzzy twin support vector machine based on information entropy for class imbalance learning. Neural Comput. Appl. 1–12 (2018). https://doi.org/10.1007/s00521-018-3551-9
19. Tanveer, M., Shubham, K.: Smooth twin support vector machines via unconstrained convex minimization. Filomat **31**(8), 2195–2210 (2017). https://doi.org/10.2298/FIL1708195T

Real-time RADAR and LIDAR Sensor Fusion for Automated Driving

Rohini Devagiri, Nalini C. Iyer and Shruti Maralappanavar

Abstract In this paper, an attempt to detect obstacle using fusion of RADAR and LIDAR data is being made using Kalman filter. Kalman filter is the preferred algorithm for multi-sensor data fusion as it is computationally efficient compared to deep learning techniques. The algorithm emphasizes on widening the field of view in comparison with individual sensors and corrects failure of either of the sensors. Observation and measurement noise covariance data for Kalman filter is computed using the operating range of sensors and real-time sensor data, respectively. The proposed fusion algorithm has been tested on a real vehicle and verified for different region of operations of both RADAR (AWR1642) and LIDAR (CE30) sensors and the comparison is reported.

Keywords Multi-sensor data fusion · Kalman filter · State estimation · Prediction · Correction

1 Introduction

Obstacle detection is one of the important components of autonomous vehicle as it helps systems to locate the obstacles and to avoid the possible collisions. It is achieved by using various sensors like RADAR, LIDAR, and camera. Each of these sensors has different operating range, accuracy, and precision. RADAR sensor is used for detection of obstacles in a very long range while LIDAR is used for the detection of obstacles in a smaller range. In order to obtain the reliable distance of obstacles present around the vehicle fusion of multiple sensor data is required. Multi-sensor

R. Devagiri · N. C. Iyer (✉) · S. Maralappanavar
School of Electronics and Communication Engineering, KLE Technological University,
Hubli, India
e-mail: nalinic@kletech.ac.in

R. Devagiri
e-mail: rohinidevagiri@gmail.com

S. Maralappanavar
e-mail: shruti_m@kletech.ac.in

© Springer Nature Singapore Pte Ltd. 2020   137
S. Agarwal et al. (eds.), *Machine Intelligence and Signal Processing*,
Advances in Intelligent Systems and Computing 1085,
https://doi.org/10.1007/978-981-15-1366-4_11

data fusion is the technology concerned with the combination of how to combine data from multiple sensors in order to make inferences about the physical event, activity, or a situation.

A brief of the different methods or techniques employed for multi-sensor data fusion as reported by various authors is summarized below. Y. A. Vershinin [1] have proposed a methodology where continuous-time decentralized Kalman filters are configured as fusion units for each sensor, which is considered as a node and the data is communicated between the nodes. This avoids the need of a centralized system and as its failure leads to the degradation in performance of the complete system. In this approach, even though there is a failure in any of the subsystems, it will not adversely affect the performance of complete system. The drawback of this approach is that the synchronization between the nodes must be established, failing to which the consistency of data is lost and results in erroneous results.

On the other hand, Giannì et al. [2] use single Kalman filter to fuse the data obtained by the different sensors like RADAR, LIDAR, and SONAR. The sensor data is preprocessed using the rule of sieve which filters the distances measured beyond the operating range of sensors. Later, this preprocessed data is input to Kalman filter which is configured as fusion device.

Further Ramli et al. [3] used the fusion of camera and LIDAR sensors to detect the obstacles around a small unmanned aerial vehicle (UAV). This is a pipelined approach where LIDAR sensor is initially used to obtain the distances of obstacles from UAV and later, camera is instructed to capture the image. To analyze this captured image, the algorithms like SURF detection and Harris corner detector are employed. The major drawback of this methodology is that there is no parallelism in the operation of the sensors and the pipelined approach adds time delay.

Also, Farias et al. [4] proposed an obstacle detection model for a robot based on fusion of sensor data and automatic calibration using artificial neural networks (ANN). A total of eight Infrared sensors and five ultrasonic sensors placed at every 45° around the robot act as input activation functions of the neural network while 20 neurons formed the hidden layer. This network results in eight output neurons which are the distances to the obstacles around the robot. The demerit of this approach includes the uncertainties introduced by the surface of the objects in case of proximity sensors and requirement of large training data. Another challenge could be the changing parameters of ANN with respect to the changing positions of sensors around the robot.

There exist some other simple techniques implemented for the fusion of multi-sensor data like the development of a mobility assistive device consisting of infrared and ultrasonic sensors which are combined to avoid false detection of obstacles as proposed by Mustapha et al. [5]. Infrared sensor is employed to detect the obstacles within 80 cm range and that of ultrasonic within 6 m range. The disadvantage of this method is that there is no comparison of data rather decision is taken only on the grounds of detection range.

Based on the above contributions proposed by the respective authors to provide optimal solutions for multi-sensor data fusion system, the gaps identified are failure in maintaining the data consistency in case of decentralized fusion approach, the added delays due to pipelined usage of multiple sensors, the requirement of large data to train the ANN model and also to calibrate the obstacle detection system and lack of proper comparison of data obtained from different sensors.

To overcome the issue of delay due to pipelined usage of multiple sensors and the huge data set required to train the ANN model, this paper proposes a novel multi-sensor data fusion system to fuse the data from RADAR and LIDAR sensors for an autonomous vehicle avoiding the practical issues in the acquisition of sensor data.

The rest of the paper is organized as follows. The overview of Kalman filter is given in Sect. 2. The proposed methodology used in multi-sensor data fusion system is discussed in Sect. 3. The experimental results are discussed in Sect. 4. The paper is summarized and concluded in Sect. 5.

2 Overview of Kalman Filter

The Kalman filter is realized as a set of mathematical equations that provide an efficient computational means to estimate the state of the process, in a way that minimizes the mean of the squared error [6]. Kalman filter consists of three stages. They are initialization of system parameters, prediction of the estimated state, and correction of estimated state. The details of each stage are briefed below.

2.1 Prediction of the Estimated State

The time update equations are responsible for projecting forward (in time) the current state and error covariance estimates to obtain the priori estimates for the next step time [6]. These time update Eqs. (1) and (2) can be also called as prediction equations.

$$\hat{x}_k = A\hat{x}_{k-1} + Bu_{k-1} \tag{1}$$

$$\hat{P}_k = A\hat{P}_{k-1}A^T + Q \tag{2}$$

where $\hat{x}_k$ is the prior state estimate, $\hat{x}_{k-1}$ is the state estimate of previous time step, u_{k-1} is the external control input to the system (if any), $\hat{P}_k$ is the priori estimated error covariance, A is the state transition matrix, B is the control matrix, and Q is the process noise covariance.

2.2   Correction of the Estimated State

The measurement update equations are responsible for the feedback for incorporating a new measurement into the priori estimate to obtain an improved posterior
estimate [6]. These measurement Eqs. (3), (4) and (5) can be also called as correction
equations as the actual measured output is fed back to the system in order to retune
the parameters of the filter.

$$K_k = \hat{P}_k H^T \left(H \hat{P}_k H^T + R \right)^{-1} \tag{3}$$

$$x_k = \hat{x}_k + K_k \left(z_k - H \hat{x}_k \right) \tag{4}$$

$$P_k = (I - K_k H) \hat{P}_k \tag{5}$$

where K_k is the Kalman filter gain, x_k is the posteriori state estimate, z_k is the
input measurement given to the system, H is the observation matrix, and R is the
measurement noise covariance.

2.3   Initialization of System Parameters

The initialization of system parameters is the crucial step in the filter design. A, B,
Q, R, and H as given in Eqs. (1), (2), (3), (4), and (5) are the parameters to be tuned.
A is a $n \times n$ state transition matrix, where n stands for number of states. A, state
transition matrix, relates the state at the previous time step to the state at the current
step. B is a $n \times 1$ control matrix which relates the control input u_k to the state x_k.
Q is process noise covariance, R is measurement noise covariance, and H is $m \times n$
matrix indicating observation model that relates the measurement z_k to the state x_k.

3   Proposed Methodology

The proposed methodology for multi-sensor data fusion in AEV includes data acquisition, system modeling, and Kalman filter as shown in Fig. 1. The distance of obstacle
as detected by LIDAR sensor and RADAR sensor is given as input to the system. Data
acquisition subsystem is responsible for synchronization of data obtained from the
sensors. This data is modeled according to the application and given as measurement
input to the Kalman filter.

   The obstacle detection for an autonomous vehicle is achieved using fusion of
data from RADAR and LIDAR sensors. The details of multi-sensor data fusion are
discussed in this section.

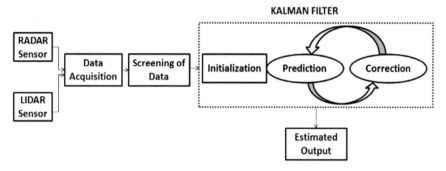

Fig. 1 The block diagram of multi-sensor data fusion system

3.1 RADAR Sensor

RADAR sensors use Frequency Modulated Continuous Waves (FMCW) to reliably detect stationary or moving obstacles. These are widely used in autonomous applications. AWR1642 Single-Chip FMWC RADAR sensor is used to measure the distance of the obstacles. It is a device capable of operating in the 76- to 80-GHz band. The device includes an ARM R4F-based processor subsystem, which is responsible for radio configuration, control, and calibration. Simple programming model changes can enable a wide variety of sensor implementation (short, mid, long) with the possibility of dynamic reconfiguration for implementing a multimode sensor. The device is configured to operate up to 50 m distance with 30° angular coverage. The device outputs the distance of obstacle.

3.2 LIDAR Sensor

LIDAR sensor calculates the distance to the object by illuminating the target with pulsed laser light and measuring the reflected pulses. CE30 is the LIDAR sensor used to measure the distance of obstacle precisely. The detection region up to 4 m depth and 62 cm width is configured. The obstacle present in this rectangular area is detected by the sensor and its précised distance is measured.

3.3 Data Acquisition

Data acquisition subsystem is responsible for synchronization of data obtained from the RADAR and LIDAR sensor. RADAR sensor measures the distance of obstacle in terms of meters while LIDAR does in terms of centimeters. The output frames per second for both sensors are different. Hence, it is essential to synchronize the data with respect to time and units. RADAR gives 4 distance values per second and LIDAR gives 10 distance values per second. Therefore, the response time of RADAR is 0.25 s while that of LIDAR is 0.1 s. To consider the rigorous changes in the position of obstacle, the response time of fusion system is scaled to 0.1 s. Though it requires 0.25 s to reflect the change in RADAR output, the previous results of it are considered with the robust results obtained from LIDAR at the rate of 0.1 s.

3.4 Screening of Data

Screening of data is responsible to configure the H matrix which indicates the observation model. Based on the requirements of application and the accuracy of data measured from the sensors, certain threshold values are set. These threshold values ensure that the erroneous data obtained from the sensors, in cases where the obstacles are out of the operating range of sensor, are sieved off.

$$H = \begin{bmatrix} h_R \\ h_L \end{bmatrix}$$

The observation matrix H is a 2×1 matrix in which h_R is the observation model parameter for RADAR sensor and h_L is the observation model parameter for LIDAR sensor.

Table 1 lists the operating range of RADAR and LIDAR sensors. These ranges are practically tested. If the output of the RADAR sensor is out of the range as mentioned in Table 1, then the value of h_R is taken to be zero otherwise it is taken to be one. Similarly, if the output of LIDAR sensor is out of the range, then the value of h_L is taken to be zero otherwise it is taken to be one. In this way, the system is modeled for every iteration or frame.

In our application, the output of the fusion is the distance of the obstacle. Hence, there is a single state and x_k is the vector having the state variables. A is a 1×1 matrix and B is taken to be zero (any value can be considered) as there is no control input u_k to the system. Q is a 1×1 matrix which might change in every step of time but is assumed to be constant. The initial estimate error covariance $\hat{P}_0$ is assumed to be any nonzero value as it will get tuned itself in the next time step.

Table 1 Operating range of sensors

Sensor	FPS	Minimum range (m)	Maximum range (m)	Angular coverage (°)
RADAR (AWR1642)	4	2	50	30
LIDAR (CE30)	10	0.1	3.5	180

$$R = \begin{bmatrix} \sigma_R & 0 \\ 0 & \sigma_L \end{bmatrix}$$

The measurement noise covariance R has to be measured practically before the operation of filter. The measurement noise covariance R is a 2×2 matrix in which σ_R is the measurement noise covariance of the RADAR sensor and σ_L is the measurement noise covariance of the LIDAR sensor. These values are calculated using the test dataset of RADAR and LIDAR output for the same obstacle. The measurement noise covariance of a sensor is the standard deviation of error between the true value and measured output. The values of σ_R and σ_L obtained are 0.04962 and 0.12452, respectively.

$$z_k = \begin{bmatrix} z_{kR} \\ z_{kL} \end{bmatrix}$$

The input measurement z_k is a 2×1 matrix in which z_{kR} is the distance of obstacle (measurement value) given by RADAR sensor and z_{kL} is the distance of obstacle (measurement value) given by LIDAR sensor.

4 Experimental Results and Discussion

The multi-sensor data fusion system for autonomous vehicle designed to obtain the better reliable distance of obstacle than individual sensors is implemented and the experimental results obtained are discussed in this section. The algorithm is tested for various cases where the obstacle is in the LIDAR region only, in the RADAR region only, and in the region of both RADAR and LIDAR are discussed in detail.

4.1 Experimental Setup

The data for fusion algorithm is acquired from RADAR (AWR1642) and LIDAR (CE30) sensors by mounting them on vehicle as shown in Fig. 2. The operating regions of RADAR and LIDAR sensors are as shown in Fig. 3. There are three different regions of operations identified. They are the RADAR region, the LIDAR region, and the region where operation of both the sensors overlaps. The data is collected considering these three regions and thus, obtained results are listed below.

Fig. 2 Sensor placement on
vehicle

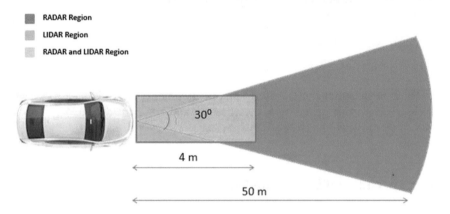

Fig. 3 Operating region of sensors

4.2 Obstacle in the RADAR Region

When the obstacle is out of the region of LIDAR sensor, then the system selects the
distance value as given by the RADAR sensor. The output of Kalman filter is similar
to that of RADAR sensor and also the true distance value of obstacle as shown in
Fig. 4.

4.3 Obstacle in the LIDAR Region

When the obstacle is out of the region of RADAR sensor, then the system selects the
distance value as given by the LIDAR sensor. The Kalman filter reduces the error in
output of LIDAR sensor and gives the distance which is approximately equal to the
true distance value of obstacle as shown in Fig. 5.

Fig. 4 Output of the system when obstacle is in only RADAR region

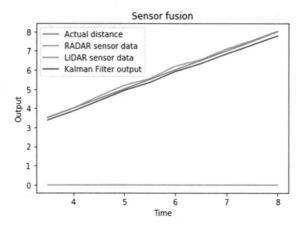

Fig. 5 The output of the system when obstacle is in only LIDAR region

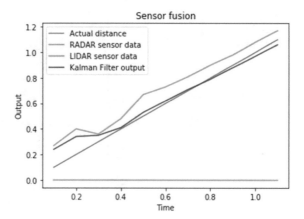

4.4 Obstacle in the Region of Both RADAR and LIDAR

When the obstacle is in the region of both RADAR and LIDAR sensor, then the system considers the distance value given by both the sensors. The Kalman filter reduces the error in output of LIDAR as well as RADAR sensor and gives the distance which is approximately equal to the true distance value of obstacle as shown in Fig. 6. Here, the values of the measurement noise covariance matrix R are responsible in reducing the error.

4.5 System Behavior with Respect to Time

As the response time of RADAR sensor is slower than that of LIDAR sensor, the rapid change in the position of obstacle may not be reflected in the output of RADAR.

Fig. 6 The output of the system when obstacle is in the region of both RADAR and LIDAR

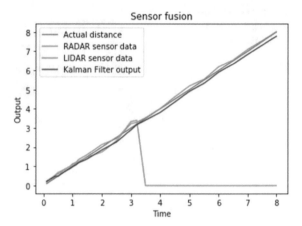

Fig. 7 The system response when obstacle changes within one second

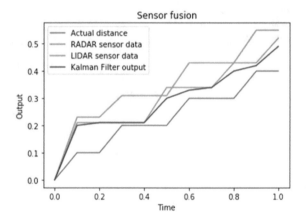

It takes 0.25 s to refresh the output buffer of RADAR sensor while it takes 0.1 s for that of LIDAR sensor. Hence, the responses of RADAR and LIDAR sensors when obstacle changes the position rapidly are different. This is overcome in multi-sensor fusion. The response time of fusion system is 0.1 s and the output of fusion is approximately equal to the true distance value of obstacle as shown in Fig. 7.

5 Summary and Conclusion

The multi-sensor fusion system using Kalman filter is successfully implemented and verified for the different region of operation of sensors. The issue of pipelined usage of multiple sensors is overcome. The field of view is widened in comparison with the field of view of individual sensors. The sensors are calibrated in order to minimize the error. The limitation of the proposed methodology is that the algorithm is dependent on the type of sensors used. Hence, the approach is specific with respect to the operating regions of the sensors. The measurement noise covariance has to be calculated practically and there is no standard formula to calculate it. In future work, the calculation of measurement noise covariance could be learned using either machine learning or deep learning techniques in order to achieve better results. The proposed algorithm can be extended to fuse the data from more than two sensors and also predict the dimensions of the obstacle.

References

1. Vershinin, Y.A.: A data fusion algorithm for multi sensor systems. In: Proceedings of the Fifth International Conference on Information Fusion. FUSION 2002 (IEEE Cat. No. 02EX5997), vol. 1, pp. 341–345. Annapolis, MD, USA (2002)
2. Giannì, C., Balsi, M., Esposito, S., Fallavollita, P.: Obstacle detection system involving fusion of multiple sensor technologies. In: ISPRS—International Archives of the Photogrammetry, Remote Sensing and Spatial Information Sciences, vol. XLII-2/W6, pp. 127–134 (2017) https://doi.org/10.5194/isprs-archives-xlii-2-w6-127-2017
3. Ramli, M.F.B., Legowo, A., Shamsudin, S.S.: Object detection technique for small unmanned aerial vehicle. In: IOP Conference Series: Materials Science and Engineering, vol. 260, p. 012040 (2017)
4. Farias, G., et al.: A neural network approach for building an obstacle detection model by fusion of proximity sensors data. Sensors (Basel, Switz) **18**(3), 683 (2018). (PMC. Web. 10 Sept 2018)
5. Mustapha, B., Zayegh, A., Begg, R.K.: Multiple sensors based obstacle detection system. In: 2012 4th International Conference on Intelligent and Advanced Systems (ICIAS 2012), pp. 562–566. Kuala Lumpur (2012)
6. Welch, G., Bishop, G.: An Introduction to the Kalman Filter. Proc. Siggraph Course. 8. (2006)

Generalizing Streaming Pipeline Design for Big Data

Krushnaa Rengarajan and Vijay Krishna Menon

Abstract Streaming data refers to the data that is sent to a cloud or a processing centre in real time. Even though we have limited exposure to such applications that can process data streams on a live basis and generate useful insights, we are still in infancy when it comes to complicated stream processing. Current streaming data analytics tools represents the third-/fourth-generation data processing capability in the big data hierarchy which includes the Hadoop ecosystem, Apache Storm™ and Apache Kafka™ and its likes, Apache Spark™ framework, and now, Apache Flink™ with its non-batch stateful streaming core. Each of these individually cannot handle all the aspects of a data processing pipeline, alone. It is essential to have good synergy between these technologies to cater to the management, streaming, processing and fault-tolerant requirements of various data-driven applications. Companies tailor their pipelines exclusively for their requirements, since making a general framework entails mammoth interfacing and configuration efforts that are not cost-effective for them. In this paper, we envision and implement such a generalized minimal stream processing pipeline and measure its performance, on some data sets, in the form of delays and latencies of data arrival at pivotal checkpoints in the pipeline. We virtualize this using a Docker™ container without much loss in performance.

Keywords Big data · Apache kafka™ · Apache Flink™ · Docker™

1 Overview

Big data analytics has become an inevitable infrastructure tool for every one generating or amassing large volumes of data at very high speed (data velocity/rate of generation). A data process is, in a sense, similar to power generation. Just like how

K. Rengarajan (✉) · V. K. Menon
Centre for Computational Engineering and Networking (CEN), Amrita School of Engineering, Amrita Vishwa Vidyapeetham, Coimbatore, India
e-mail: krushnaarengarajan@gmail.com

V. K. Menon
e-mail: m_vijaykrishna@cb.amrita.edu

© Springer Nature Singapore Pte Ltd. 2020
S. Agarwal et al. (eds.), *Machine Intelligence and Signal Processing*,
Advances in Intelligent Systems and Computing 1085,
https://doi.org/10.1007/978-981-15-1366-4_12

energy can be harnessed only in certain forms such as heat or electricity, data stores information (similar to energy) which could be understood or extracted only when the data is aptly transformed in certain ways. Raw data is seldom found useful in any way. Traditional data processing was done on data that is collected, stored and catalogued. Data scientists study this data and prescribe the required transformations and computations in order to yield useful scientific or business insights. These analyses were typically done on smaller quantum of data; now, data is produced in huge volumes and the rate of generation of data is beyond exponential in most fields of engineering, science and business [1–5]. Shear high volume and velocity itself make any type of traditional data processing impossible; this high growth of data can be attributed to a high increase in data recorders such as smartphones, sensors, social media platforms, etc. This is where we get big data (BD) from. BD process has newer dimensions never was found necessary before, such as resource management, data management, high dimensionality and fault tolerance. Hence, we require dedicated data pipelines that cater to all the above needs efficiently. It is necessary to understand the current requisites for BD processes so that we can see how it differs from traditional data processing. Real-time stream processing is an evolving aspect of BD, which has found application in almost all aspects of engineering, business and healthcare. The data will stream as raw bytes or in structured frames using a communication protocol, or simply as strings through TCP Sockets, from a data source to a subscriber application. The application can, in real time, apply transformations or computations to it, in order to generate insights or to forecast future values. Streaming analytics is required, where the insight has to be trendy and immediate (real time or near real time) [6–9] and aims to process anything from tens of thousands of events per second to millions per second, depending on the requirement, the nature of the data and the target consumer. There are many benchmarked Apache-hosted software platforms for data pipelining and real-time data analytics.

This paper aims to show an implementation perspective of a generalized pipeline with data resource management, and real-time processing; an end-to-end streaming data analytics experiment is demonstrated. The complete infrastructure is tested on native standalone platforms and also deployed on a Docker™ container. Further, we test the pipeline with different data sets and show latency in dataflow for a few data sets using Apache Kafka™. For on-stream real-time computations (micro-service implementations), we use Apache Flink™, a fourth-generation non-batch stream processing framework.

1.1 Generalizing Big Data Pipelines

Like any traditional data processing systems, BD has mostly the same basic processing types. It is at times difficult to segregate them on real analytics task; any BD pipeline will have them indiscriminately and multiple times for certain purposes, as part of the analytics being performed. In this section we are trying to put more clarity on how some BD requisites, can map to the types of processes The basic types of data processes include *Transaction, Batch, Stream, Warehouse and Graph* [7, 10, 11]. It would be interesting to observe, how the requisites of BD processes maps to these fundamental types of data processing. It helps better visualize and generalize a BD pipeline. The following are the requisites we would consider predominant in any BD pipeline.

1. Computation
2. Shuffle
3. Fault tolerance
4. Data management
5. Resource management
6. Stateful access.

2 Data Processing Types

a. **Transaction Processing in BD Pipeline**.

A transaction is a data-driven event, involving a series of database modifications or 'commits' in a specific order. These set of data base changes are said to be 'atomic', i.e. either all of them complete or none should complete. The computations involved are limited to validation of the current transaction being committed as to the source of the event, the destination, the inventory balance and other constrains such max and min commodity counts, etc. If at all any fault might occur, it will only result in roll back of the committed records and will not affect any prior or follow transactions. The data is of elementary type and no sophisticated data or resource management efforts are needed. The committed transactions cannot be changed but a new nullifying transaction may be added for reversal, during audits.

b. **BD Batch Processing Tasks**

This is the typical MapReduce process type. Batches of data records are transformed into different 'shapes' [12–16] to facilitate information extraction. It is also the most common type of process found in BD pipelines. However, these are in no sense trivial; the 'map' can have very hard computations. The basic premise is that successive records are completely data independent; this is true for any 'map' task. A 'batch' is a data set abstraction like an HDFS block, an RDD [15, 17], a data set [15, 17–19] or data frame [18, 19]. Faults in batch processing occur solely due to data loss and/or task failure. In Hadoop ecosystem, redundancy is the main key in tolerating data loss. Tasks that fail are just relaunched. However, in the new generation technologies such as Spark, the in-memory data representations are simply reconstructed, if lost, from a checkpoint. This is termed as lineage. Batch tasks are fully independent and require task brokers, schedulers and hypervisors to pool and manage them efficiently when large amounts of data are involved.

c. **Real-time Stream Processing**

Real-time data processing has been happening in many safety–critical applications for decades. Most of the results are used for control and navigation. When it comes to BD, real-time processing is what is generally termed as stream processing [15, 20–22]. The third- and fourth-generation BD tools [23, 24] will allow us to directly apply map and reduce computation on live data streams. Streams are a different type of data abstractions, yet we can perform complete batch type processes on them with maps and reductions. However, most streaming tasks have to complete in real time, i.e. they have a very close dead line after which these results will not be useful. Here, the prudence is to keep the batch small that can be collected and completely processed within a few $100 \mu s$ and the results provided. This is often referred to as 'micro-batching' [15, 23, 25]. Here, data stream window can only be time bound due to the floating batch size (data might not come at the same speed). If the frequency of data flow is 50 Hz (50 data per second), then a window of 1 s expects roughly 50 ± 20, say, as the range of micro-batch sizes [15, 23, 26]. It also incurs latency (due to micro-batching) making the computations eventually near real time. Apache Flink™ has the capability to stream without micro-batching and is completely stateful; this can perform complex computations in real time as well as have (fixed) size-based stream windows. Despite the same premise on data independence as with 'map' tasks (refer previous section), streams are much more efficient at computing even auto correlated models which violate the above independence criteria.

d. **Warehousing in BD Pipeline**

Data might be acquired from different sources and synchronized based on a common field. The best contemporary example is phasor measurement data on power grid transmission lines, which are synchronized on GPS clock signal. This timestamp serves as the key for combining data from several phasor measurement units (PMUs) to a single row giving the opportunity to do further batch process on it to get insights into system dynamics. The main computation here is a join query that will combine data sets or data streams (static and dynamic data) to a common view over a synchronized key. Traditionally, this is done to generate static tables in a database and was called data warehousing. In a BD pipeline, however, static views are less preferred and in-memory dynamic views are mostly used. The warehouse joins do not stand separate and is a part of mostly all batch tasks. The distribution of data is crucial to efficient joins. Joins are performed locally in each node and entails limited network shuffles. This can change in some cases, especially when the data is not properly distributed within the cluster, such as the case, when disparate data sets reside on different nodes; the data need to be brought to a common node in order to generate a joined view or data set. Joins are simple operations that do not create any state changes and can persist as long as they are needed.

e. **Graph Processing Tasks**

A Graph is a representation of measured dependencies between nodal entities, in this case between various data. Of all the processes on a BD pipeline, a graph process is unique in the way that it strives on pairwise dependencies in data sets. Data is interpreted as nodes or vertices which are connected to each other by an edge weighted by the pairwise relationship between the two data. The ideal example is the Page-Rank algorithms that rank web pages (vertices) by their popularity, considering the incoming and outgoing links (edges) they have. There is no other BD task that considered relationships between data as the premise for computations. All major BD frameworks have a graph processing module like Giraph™ for Hadoop™ or GraphX™ for Spark™. This is a computation and shuffle-intensive process. Graph computations are modelled using MapReduce tasks but entails a whole lot of indexing and internal joins that help to capture the pairwise dependencies as edges. Table 1 gives a detailed tabulation of the BD processes and their natures as pertaining to the aspects listed in Sect. 1.1.

3   Available Frameworks

Since the advent of BD era many frameworks have been developed private and public alike. The most popular among publically available frameworks was Apache Hadoop™. Typically, it represents the first generation of all BD frameworks and performed batch processing over large amount of data store in a distributed file system called Hadoop Distributed File-Systems (HDFS). Hadoop implemented the famous MapReduce paradigm proposed by Google [12] in a publically available API. However, Hadoop did have any specific data abstraction than HDFS that allowed to cache data in-memory. With sufficiently large cluster or cloud Hadoop was able to perform amicably. Hadoop was unable to do iterative task that required saving intermediate results, for further processing. Any attempt in this direction resulted in excessive disc overheads. Furthermore, task that requires real-time stream processing was not possible on Hadoop. This required a separate type processing core that supported stream processing; fundamentally different from batch processing hence frameworks such as Apache Storm™ and Apache Samza™ where developed. Both these frameworks are considered to be in second generation among all of BD tool evolution. However, stream processing does not replace batch processing, and hence, it was necessary that a cluster be parallelly configured with both types of frameworks. Unification of batch and stream processing was in evitable as the requirement for both started coexisting. Hence Apache Spark™ (third generation BD framework) gained momentum and became one of the most active in Apache community. The real contribution is a data abstraction called Resilient Distributed Datasets(RDD). RDDs are in-memory data structures that can fully or partially be persistent in the RAM or the disc simultaneously. RDDs have no or very little disc latency and can accelerate map-reduce jobs from 10 to 100 times faster than Hadoop. Spark also introduced Scala as its natural interface language, which enables us to code very precise map-reduce tasks. Spark supports another abstraction called Dstreams with which real-time stream processing that can be done with ease. However, the stream is implemented as micro-batches. A newer framework Apache Flink™ removed

Table 1 Comparison of various BD processing techniques

	Transaction	Batch	Stream	Warehouse	Graph
Computation	Very less mostly validation	Map and reduce tasks	Map and reduce tasks window task	Join queries	Map and reduce tasks
Shuffle	No shuffle	Shuffle-intensive	Minimal shuffle	No or limited shuffle	Shuffle-intensive
Fault tolerance	Atomic consistency	Redundant data or lineage	Lineage	Lineage	Lineage
Data management	No data management	Disc/memory serialization	In-memory data structures	In-memory data frames (views)	In-memory data structures
Resource management	No resource management	NA/hypervisors	Task brokers/hypervisors	Data brokers/hypervisors	Task brokers/hypervisors
Stateful access	Only within a transaction	Only within a batch	Yes	No state changes	Yes
Persistence & iteration	No	Partially iterable	Fully iterable	Fully iterable	Fully iterable

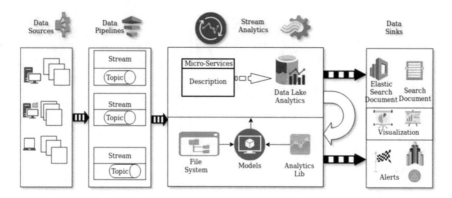

Fig. 1 Apache Flink™ concept of stateful computation

micro-batching with a streaming core as its main data abstraction. Apache Flink™ is capable of stateful streams and performing real-time analytics on live streams; it represents the fourth-generation BD tool (Fig. 1; Table 2).

Message queuing framework form an integral part any data pipeline that work as conduit and as micro-service serving data from source producer to target consumer on demand. There is multitude of queuing framework's and chosen according to the data and response required. Our generalised pipeline uses one data broker and another computation and streaming framework; Apache Kafka™ is used for the former and both Apache Spark™ and Apache Flink™ are used for the later. The final API will supports Apache Spark™ and Apache Flink™ for data processing and Apache Kafka™ and ActiveMQ™ as data brokers (Fig. 2; Table 3).

4 Experimental Pipeline and Results

The experimental pipeline ingests different types of data, pushes it to analytics engine and where transformations and computations are performed before reaching a target micro-service. The purpose is to measure the latency in data delivery and weather it can be classified as real-time delivery. The experiment is repeated multiple times and the average latency is reported. There are two pipelines evaluated; one on normal Linux host system, second using a Docker container.

Docker™ uses the concept of OS-level-virtualization and the Docker container is a packaging mechanism for orchestrated software stacks to be easily deployed anywhere without any library dependencies. Docker image includes APIs with all its dependencies and libraries as a single Docker ship (Fig. 3). The pipeline architecture is as shown in Fig. 4; the results of the latency computations are given in Table 4.

Table 2 Evolution of data analytics engine

Generation	1st generation	2nd generation	3rd generation	4th generation
Technology	APACHE HADOOP™	APACHE TEZ™	APACHE SPARK™	APACHE FLINK™
Processing types	Batch	Batch interactive	Batch interactive **Near-real**-time streaming **Iterative** processing	Hybrid (Streaming + Batch) Interactive **Real-time** streaming **Native-iterative** processing
Transformations	MapReduce	**D**irect **A**cyclic **G**raphs (DAG) Data flows	RDD: **R**esilient **D**istributed **D**atasets	**Cyclic data flows**

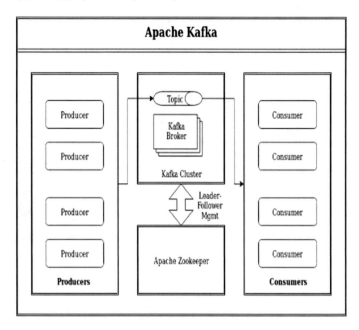

Fig. 2 Apache kafka™ producer–consumer concept

Table 3 Comparison between message queues

	Apache ActiveMQ™	RabbitMQ™	ZeroMQ™	Apache Kafka™	IronMQ™	Apache Qupid™
Broker less/decentralized	No	No	Yes	Distributed	Distributed & cloud-based	No
Language support	C, C++, JAVA, others	C, C++, JAVA, others	C, C++, JAVA, others	C, C++, JAVA, others	C, C++, JAVA, others	C, C++, JAVA, others

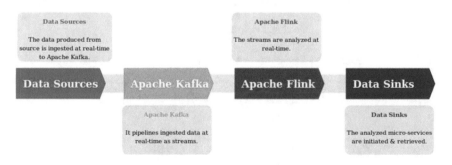

Fig. 3 Generic experimental pipeline

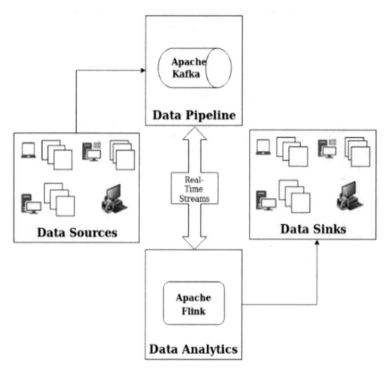

Fig. 4 Architecture

Table 4 Experimental results

	Type of data	Data sources	Computation performed	Data rate kB/sec	Latency per record (ms)	Latency per record (docker) (ms)
JMETER	Numeric	Generated data with floating values	Variance and standard deviation	382	50.1047	65.3665
WIKI SAMPLE	Textual	Sample of wiki data	Word count	624	51.9751	65.3665
SEVER LOGS	Categorical	Server filter query recorded logs	Regex	222	48.8610	51.4221
NSE SAMPLE	Categorical	Sample stock data	ARMA forecasting	64	51.1252	53.6154

5 Conclusion

Our experiment modelled popular aspects of a generic BD stream pipeline. We have considered four different types of data and computations. The first type of data is numeric; floating value generated using Apache Jmeter™. The on-stream computation was computing variance and standard deviation on a window of 1000 values each. The second type of data came from sample of Wikipedia where the computation was to count unique set of word per window of 1000 lines. The third type of data were server logs where Regex based pattern matching is done on 1000 logs each.

The final type of data was stock market data where forecasting is done on window of 50 values and 3 companies using Auto Regressive Moving Average. All latencies is recorded and averaged over multiple execution runs to get the final latency value in each case. In all the cases, we have observed the latency 48–66 ms which is well the real-time response frame. The latencies for normal Linux Host and Docker are separately tabulated; as expected Docker lags behind the host marginally due to extra software overhead.

References

1. Villars, R.L., Olofson, C.W., Eastwood, M.: Big Data: What It Is and Why You Should Care. AMD, USA (2011)
2. Chen, H., Chiang, R.H., Storey, V.C.: Business intelligence and analytics: from big data to big impact. Management information systems research center, University of Minnesota. MIS Quart. **36**(4), 1165–1188 (2012). (13-10-2018 07:18 UTC)
3. Ming, Z., Luo, C., Gao, W., Han, R., Yang, Q., Wang, L., Zhan, J.: BDGS: a scalable big data generator suite in big data benchmarking. In: Lecture Notes in Computer Science Book Series. LNCS, vol. 8585 (2013)
4. Khan, N., Yaqoob, I., Hashem, I.A.T., Inayat, Z., Ali, M., Kamaleldin, W., Gani, A.: Big data: survey, technologies, opportunities, and challenges. Sci. World J., Article ID 712826, 18 (2014)
5. Chen, C.P., Zhang, C.Y.: Data-intensive applications, challenges, techniques and technologies: a survey on Big Data. Inf. Sci. **10**(275), 314–347 (2014)
6. Fu, T.Z.J., et al.: DRS: dynamic resource scheduling for real-time analytics over fast streams. In: 2015 IEEE 35th International Conference on Distributed Computing Systems, 23 July 2015
7. Ellis, B.: Real-Time Analytics: Techniques to Analyze and Visualize Streaming Data. Wiley, Indianapolis (2014)
8. Gandomi, A., Haider, M.: Beyond the hype: big data concepts, methods, and analytics. Int. J. Inf. Manag. **35**(2), 137–144 (2015)
9. Bifet, A., Holmes, G., Pfahringer, B., Read, J., Kranen, P., Kremer, H., Seidl, T.: MOA: A real-time analytics open source framework. In: Lecture Notes in Computer Science Book Series. LNCS, vol. 6913, pp. 617–620 (2011)
10. Stonebraker, M., Çetintemel, U., Zdonik, S.: The 8 requirements of real-time stream processing. ACM SIGMOD Rec. **34**(4), 42–47 (2005)
11. Madden, S.: From Databases to Big data. IEEE Internet Computing: Massachusetts Institute of Technology, IEEE Internet Computing (2012)
12. Dean, J., Ghemawat, S.: MapReduce: simplified data processing on large clusters issue: 1958–2008. In: Communications of the ACM-50th Anniversary, pp. 107–113
13. Boykin, O., et al.: Summingbird: a framework for integrating batch and online MapReduce computations. Proc. VLDB Endow. **7**(13), 1441–1451 (2014)

14. Morton, K., Balazinska, M. Grossman, D.: ParaTimer: a progress indicator for MapReduce DAGs. In: Proceedings of the 2010 International Conference on Management of data, SIGMOD 10, pp. 507–518. ACM SIGMOD, Indianapolis (2010)
15. Patel, K., Sakaria, Y., Bhadane, C.: Real time data processing frameworks. Int. J. Data Mining Knowl. Manag. Process (IJDKP) 5(5), 49–63 (2015)
16. Chen, Y., et al.: Energy efficiency for large-scale MapReduce workloads with significant inter-active analysis. In: Proceedings of the 7th ACM European Conference on Computer Systems, EuroSys 12, pp. 43–56. ACM, Bern, Switzerland (2012)
17. Samadi, Y., Zbakh, M., Claude, Tadonki, C.: Comparative study between Hadoop and Spark based on Hibench benchmarks. In: 2016 2nd International Conference on Cloud Computing Technologies and Applications (CloudTech). IEEE, Marrakech, Morocco (2016)
18. Mozafari et al.: Snappy Data: Streaming, Transactions, and Interactive. SIGMOD'16. ACM, San Francisco, CA, USA (2016)
19. Mozafari, B., Menon, S., Ramnarayan, J.: SnappyData: A unified cluster for streaming. In: 8th Biennial Conference on Innovative Data Systems Research (CIDR'17). Creative Commons Attribution, Chaminade, California, USA (2017)
20. Kamburugamuve, S., Fox, G.C.: Survey of Distributed Stream Processing. Indiana University, Bloomington, IN, USA (2016)
21. Lucknow, A., Chantzialexiou, G., Jha, S.: Pilot-streaming: a stream processing framework for high-performance computing. In: 27th International Symposium on High-Performance Parallel and Distributed Computing (HPDC'18). ACM, Tempe, Arizona, USA (2018)
22. Jayanthi, D., Sumathi, G.: A framework for real-time streaming analytics using machine learn-ing approach. In: Sriperumbudur: Compusoft: An International Journal of Advanced Computer Technology. Proceedings of National Conference on Communication and Informatics-2016, pp. 85–92 (2016)
23. Marcu, O.C., Costan, A., Antoniu, G., Pérez-Hernández, M.S.: Spark Versus Flink: Under-standing Performance in Big Data Analytics Frameworks. European Union Under the Marie Sklodowska-Curie Actions
24. Carboney, P., Katsifodimos, A., Ewenz, S., Markl, V., Haridi, S., Tzoumas, K.: Apache Flink™: Stream and Batch Processing in a Single Engine, pp. 28–38. IEEE Computer Society Technical Committee on Data Engineering (2015)
25. Salloum, S., Dautov, R., Chen, X., Peng, P.X.: Big data analytics on Apache Spark. Int J Data Sci Anal 1, 145–164 (2016). (Springer)
26. Shyam, R., HB, B.G., Kumar, S., Poornachandran, P.: Apache Spark a Big Data Analytics Platform for Smart Grid, pp. 171–178. Elsevier Ltd. (2015)

Adaptive Fast Composite Splitting Algorithm for MR Image Reconstruction

Raji Susan Mathew and Joseph Suresh Paul

Abstract Reconstruction of an image from undersampled k-space data using conventional sparsity priors use any of the pre-specified transforms like wavelet, which utilizes the global sparsity of images. Alternatively, methods such as block-matching and 3D filtering (BM3D) utilize the non-local sparsity of images, resulting in better preservation of finer details. However, use of either of the methods cannot utilize global and non-local sparsity at the same time. In this paper, we propose an efficient combination of the global and non-local sparsity realized using wavelet and BM3D, implemented using the fast composite splitting denoising (FCSD) algorithm. Furthermore, we use an adaptive regularization parameter for the wavelet to yield the best achievable performance using adaptive thresholded Landweber (ATL). The proposed adaptive fast composite splitting denoising algorithm (AFCSD) combines the advantages of both BM3D with adaptive TL through the use of FCSD and thus exploits the self-repetitive structure in images by enforcing non-local sparsity as well as entire consistency of the image. Experimental results reveal that the proposed method is superior to any of the individual methods in terms of reconstruction accuracy and computation complexity. Additionally, when used as an auxiliary constraint in dictionary learning (DL), this can result in further improvements in reconstruction with significant reduction in the computation time as compared to a conventional DL reconstruction.

Keywords Adaptive fast composite splitting denoising(AFCSD) · Adaptive thresholded Landweber (ATL) · Block-matching and 3D filtering (BM3D) · Threshold · Wavelet

R. S. Mathew (✉) · J. S. Paul
Medical Image Computing and Signal Processing Lab, Indian Institute of Information Technology and Management, Trivandrum, Kerala, India
e-mail: rajisusan.res15@iiitmk.ac.in

J. S. Paul
e-mail: j.paul@iiitmk.ac.in

© Springer Nature Singapore Pte Ltd. 2020
S. Agarwal et al. (eds.), *Machine Intelligence and Signal Processing*,
Advances in Intelligent Systems and Computing 1085,
https://doi.org/10.1007/978-981-15-1366-4_13

161

1 Introduction

Longer scan times during magnetic resonance (MR) data acquisition demands the development of sophisticated techniques that can accelerate the scanning process by reducing the number of k-space measurements without compromising the image quality. The image recovery using compressed sensing (CS) principles has achieved the attention of the scientific community as the reconstructions relying on the sparsity assumptions can exhibit significant improvements in the reconstruction of the MR image from a fewer number of measurements. Due to the acquisition noise as well as the inherent ill-conditioning of the underlying Fourier transform operator, the image reconstruction problem becomes ill-posed. This requires employment of an accurate image prior model to obtain stable solutions with high image quality images. Although many regularization models have been proposed in the literature [1], currently researchers focus on transform domain sparsity-based priors [2–5].

An optimal sparse representation of images is very crucial for high-quality sparsity-based image recovery. Pre-specified transforms like wavelets [6–8], contourlets [9], and curvelets [10] can yield recovered images with efficient utilization of the global sparsity of the images. On the other hand, methods which utilizes the self-repetitive structure in images such as block-matching and 3D filtering (BM3D) exploit the non-local sparsity of images [11]. Even though the preservation of finer details is achieved through the use of non-local sparse regularization, it has been shown to yield noise-like artifacts in piece-wise smooth regions [12]. However, such a method cannot simultaneously utilize the global and non-local sparsity.

To overcome this, we propose an efficient combination of the global and non-local sparsity realized using wavelet and BM3D to be implemented as a fast composite splitting denoising approach (FCSD) [13]. Furthermore, we use an adaptive regularization parameter for the wavelet to yield the best achievable performance using adaptive thresholded Landweber (ATL). The proposed adaptive fast composite splitting denoising algorithm (AFCSD) combines the advantages of both BM3D with adaptive TL through the use of FCSD and thus exploits the self-repetitive structure in images by enforcing non-local sparsity as well as entire consistency of the image. In an FCSD framework, the composite regularization problem is decomposed into simpler subproblems, and solving each of them individually by thresholding methods. Experimental results reveal that such a framework is superior to any of the individual methods in terms of reconstruction accuracy, and computation complexity.

This paper is organized into four sections. In the Theory section, we give a brief overview of the reconstruction model, the proposed AFCSD algorithm with its application to dictionary learning (DL) approaches. Method sections describe the implementation details of the proposed approach along with the algorithmic procedure and Results section includes illustrations of the AFCSD applied to different datasets. The paper concludes with a discussion on the implications of the proposed method and relative advantages.

2 Theory

2.1 Reconstruction Model

The undersampled k-space representation K_u of an image U is given by

$$K_u = F_u U, \tag{1}$$

The recovery of an MR image using the compressive sensing theory is an underdetermined problem. This implies that there exists more than one solution for the same optimization problem. With the prior information that the signal is sparse in some transform domain, the original signal can be recovered by minimization of the optimization problem,

$$\min_U \| \Psi U \|_1 \text{ subject to} \frac{1}{2} \| K_u - F_u U \|_2^2 \le \varepsilon, \tag{2}$$

where Ψ is the sparsifying transform operator with Ψ' denoting its adjoint operator. Although various methods [14–17] can be used to solve the aforementioned problem, the simplest minimization strategy is that of iterative shrinkage–thresholding algorithm (ISTA) [5]. ISTA is also referred to as thresholded Landweber (TL) as it involves a Landweber update step followed by hard or soft thresholding. The update step gives

$$\widehat{w}^{(k)} = w^{(k-1)} + \Psi\left(F_u'\left(K_u - F_u\left(\Psi' w^{(k-1)}\right)\right)\right), \tag{3}$$

where $w^{(k-1)}$ represents the wavelet coefficients of the recovered image in the $(k-1)'$th iteration, and $\varepsilon_{res}^{(k)} = \Psi\left(F_u'\left(K_u - F_u\left(\Psi' w^{(k-1)}\right)\right)\right)$ denotes the consistency error. Following the Landweber update, the wavelet coefficients are soft thresholded to get the current iterate as

$$w^{(k)} = \mathcal{T}_{\alpha^{(k-1)}}\left(\widehat{w}^{(k)}\right), \tag{4}$$

where $\mathcal{T}_\alpha$ denotes a component-wise application of the soft thresholding function defined as

$$\left(w^{(k)}\right)_i = \begin{cases} \left(\widehat{w}^{(k)}\right)_i - \alpha \dfrac{\left(\widehat{w}^{(k)}\right)_i}{\left|\left(\widehat{w}^{(k)}\right)_i\right|} , if \left|\left(\widehat{w}^{(k)}\right)_i\right| > \alpha \\ 0, \qquad\qquad\qquad if \left|\left(\widehat{w}^{(k)}\right)_i\right| \le \alpha. \end{cases} \tag{5}$$

Those coefficients with values lesser than the threshold value are set to 0, whereas the threshold value is subtracted from those coefficients with values larger than the threshold. This is based on the assumption that the threshold corresponds to the noise variance of the realizations of the transform domain coefficients. If the measurements are sparse, we can recover the original signal by minimizing the corresponding l_1 problem.

2.2 BM3D for Non-local Sparsity

BM3D is a non-local image modeling method that uses high-order adaptive group-wise models. A detailed description on BM3D can be found in [11]. The basic concept of the BM3D modeling can be explained based on three steps. In the first step, similar image blocks are collected in patches. The blocks are arranged together to form 3D arrays and decorrelated using an invertible 3D transform. In the second step, 3D group spectra obtained in the first step are filtered by hard thresholding. In the third step, the hard thresholded spectra are inverted to provide the estimate of each block in the group. Finally, the block-wise estimates are mapped to their original positions and the restored image is computed as a weighted average of all the block-wise estimates. The block-wise operation, by default, imposes a localization of the image on small pieces. Higher level of sparse signal representation as well as lower complexity of the model can be attained through the use of joint 3D group-wise transforms compared to 2D block-wise transforms. As the total number of group-wise spectrum components involved is much large in comparison to the image size, the resultant data approximation is overcomplete or redundant. This redundancy is the primary factor for the effectiveness of the BM3D modeling.

2.3 Proposed Adaptive Composite Model

In this paper, we propose an efficient algorithm for image restoration using the adaptive fast composite splitting denoising (AFCSD) in which the wavelet threshold is updated in each iteration. This composite model yields faster convergence of steady-state errors utilizing both the global sparsity and local sparsity of images. FCSD minimizes a linear combination of three terms corresponding to a least-square data fitting, wavelet-based l_1-norm regularization together with BM3D-based non-local sparsity enforcing term. The original problem is decomposed into l_1 and with BM3D based non-local sparsity enforcing subproblems, respectively. The soft thresholding parameter in the l_1- norm regularization is adapted in each step using the algorithm proposed by Mathew and Paul in [18]. Accordingly, the threshold in each iteration is updated as

$$\alpha^{(k)} = \frac{\mathbb{E}\left(\left|\mathcal{E}_{res}^{(k)}\right|\right)}{\Phi\left(\delta_1^{(k)}\right) + \left(\mathbb{E}\left(\left|\mathcal{E}_n^{(k-1)}\right|\right)\right)/\alpha^{(k-1)}}, \tag{6}$$

where $\mathbb{E}(\cdot)$ denote the expectation of the corresponding random variable, $\mathcal{E}_n^{(k-1)}$ is called the sparse approximation error computed as the difference between the thresholded and unthresholded wavelet coefficients, and $\Phi(\cdot)$ an increasing function of $\delta_1^{(k)}$ satisfying conditions for a selection $\Phi(\cdot)$ as described in [18]. Here, $\delta_1^{(k)}$ is called the discrepancy level given by

$$\delta_1^{(k)} \triangleq \left| \left\| \mathcal{E}_{res}^{(k)} \right\|_1 - \left\| \mathcal{E}_n^{(k-1)} \right\|_1 \right|. \tag{7}$$

The weighted average of solutions obtained from two subproblems: BM3D and adaptive ISTA yield the recovered image in an iterative framework. The relative decrease in l_2-norms of the reconstructed images is then used to define a stopping criterion as

$$\frac{\left\| U^{(k)} - U^{(k-1)} \right\|_2}{\left\| U^{(k-1)} \right\|_2} \leq \text{tol}. \tag{8}$$

In our study, we have set a tolerance level of $1.0e - 4$, as tolerance values less than $1.0e - 4$ it did not result in any perceptible improvement in visual quality. The algorithmic steps for AFCSD are enclosed below in Algorithm (A1).

Algorithm-1 (A1): AFCSD Initialization
  Initialize $r^{(0)} = 0$, $U^0 = 0$, $t^{(0)} = 1$, α, γ
  Input the zero-filled k-space data K_u
  Main Iteration

(1) Iterate for $k = 1, \ldots, \text{MaxIter}$:

  (i) $U_g = r^{(k)} + F_u'\left(K_u - F_u\left(U^{(k-1)}\right)\right)$

  (ii) $U_1 = \min_U \left\{ 2\alpha \left\| \Psi U^{(k-1)} \right\|_1 + \left(\left\| U^{(k-1)} - U_g \right\|_2 \right)^2 / 2 \right\}$

  (iii) $U_2 = \min_U \left\{ 2\gamma \left\| U^{(k-1)} \right\|_{NL} + \left(\left\| U^{(k-1)} - U_g \right\|_2 \right)^2 / 2 \right\}$

  (iv) $U^{(k)} = (U_1 + U_2)/2$

  (v) $t^{(k+1)} = \frac{1+\sqrt{1+4t^{(k)}}}{2}$

  (vi) $r^{(k+1)} = U^k + \frac{(t^{(k)}-1)}{t^{(k+1)}}(U^{(k)} - U^{(k-1)})$

(2) Repeat step 1) until $\frac{\left\| U^{(k)} - U^{(k-1)} \right\|_2}{\left\| U^{(k-1)} \right\|_2} \leq tol.$

  end
  NL stands for non-local wavelet norm. Here, l_1-norm of the wavelet coefficients is computed by stacking similar patches to 3D-group. The 2D-wavelet transform is first calculated on each patch and 1D-wavelet transform is then computed along the third direction.

3 Results

We used brain images (a) and (b) in Fig. 1 shared at the Web site http://www.quxiaobo.
org/csg_software_en.html to demonstrate our results. The corresponding fully sampled k-spaces are retrospectively undersampled using a sampling mask (shown in
either (c) and (d)). The sampling mask in (c) was generated using 30% acquired
samples, with 1.55% of samples fully acquired from the k-space center, whereas the
radial sampling mask in (d) is generated using 80 spokes with 256 samples along
each spoke and the angles selected using golden ratio [19]. The reconstruction errors
are computed as the relative l_2-norm error (RLNE) [20–22] defined as

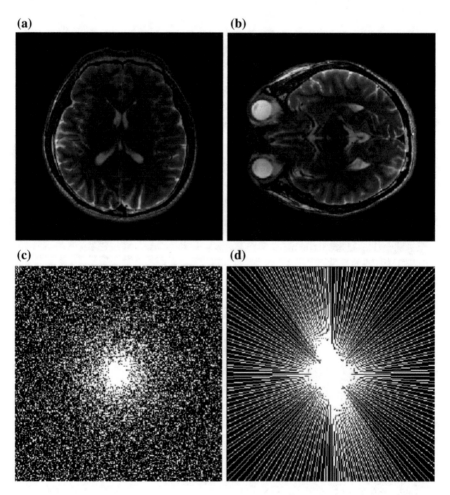

Fig. 1 **a–b** Sample MR images, **c** random sampling mask, and **d** radial sampling mask

$$RLNE = \|U - U_{\text{ref}}\|_2 / \|U_{\text{ref}}\|_2, \tag{9}$$

where U_{ref} and U denote the ground truth image and reconstructed image, respectively. The algorithms are implemented using MATLAB (The Mathworks, Nattick, MA) on a PC with Intel Xeon 2.4 GHz processor and 16 GB of RAM running Windows 7 OS. The adaptation function used is $\Phi\left(\delta_1^{(k)}\right) = \delta_1^{(k)}$. For comparison of the proposed AFCSA algorithm, we have used thresholded Landweber (TL), adaptive thresholded Landweber (ATL), and BM3D; BM3D is implemented using the MATLAB script shared at (BM3D http://www.cs.tut.fi/foi/GCF-BM3D/) [11].

Figures 2 and 3 show reconstructed images using mask (c) and (d), respectively. Top and bottom panel show reconstructed and difference images, respectively.

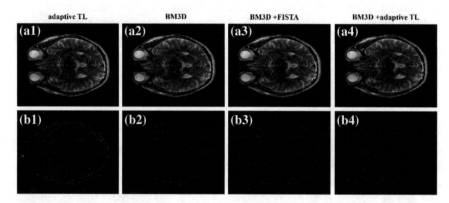

Fig. 2 Reconstructions using adaptive TL, BM3D, and FCSD with BM3D and FISTA and FCSD with adaptive TL and BM3D for mask shown in (**c**). The difference images are shown in the bottom panel. The RLNE values are 0.0637, 0.0562, 0.0453, and 0.0403, respectively

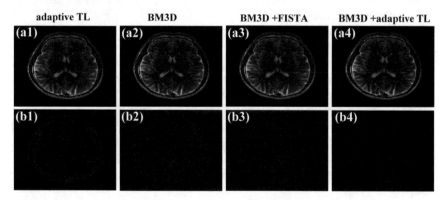

Fig. 3 Reconstructions using adaptive TL, BM3D, and FCSD with BM3D and FISTA and FCSD with adaptive TL and BM3D for mask shown in (**d**). The difference images are shown in the bottom panel. The RLNE values are 0.0556, 0.0476, 0.0321, and 0.0210, respectively

Although the finer details are preserved well using non-local sparse regularization, it results in noise-like artifacts in the piece-wise smooth regions. The l_1-norm-based sparse regularization can maintain the consistency of the entire image compared to BM3D that uses soft thresholding on 3D image blocks by stacking similar patches. The advantages of these two approaches can be obtained by combining BM3D with adaptive TL through the use AFCSD.

3.1 Application to Dictionary Learning

When the AFCSD algorithm used as an additional constraint in DLMRI [23, 24], it is shown to result in reduction in RLNE along with accelerated convergence of the algorithm in highly undersampled datasets. Figure 4 shows reconstructions obtained with the DL algorithm compared with the results of conventional DL at 20% undersampling in dataset 3.

4 Discussion

In this paper, a fast adaptive composite splitting strategy for MR image reconstruction is proposed so as to achieve faster reduction of steady-state reconstruction errors and convergence to the minimum possible steady-state error. The proposed scheme provides an efficient method to combine the global and non-local sparsity simultaneously. Here, a combination of wavelet and BM3D are implemented using FCSA. Additionally, an adaptive regularization parameter for the wavelet is used to yield the best achievable performance using ATL. Further improvements can be achieved in the reconstruction by using the AFCSD reconstructed image as an auxiliary constraint in DL-based reconstruction.

The accuracy of regularized output depends on the selection of appropriate regularization parameter. In this context, the optimum parameter controls the degree of regularization and thus determines the compromise between SNR and artifacts. While under-regularization leads to residual noise or streaking artifacts in the image, over-regularization removes the image features and enhances aliasing. To account for this, an iteration-dependent threshold is used in this method.

The rate of convergence can be improved by the incorporation of two-step IS strategies like TwIST [25] and FISTA [26, 27] algorithms. This is utilized in this method with FISTA-based over-relaxation scheme. This inclusion can result in a considerable reduction of the computational cost compared to that without using FISTA.

The proposed adaptation scheme is shown to result in reconstruction error reduction together with acceleration of the compute time. Further improvements in image quality can be brought in using incorporation of TV-based sparsity also.

DL   DL with AFCSD

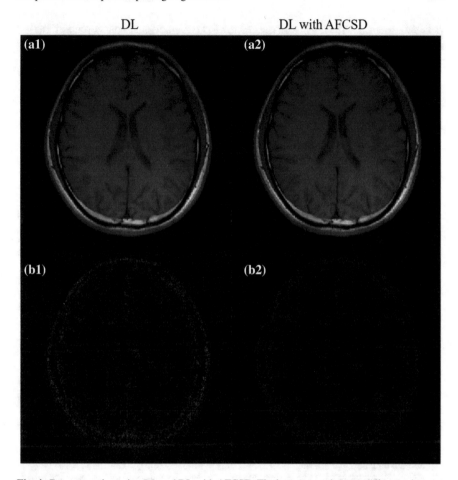

Fig. 4 Reconstruction using DL and DL with AFCSD. The bottom panel shows difference images. RLNE values are 0.0543 and 0.0421, respectively. The corresponding CPU Time are 62.5 s and 41.3 s, respectively

5   Conclusion

In this paper, a combination of wavelet and non-local sparsity is implemented using the fast composite splitting approach. An adaptive regularization parameter is used for the wavelet to yield the best achievable performance using ATL. This method exploits the self-repetitive structure in images by enforcing non-local sparsity as well as entire consistency of the image. Numerical results reveal that the proposed method is superior to any of the methods performed individually in terms of reconstruction accuracy and computation complexity.

Acknowledgements The authors are thankful to Maulana Azad national fellowship by UGC and planning board of Govt. of Kerala (GO(Rt)No.101/2017/ITD.GOK(02/05/2017)) for the financial assistance.

References

1. Tikhonov, A.N., Arsenin, V.I.: Solutions of Ill-Posed Problems. Winston distributed by Halsted Press, Washington/New York (1977)
2. Ramlau, R., Teschke, G.: Tikhonov replacement functionals for iteratively solving nonlinear operator equations. Inverse Probl. **21**(5), 1571–1592 (2005)
3. Ramlau, R., Teschke, G.: A Tikhonov-based projection iteration for nonlinear ill-posed problems with sparsity constraints. Numer. Math. **104**(2), 177–203 (2006)
4. Anthoine, S.: Different wavelet-based approaches for the separation of noisy and blurred mixtures of components, Application to astrophysical data. Ph.D. thesis, Princeton University (2005)
5. Daubechies, I., Defrise, M., De Mol, C.: An iterative thresholding algorithm for linear inverse problems with a sparsity constraint. Comm. Pure Appl. Math. **57**(11), 1413–1457 (2004)
6. Chambolle, A., De Vore, R.A., Lee, N.-Y., Lucier, B.J.: Nonlinear wavelet image processing: variational problems, compression, and noise removal through wavelet shrinkage. IEEE Trans. Image Process. **7**(3), 319–335 (1998)
7. Fornasier, M., Rauhut, H.: Iterative thresholding algorithms. Appl. Comput. Harmon. Anal. **25**(2), 187–208 (2008)
8. Figueiredo, M.A., Nowak, R.D.: An EM algorithm for wavelet-based image restoration. IEEE Trans. Image Process. **12**(8), 906–916 (2003)
9. Qu, X., Zhang, W., Guo, D., Cai, C., Cai, S., Chen, Z.: Iterative thresholding compressed sensing MRI based on contourlet transform. Inverse Probl. Sci. Eng. **18**(6), 737–758 (2010)
10. Candes, E., Demanet, L., Donoho, D., Ying, L.: Fast discrete curvelet transforms. Multiscale Model. Simul. **5**(3), 861–899 (2006)
11. Dabov, K., Foi, A., Katkovnik, V., Egiazarian, K.: Image denoising by sparse 3-D transform-domain collaborative filtering. IEEE Trans. Image Process. **16**(8), 2080–2095 (2007)
12. Knaus, C., Zwicker, M.: Dual-domain filtering. SIAM J. Imaging Sci. **8**(3), 1396–1420 (2015)
13. Huang, J., Zhang, S., Metaxas, D.: Efficient MR image reconstruction for compressed MR imaging. Med. Image Anal. **15**(5), 670–679 (2011)
14. Hosseini, M.S., Plataniotis, K.N.: High-accuracy total variation with application to compressed video sensing. IEEE Trans. Image Process. **23**, 3869–3884 (2014)
15. Ling, Q., Shi, W., Wu, G., Ribeiro, A.: DLM: decentralized linearized alternating direction method of multipliers. IEEE Trans. Signal Process. **63**, 4051–4064 (2015)
16. Yin, W., Osher, S., Goldfarb, D., Darbon, J.: Bregman iterative algorithms for l1-minimization with applications to compressed sensing. SIAM J. Imaging Sci. **1**, 143–168 (2008)
17. Qiao, T., Li, W., Wu, B.: A new algorithm based on linearized Bregman iteration with generalized inverse for compressed sensing. Circuits Syst. Signal Process. **33**, 1527–1539 (2014)
18. Mathew, R.S., Paul, J.S.: Sparsity promoting adaptive regularization for compressed sensing parallel MRI. IEEE Trans. Comput. Imaging **4**(1), 147–159 (2018)
19. Feng, L., et al.: Golden-angle radial sparse parallel MRI: combination of compressed sensing, parallel imaging, and golden angle radial sampling for fast and flexible dynamic volumetric MRI. Magn. Reson. Med. **72**(3), 707–717 (2014)
20. Liu, Y., Zhan, Z., Cai, J.-F., Guo, D., Chen, Z., Qu, X.: Projected iterative soft-thresholding algorithm for tight frames in compressed sensing magnetic resonance imaging. IEEE Trans. Med. Imaging **35**(9), 2130–2140 (2016)

21. Qu, X., et al.: Undersampled MRI reconstruction with patch-based directional wavelets. Magn. Reson. Imaging **30**(7), 964–977 (2012)
22. Qu, X., Hou, Y., Lam, F., Guo, D., Zhong, J., Chen, Z.: Magnetic resonance image reconstruction from undersampled measurements using a patch-based nonlocal operator. Med. Image Anal. **18**(6), 843–856 (2014)
23. Ravishankar, S., Bresler, Y.: MR image reconstruction from highly undersampled k-space data by dictionary learning. IEEE Trans. Med. Imag. **30**(5), 1028–1041 (2011)
24. Bresler, Y.: DLMRI-Lab. http://www.ifp.illinois.edu/~yoram/Software.html
25. Bioucas-Dias, J.M., Figueiredo, M.A.: A new TwIST: two-step iterative shrinkage/thresholding algorithms for image restoration. IEEE Trans. Image Process. **16**(12), 2992–3004 (2007)
26. Beck, A., Teboulle, M.: A fast iterative shrinkage-thresholding algorithm for linear inverse problems. SIAM J. Imaging Sci. **2**(1), 183–202 (2009)
27. Beck, A., Teboulle, M.: Fast gradient-based algorithms for constrained total variation image denoising and deblurring problems. IEEE Trans. Image Process. **18**(11), 2419–2434 (2009)

Extraction of Technical and Non-technical Skills for Optimal Project-Team Allocation

Kanika Bhatia, Shampa Chakraverty, Sushama Nagpal, Amit Kumar, Mohit Lamba and Anupam Poddar

Abstract Project portfolio management (PPM) is a crucial subject matter in academics that needs to be explored at skill level for both the projects and the students. Skill matching can play a major role in not only determining the requirements of the project but also utilizing the skills of the students. When it comes to skills one should focus on the technical skills but non-technical skills are equally important to be taken into account while performing matching. Non-technical skills are needed to be acquired by the student as a part of social and work ethics. In this paper, we have proposed a model which performs both skill extraction and skill matching of the projects to a team of students. The skill extraction involves both technical and non-technical skill extraction while for skill matching. Formal concept analysis (FCA) and project-oriented stable marriage algorithm have been employed. This model matches a project to a team of students while analyzing both technical and non-technical skills required for the project and of the student.

Keywords Naïve Bayes classification · Wiki-scraping · Natural language toolkit · Formal concept analysis · Project-oriented stable marriage algorithm

1 Introduction

Project portfolio management (PPM) to ensure resource availability and sustainability has become a crucial task in the corporate industry. The demand of employment to the candidates is based on the skill requirements of the company [1, 2].

Undergraduate courses like B.Tech., M.Tech., MBA, etc., curriculum keeps on updating to make sure that students understand the growing demands of the world. Projects like capstone, minor, major, B.Tech., final-year projects, etc., are part of the curriculum which helps them get glimpse of the real-life projects. These projects are no longer the theory or simple practical ones like they used to be 15–20 years back.

K. Bhatia (✉) · S. Chakraverty · S. Nagpal · A. Kumar · M. Lamba · A. Poddar
Department of Computer Engineering, Netaji Subhas University of Technology,
Sector-3, Dwarka, New Delhi 110078, India
e-mail: k.kanikabhatia15@gmail.com

© Springer Nature Singapore Pte Ltd. 2020
S. Agarwal et al. (eds.), *Machine Intelligence and Signal Processing*,
Advances in Intelligent Systems and Computing 1085,
https://doi.org/10.1007/978-981-15-1366-4_14

They have evolved to a massive level that are getting used and adapted, and some are even funded by government or private industry where a whole research team is formed to carry out every task in an organized manner [3]. Skill-based PPM should be introduced and researched in academic domain as well.

While employing skill-based PPM, it needs to be taken into the account that most of the time only technical skills of the project are focused upon, while non-technical skills requirement of the project is always left out presumed to be of not much importance [4]. The lack of these non-technical skills could lead to failure of the project. The workplace be academic or corporate require some crucial non-technical skills that hold equal importance, especially while working in a team or a group with others. Non-technical or generic or soft or people or impersonal or social or managerial skills are the skills required to work in an organized environment. Soft skills include good managerial and clear communication skills to carry out a task in an organized manner. Teamwork, leadership skills, and decision-making skills provide growth and development of the students, making them ready to go into the real world [5, 6].

This paper has been divided into five sections as follows. In the second section, prior work pertaining to techniques and approaches for performing skill matching or job matching in the corporate industry has been discussed. In the third section, a model has been proposed which performs skill extraction and skill matching of the projects to a group of students. While in the fourth section, the experimented results and discussions have been carried out and finally conclusion has been drawn in the fifth section.

2 Prior Work

Over the decade, a lot of research has been conducted on job matching of the profiles with the candidates in the corporate industry. Hexin et al. [7] have proposed a recruitment framework which performs matching between employer's requirement and job-seeking candidate's skills by employing semantic similarity matching but fails when it comes to larger ontologies. Marcarelli et al. [8] have highlighted in their paper how there is a need to bridge the gap between industry needs and student skills by introducing more technology courses to increase student skills. Hexin et al. [4] have proposed electronic job recruitment as a service platform for the same but fails to incorporate non-technical skills of a candidate. Matusovich et al. [9] have researched how non-technical skills like teamwork and communication are still at top-most priority in jobs for engineering graduates and the faculties need to update the learning mechanisms to impart these skills. Chandsarkar et al. [10] have employed an indexing method based on signature graphs to provide the closest match of candidate for a job if exact match is not available, but it fails to incorporate scalability of the data and does not work on soft or logical skills. Corde et al. [11] have presented a bird mating optimization method for one-to-n skill matching for a job description from n CVs. Shankaraman et al. [12] have proposed a system which not only generates skill set for a student but also generates recommended jobs for the student in ICT sector. Alimam et al. [13] have proposed an automated guidance system to match students to their appropriate career path using a skill-based model which can be employed for counselling of students. Maurya et al. [14] have explored unsupervised Bayesian matching to match member's profiles to job postings on LinkedIn. Pisalyon et al. [15] have presented a methodology using FINDEREST model which explores the personal skills from social media use behaviour and recommends the fields of study. We have earlier proposed a skill matcher for PPM to optimally match students with projects as per technical skills requirement [16]. However, this work did not take into account non-technical extraction which we now incorporate.

3 Model

The proposed model consists of two phases, namely skill extraction and skill matching. The skill extraction phase is responsible for extraction of both technical and non-technical skills required for fulfilment of a particular project. On the other hand, the skill matching phase is responsible for matching of project to a team of students keeping in mind the skills required for that particular project and student's skills (Fig. 1).

3.1 Skill Extraction

Skill extraction phase is not only responsible for extraction of technical skills required for a project but also extracts the non-technical skills that are required for that project. Thus, we have subdivided this phase into two stages, namely technical skill extraction and non-technical skill extraction (Fig. 2).

Technical Skill Extraction
Technical skill extraction has been taken from our previous work which first involves

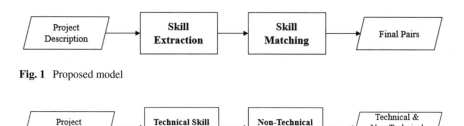

Fig. 1 Proposed model

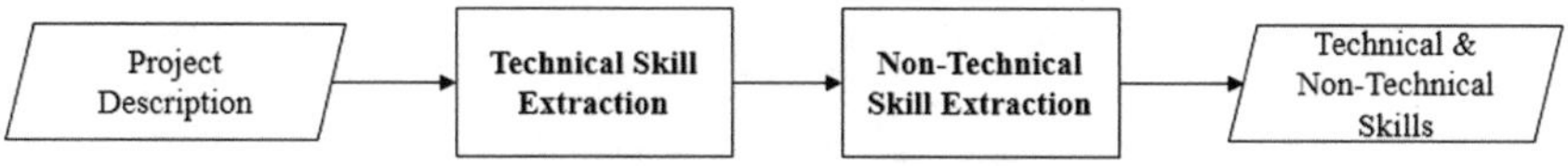

Fig. 2 Skill extraction

training module after which we run it on the trained module. First, we perform segregation of the data of project descriptions and their keywords into different text files. Then, we perform pre-processing where we determine TF values using natural language processing toolkit (NLTK) in python. Finally, naive Bayes classification is performed with respect to binary relevance [16].

Now, the technical skill extraction training module is ready to be tested. Thus, when a new project description is subjected to the trained module, it extracts technical skills required for that particular project. The technical skills obtained from this stage are further utilized for extraction of non-technical skills.

Non-technical Skill Extraction
The database [17] is a dataset consisting of key skills and their subskills that have been utilized for creation of three dictionaries, namely, Dictionary_1, Dictionary_2, and Dictionary_3 as shown in Fig. 3. Dictionary_1 is a dictionary having key skills associated with their subskills, which is further utilized to construct. Dictionary_2 consisting of subskills associated with their key skills and this is further used to create Dictionary_3. Dictionary_3 is the third dictionary consisting of sub-subskills associated with their subskills.

After the creation of all the three dictionaries, we are ready to extract non-technical skills required by the project. The new project description gets tokenized and technical skills obtained for the same are further utilized for extraction after subjecting to wiki-scraping as depicted in Fig. 4. First, we match every word with Dictionary_3 to obtain subskills and finally with Dictionary_2 to obtain the key skills. These key skills are the non-technical skills required for the fulfilment of the project.

Fig. 3 Creation of dictionaries

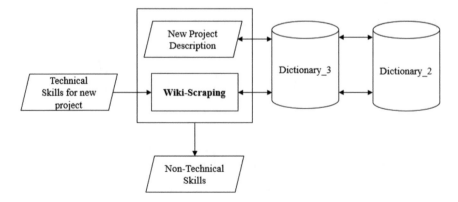

Fig. 4 Non-technical skill extraction

3.2 Skill Matcher

The technical and non-technical skills obtained from the first phase is later on subjected to the second stage required for completion of the project, we fed the data in the form of tables where rows and columns are for projects and skills, respectively.

FCA
The tables are in the form of binary values that are further converted to cross tables. These cross tables are fed to FCA which first produces formal concepts that later on form the nodes in the concept lattices. These concept lattices are converted into matrices and affinity matrix, and later on preference matrices are obtained from them. The detail description has been included in our previous work "Project Portfolio Management in Education Domain using Skill Matcher Model" [16].

Project-Oriented Stable Marriage Algorithm
The preference matrices are further subjected to extended stable marriage algorithm to find the final pairs of projects and students. The detail description has been included in our previous work "Project Portfolio Management in Education Domain using Skill Matcher Model" [16].

4 Results and Discussion

4.1 Skill Extraction

A new project description is fed to the skill extraction phase as an input as shown in Fig. 5 which first undergoes technical skill extraction stage. The input data is subjected to trained module of technical extraction which after performing naïve Bayes classification produces technical skills as shown in Fig. 6.

```
This work proposes an intelligent learning diagnosis system that supports a
Web-based thematic learning model, which aims to cultivate learners' ability
of knowledge integration by giving the learners the opportunities to select
the learning topics that they are interested, and gain knowledge on the specific
topics by surfing on the Internet to search related learning courseware and
discussing what they have learned with their colleagues. Based on the log files
that record the learners' past online learning behavior, an intelligent diagnosis
system is used to give appropriate learning guidance to assist the learners in
improving their study behaviors and grade online class participation for the
instructor. The achievement of the learners' final reports can also be predicted
by the diagnosis system accurately. Our experimental results reveal that the
proposed learning diagnosis system can efficiently help learners to expand their
knowledge while surfing in cyberspace Web-based "theme-based learning" model.
```

Fig. 5 Project description

```
----------------Technical skills---------------
Cyberspace
Machine learning
Learning systems
Support vector machines
```

Fig. 6 Technical skills

```
                                              ------------------------------------
                                                          -- Others --
                                              ------------------------------------
                                              applying innovation techniques ,
----------------NON - Technical skills----    governance ,
                                              brand management ,
--------------------------------              benefits management ,
            -- Major skills --                program lifecycle ,
--------------------------------              marketing strategy & planning ,
                                              scope management ,
core project management skills ,              strategic program management ,
core executive leadership skills ,            influencing to negotiate ,
mba skills ,                                  business & product development ,
soft skills ,                                 human resources ,
communication skills ,                        international business ,
technical skills ,                            market research ,
selling skills ,                              finance ,
professional skills ,                         soft skills ,
--------------------------------              integration management ,
            -- Management --                  selling skills ,
--------------------------------              core executive leadership skills ,
scope management ,                            decision making ,
core project management skills ,              information modeling & design ,
strategic program management ,                mba skills ,
integration management ,                      artistic abilities ,
brand management ,                            emotional intelligence ,
benefits management ,                         visual abilities ,
management ,                                  communication skills ,
                                              technical skills ,
                                              writing ,
                                              professional skills ,
```

Fig. 7 Non-technical skills

After we obtain technical skills, the technical skills and the project description are fed to non-technical skill extraction stage as an input. In this stage, wiki-scraping is performed, which then extracts non-technical skills required for the fulfilment of the project from the database as shown in Fig. 7.

4.2 Skill Matcher

We have taken a small dataset of 10 students and 10 projects. For projects, we employed the proposed model and for students we have taken a data of 10 students a set of skills. After the skill extraction phase, technical and non-technical skills of projects and students are fed to skill matching phase in the form of tables with rows as students/projects and columns as skills having binary values 1/0, for presence and absence of skills.

FCA

Tables with binary values are then converted to cross tables where '1' is replaced with 'X' and '0' with black space. These cross tables are then fed to formal context analysis phase. In FCA, first, the formal concepts are formed in the form of extent and intent. The formal concepts are formed for student cross table with students as extent and skills as intent as shown below.

$c0 = ()$ > ('data integration', 'frequent patterns', 'supervised learning', 'neural chips', 'core memory', 'manufacturing systems', 'time management', 'core management skills', 'entrepreneurial skills', 'communication skills')

$c1$ = ('student1',) > ('frequent patterns', 'supervised learning', 'manufacturing systems', 'core management skills', 'communication skills')

$c2$ = ('student3',) > ('frequent patterns', 'entrepreneurial skills', 'communication skills')

$c3$ = ('student5',) > ('data integration', 'frequent patterns', 'core memory', 'manufacturing systems', 'communication skills')

$c4 = ('student6',)$ > ('core memory', 'time management')

$c5 = ('student8',)$ > ('data integration', 'supervised learning')

$c6$ = ('student1', 'student5') > ('frequent patterns', 'manufacturing systems', 'communication skills')

$c7$ = ('student1', 'student10') > ('supervised learning', 'core management skills', 'communication skills')

$c8 = ('student5', 'student6')$ > ('core memory',)

$c9 = ('student6', 'student9')$ > ('time management',)

$c10 = ('student1', 'student3', 'student5')$ > ('frequent patterns', 'communication skills')

$c11$ = ('student1', 'student7', 'student10') > ('core management skills',)

$c12 = ('student2', 'student5', 'student8')$ > ('data integration',)

$c13 = ('student1', 'student3', 'student5', 'student10')$ > ('communication skills',)

$c14$ = ('student1', 'student4', 'student8', 'student10') > ('supervised learning',)

$c15$ = ('student1', 'student2', 'student3', 'student4', 'student5', 'student6', 'student7', 'student8', 'student9', 'student10') > ()

Similarly, we are provided with formal concepts for projects which has project as extents and skills as intents as shown below.

$c0$ = () > ('data integration', 'frequent patterns', 'supervised learning', 'neural chips', 'core memory', 'manufacturing systems', 'time management', 'core management skills', 'entrepreneurial skills', 'communication skills')

$c1$ = ('project1',) > ('supervised learning', 'neural chips', 'core management skills', 'communication skills')

$c2$ = ('project5',) > ('manufacturing systems', 'communication skills')

$c3$ = ('project6',) > ('frequent patterns',)

$c4$ = ('project7',) > ('data integration', 'neural chips', 'entrepreneurial skills', 'communication skills')

$c5$ = ('project10',) > ('data integration', 'supervised learning', 'core memory', 'time management', 'core management skills', 'communication skills')

$c6$ = ('project1', 'project7') > ('neural chips', 'communication skills')

$c7$ = ('project1', 'project10') > ('supervised learning', 'core management skills', 'communication skills')

$c8$ = ('project2', 'project10') > ('data integration', 'time management')

$c9$ = ('project3', 'project10') > ('data integration', 'supervised learning', 'core memory')

$c10$ = ('project4', 'project7') > ('entrepreneurial skills',)

$c11$ = ('project5', 'project9') > ('manufacturing systems',)

$c12$ = ('project7', 'project10') > ('data integration', 'communication skills')

$c13$ = ('project1', 'project3', 'project10') > ('supervised learning',)

$c14$ = ('project2', 'project8', 'project10') > ('time management',)

$c15$ = ('project1', 'project5', 'project7', 'project10') > ('communication skills',)

$c16$ = ('project2', 'project3', 'project7', 'project10') > ('data integration',)

$c17$ = ('project1', 'project2', 'project3', 'project4', 'project5', 'project6', 'project7', 'project8', 'project9', 'project10') > ()

After obtaining formal concepts for students and projects, concept lattices are formed for both student and project as shown in Fig. 8. The formal concepts form the nodes of the concept lattices depicting specialization at the top and generalization of formal concepts at the bottom.

After obtaining the concept lattices, student concept matrix and project concept matrix are obtained from them where rows signify student concept/project concept and columns signify skills. These concept matrices are later on used to provide affinity matrix after employing affinity algorithm, which is further used to obtain preference matrices for both student and project depicting their preferences of projects and students, respectively, as shown in Fig. 9.

Project-Oriented Stable Marriage Algorithm

After obtaining the preference matrices, extended stable matching algorithm is employed that provides us the final stable pairs. The given below data shows us the team of students required for a particular project matched as per the student's skills and the skills required for completion of the project for its completion as per the skills required by it.

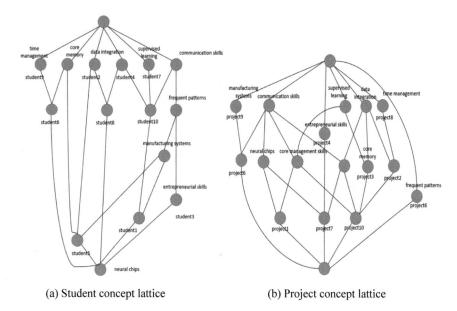

(a) Student concept lattice   (b) Project concept lattice

Fig. 8 **a** Student concept lattice. **b** Project concept lattice

0 [6, 0, 1, 2, 10, 12, 14, 4, 5, 11, 8, 9, 13, 15, 3, 7]
1 [2, 9, 14, 1, 3, 5, 11, 6, 10, 12, 13, 15, 0, 7, 8, 4]
2 [1, 11, 2, 8, 10, 14, 15, 3, 4, 5, 7, 6, 9, 12, 13, 0]
3 [13, 7, 2, 8, 9, 10, 12, 14, 15, 1, 4, 5, 11, 6, 0, 3]
4 [8, 12, 15, 7, 11, 2, 6, 9, 10, 13, 14, 0, 1, 3, 4, 5]
5 [1, 2, 10, 14, 5, 11, 6, 9, 12, 13, 15, 0, 3, 7, 8, 4]
6 [6, 0, 12, 14, 1, 4, 5, 11, 2, 8, 9, 10, 13, 15, 3, 7]
7 [2, 8, 9, 10, 12, 13, 14, 15, 1, 5, 7, 11, 6, 0, 3, 4]
8 [13, 7, 2, 9, 10, 12, 14, 15, 1, 5, 11, 6, 8, 0, 3, 4]
9 [2, 14, 1, 5, 11, 6, 9, 10, 12, 13, 15, 0, 3, 7, 8, 4]
10 [2, 6, 9, 10, 12, 13, 14, 15, 0, 1, 5, 7, 11, 8, 3, 4]
11 [15, 7, 11, 2, 8, 9, 10, 12, 13, 14, 1, 3, 5, 6, 0, 4]
12 [14, 1, 5, 11, 2, 6, 9, 10, 12, 13, 15, 0, 3, 7, 8, 4]
13 [12, 2, 6, 8, 9, 10, 13, 14, 15, 0, 1, 5, 7, 11, 3, 4]

0 [6, 0, 10, 12, 13, 4, 9, 1, 5, 7, 8, 11, 3, 2]
1 [5, 12, 9, 0, 1, 2, 6, 7, 8, 10, 11, 13, 3, 4]
2 [9, 1, 5, 7, 8, 10, 11, 12, 13, 3, 4, 0, 2, 6]
3 [1, 11, 12, 4, 9, 2, 5, 6, 7, 8, 10, 13, 3, 0]
4 [6, 3, 4, 0, 2, 7, 8, 10, 11, 12, 13, 9, 1, 5]
5 [12, 9, 1, 5, 6, 7, 8, 10, 11, 13, 3, 4, 0, 2]
6 [6, 0, 10, 12, 13, 4, 9, 1, 5, 7, 8, 11, 3, 2]
7 [8, 11, 3, 4, 7, 10, 12, 13, 9, 1, 2, 5, 6, 0]
8 [4, 7, 11, 13, 3, 2, 6, 8, 10, 12, 9, 0, 1, 5]
9 [1, 7, 8, 10, 11, 12, 13, 3, 4, 9, 5, 6, 0, 2]
10 [5, 7, 8, 10, 11, 12, 13, 3, 4, 9, 0, 1, 2, 6]
11 [11, 12, 4, 9, 1, 2, 5, 6, 7, 8, 10, 13, 3, 0]
12 [13, 4, 6, 7, 8, 10, 11, 12, 3, 9, 0, 1, 5, 2]
13 [8, 3, 7, 10, 11, 12, 13, 4, 9, 1, 5, 6, 0, 2]
14 [12, 9, 1, 5, 6, 7, 8, 10, 11, 13, 3, 4, 0, 2]
15 [11, 4, 7, 8, 10, 12, 13, 3, 9, 1, 2, 5, 6, 0]

(a) Project Preference Matrix   (b) Student Preference Matrix.

Fig. 9 **a** Project preference matrix. **b** Student preference matrix

p0 (6 , 0.8) >{s0,s2,s3,s4,s7,s9,}
p1 (5 , 1.0) >{s1,s4,s5,s7,s8,}
p2 (6 , 1.0) >{s0,s1,s3,s4,s7,s9,}
p4 (4 , 1.0) >{s0,s9,s2,s4,}
p3 () > {}
p5 (6 , 1.0) >{s0,s2,s4,s5,s6,s9,}
p6 (6 , 0.8) >{s0,s1,s2,s4,s7,s9,}
p7 (2 , 1.0) >{s8,s5,}
p8 () > {}
p9 (9 , 1.0) >{s0,s1,s2,s3,s4,s5,s7,s8,s9,}

Similarly, we are also able to obtain the data on students pertaining to the projects they are required and matched on the basis of the skills.

s0 (6 , 1.0) >{p0,p2,p4,p5,p6,p9,}
s1 (4 , 1.0) >{p1,p2,p6,p9,}
s2 (5 , 1.0) >{p0,p4,p5,p6,p9,}
s3 (3 , 1.0) >{p0,p9,p2,}
s4 (7 , 1.0) >{p0,p1,p2,p4,p5,p6,p9,}
s5 (4 , 1.0) >{p1,p5,p9,p7,}
s6 (1 , 0.0) >{p5,}
s7 (5 , 1.0) >{p0,p1,p2,p6,p9,}
s8 (3 , 1.0) >{p1,p9,p7,}
s9 (6 , 1.0) >{p0,p2,p4,p5,p6,p9,}

Since students are matched to quite a large number of projects depending upon their skill requirement, it will not be feasible to assign them to every project. To deal with this, we have limited the count of projects to which a student can work at the same time interval. This will increase the feasibility of students making them focus on only two projects simultaneously rather than on five or six projects as shown below. This could lead to the possibility that some projects are not assigned with a team of students like project 3 and project 8.

p0 (1 , 0.8) >{s9,}
p1 (2 , 1.0) >{s1,s8,}
p2 (3 , 1.0) >{s3,s4,s7,}
p3 () > {}
p4 (1 , 1.0) >{s0,}
p5 (2 , 1.0) >{s2,s6,}
p6 (2 , 0.8) >{s1,s2,}
p7 (2 , 1.0) >{s8,s5,}
p8 () > {}
p9 (6 , 1.0) >{s0,s3,s4,s5,s7,s9,}

s0 (2 , 0.8) >{p4,p9,}
s1 (2 , 1.0) >{p1,p6,}
s2 (2 , 1.0) >{p5,p6,}
s3 (2 , 1.0) >{p9,p2,}
s4 (2 , 0.6) >{p2,p9,}
s5 (2 , 1.0) >{p9,p7,}
s6 (1 , 0.0) >{p5,}
s7 (2 , 1.0) >{p2,p9,}
s8 (2 , 1.0) >{p1,p7,}
s9 (2 , 1.0) >{p0,p9,}

5   Conclusion

Skill-based project portfolio management (PPM) in academics should not only take technical skills in context but also the non-technical (or soft) skills required for the fulfilment of the project. This paper proposes a model which consists of both skill extraction and skill matching of a project to a team of students as per the skills requirement of the project. The skill extraction phase extracts both technical and non-technical skills from the project description. After extraction of skills, skill matching is performed by employing formal context analysis (FCA) and project-oriented stable marriage algorithm. This model provides us information by providing final pairs of projects and students. Number of students required in a project and number of projects a student can take part in as per skills. We have extended our previous work by adding non-technical extraction of the skills required in a project and further limit has been put to the number of projects a student can get involved in simultaneously to make it feasible.

References

1. Guoying, Z.: Analysis of strategic decision-making of company's project investment based on portfolio theory. In: 2009 16th International Conference on Industrial Engineering and Engineering Management, pp. 118–122 (2009)
2. Bode-Greuel, K.M., Nickisch, K.J.: Value-driven project and portfolio management in the pharmaceutical industry: drug discovery versus drug development—commonalities and differences in portfolio management practice. J. Commer. Biotechnol. **14**, 307–325 (2008)
3. Guo, P., Liang, J.J., Zhu, Y.M., Hu, J.F.: R&D project portfolio selection model analysis within project interdependencies context. In: 2008 IEEE International Conference on Industrial Engineering and Engineering Management, pp. 994–998 (2008)
4. Hexin, L., Bin, Z.: Elastic information matching technology and its application in electronic recruitment. In: 2010 IEEE 11th International Conference on Computer-Aided Industrial Design & Conceptual Design 1, vol. 2, pp. 1582–1585 (2010)
5. Chatterjee, M.: Imparting communication skills in technical courses: a paradigm shift. In: 2013 Nirma University International Conference on Engineering (NUiCONE), pp. 1–5 (2013)
6. Matturro, G., Raschetti, F., Fontán, C.: Soft skills in software development teams: a survey of the points of view of team leaders and team members. In: 2015 IEEE/ACM 8th International Workshop on Cooperative and Human Aspects of Software Engineering, pp. 101–104 (2015)
7. Lv, H., Zhu, B.: Skill ontology-based semantic model and its matching algorithm. In: 2006 7th International Conference on Computer-Aided Industrial Design and Conceptual Design, pp. 1–4 (2006)
8. Marcarelli, C.J., Carter, L.J.: Work in progress-bridging the technology gap: an analysis of industry needs and student skills. In: 2009 39th IEEE Frontiers in Education Conference, pp. 1–2 (2009)
9. Matusovich, H.M., Paretti, M., Motto, A., Cross, K.: Work in progress—Match or mismatch? the teaching and learning of teamwork and communication skills. In: 2011 Frontiers in Education Conference (FIE), pp. T2C-1 (2011)
10. Chandsarkar, R., Shankarmani, R., Gharpure, P.: Information retrieval system: For skill set improvement in software projects. In: 2014 International Conference on Circuits, Systems, Communication and Information Technology Applications (CSCITA), pp. 277–282 (2014)
11. Corde, S., Chifu, V.R., Salomie, I., Chifu, E.S., Iepure, A.: Bird Mating Optimization method for one-to-n skill matching. In: 2016 IEEE 12th International Conference on Intelligent Computer Communication and Processing (ICCP), pp. 155–162 (2016)
12. Shankaraman, V., Gottipati, S.: Mapping information systems student skills to industry skills framework. In: 2016 IEEE Global Engineering Education Conference (EDUCON), pp. 248–253 (2016)
13. Alimam, M.A., Seghiouer, H., Alimam, M.A., Cherkaoui, M.: Automated system for matching scientific students to their appropriate career pathway based on science process skill model. In: 2017 IEEE Global Engineering Education Conference (EDUCON), pp. 1591–1599 (2017)
14. Maurya, A., Telang, R.: Bayesian multi-view models for member-job matching and personalized skill recommendations. In: 2017 IEEE International Conference on Big Data (Big Data), pp. 1193–1202 (2017)
15. Pisalayon, N., Sae-Lim, J., Rojanasit, N., Chiravirakul, P.: FINDEREST: Identifying personal skills and possible fields of study based on personal interests on social media content. In: 2017 6th ICT International Student Project Conference (ICT-ISPC), pp. 1–4 (2017)
16. Bhatia, K., Chakraverty, S., Nagpal, S., Kumar, A.: Project portfolio management in education domain using skill matcher model. In: 2019 International Conference on Emerging trends in Information Technology (2019) (presented)
17. Mar, A.: Skills inventory (2013). https://training.simplicable.com/training/new/skills-inventory. Accessed 10 Mar 2019

Modified Flower Pollination Algorithm for Optimal Power Flow in Transmission Congestion

Rupali Parmar, Sulochana Wadhwani and Manjaree Pandit

Abstract Artificial intelligence (AI) is an attractive and popular paradigm for providing any machine/system, the ability to carry out tasks in a 'smart' way. Population-based metaheuristics utilize the intelligence available in nature to search for optimal solutions of complex problems. Randomization prevents the search from being trapped at local optima. This paper presents a model based on recently developed modified flower pollination algorithm (MFPA) to solve the problem of transmission congestion management (CM) in competitive electricity market by real power rescheduling of generating units. The performance of the proposed algorithm is tested for single line outage, increased demand and variation in line power limits using modified IEEE 30 bus and IEEE 57 bus test systems. The results are compared with basic FPA, particle swarm optimization (PSO), random search method (RSM) and simulated annealing (SA).

Keywords Modified flower pollination algorithm (MFPA) · Congestion management (CM) · Levy flight · Optimal power flow · Restructuring · Price bids · Line power limits

1 Introduction

Perpetuation of power system in operational synchronism becomes relatively complex in open-access electricity market. As all players focus on enhancing the profit, their transactions frequently make the system operate beyond its limits [1].

A transmission network is said to be congested, when there is a difference between scheduled real power flow and actual real power flow capability with all constraints in

R. Parmar (✉) · S. Wadhwani · M. Pandit
Madhav Institute of Technology and Science, Gwalior, Madhya Pradesh, India
e-mail: rupaliparmar@gmail.com

S. Wadhwani
e-mail: dr.s.wadhwani@gmail.com

M. Pandit
e-mail: drmanjareep@gmail.com

© Springer Nature Singapore Pte Ltd. 2020
S. Agarwal et al. (eds.), *Machine Intelligence and Signal Processing*,
Advances in Intelligent Systems and Computing 1085,
https://doi.org/10.1007/978-981-15-1366-4_15

limits [2]. In restructured environment, the management of congestion and providing reliable and secure power flow is a challenge for independent system operator (ISO). Congestion may push the system into cascading outages and even grid failure.

Through market-based methods, like zonal pricing, generation rescheduling, countertrade and load curtailment, congestion may be relieved. As a consequence of congestion removal, increased electricity price are to be borne by retailer and end user [3].

Different congestion management (CM) techniques for different types of markets around the globe have been explicitly mentioned in Ref. [4–8]. Reference [9] manifests various power system operation phenomena including effects of contingencies on system operation and control.

In Ref. [10], author has presented a strategy to choose generators for rescheduling based on sensitivity to overloaded line and bids submitted by generating units. Sujata and Kamaraj [11] have implemented differential evolution optimization to tackle congestion and minimize rescheduling cost. Kumar and Srivastava [12] discussed a method for managing congestion using AC power transmission congestion distribution factor.

Gope et al. [13] have proposed lion optimization algorithm for CM using optimal rescheduling of real power generation. Authors have proposed CM through demand response program in Ref. [14]. An improved PSO [15] and damping inertia weight PSO [16] is used for the efficient removal of congestion. PSO with improved time-varying acceleration coefficients (ITVAC) is proposed to manage congestion in competitive electricity market in Ref. [17]. Reference [18] presents a strategy for CM in a network including renewable energy sources (RES).

Yang [19] has proposed flower pollination algorithm (FPA). Flower pollination algorithm (FPA) derives motivation from pollination pattern in flowering plants. Author has validated the technique on ten mathematical benchmarks and compared its performance with PSO and genetic algorithm (GA) [19].

An efficient technique based on flower pollination algorithm (FPA) is proposed for managing congestion in restructured power systems and the same is tested on medium-sized and large-sized standard test systems [20].

Randomness (based on switching probability) in the selection of exploration and exploitation phase maintains solution diversity on one hand, whereas, on the other hand, it causes the FPA to lose its direction and instead of moving toward, many times it moves away from global optima [21].

Dubey et al. [21] have proposed a modified flower pollination algorithm (MFPA) in which (i) local pollination process is controlled by the user instead of random operator and (ii) formerly achieved solutions undergo an additional 'intensive exploitation' phase to further tune the optimum solution.

In this research paper, MFPA [21] has been proposed for optimizing the redispatch cost of real power in order to mitigate congestion in the transmission network. Four cases of contingency, including single line outage, increased load demand and variation in line power limits, are investigated and results are validated on modified IEEE 30 bus and modified IEEE 57 bus test systems.

2 Problem Formulation

Congestion is alleviated by altering (increasing or decreasing) the real power output of generating units. This shift in real power output incurs extra cost, which is to be minimized. The cost function may be stated as Eq. (1) [20].

Min cost

$$TC_c = \sum_{j \in N_g} \left(C_k \Delta P_{gj}^+ + D_k \Delta P_{gj}^- \right) \ \$/h \tag{1}$$

where TC_c, C_k, D_k are the gross cost incurred for altering real power output ($/h), incremental and decremental price bids submitted by GENCOs ($/MWh). ΔP_{gj}^+ and ΔP_{gj}^- are incremental and decremental shift in generated real power of units (MW), respectively.

Equality and inequality constraints to be satisfied during optimization are:

2.1 Equality Constraints

$$P_{gk} - P_{dk} = \sum_{j=1}^{N_b} |V_j| |V_k| |Y_{kj}| \cos(\delta_k - \delta_j - \theta_{kj}); \tag{2}$$

$$Q_{gk} - Q_{dk} = \sum_{j=1}^{N_b} |V_j| |V_k| |Y_{kj}| \sin(\delta_k - \delta_j - \theta_{kj}); \tag{3}$$

$$P_{gk} = P_{gk}^c + \Delta P_{gk}^+ - \Delta P_{gk}^-; \quad k = 1, 2, 3, \ldots N_g \tag{4}$$

$$P_{dj} = P_{dj}^c; \quad j = 1, 2, 3, \ldots N_d \tag{5}$$

Here, real and reactive power at bus k are denoted by P_{gk} and Q_{gk}, respectively; real and reactive power consumed at bus k are denoted by P_{dk} and Q_{dk}, respectively; bus voltage angles at bus j and bus k are denoted by δ_j and δ_k, respectively; admittance angle of line connected between bus k bus j is denoted by θ_{kj}; number of total buses, generator buses and load buses are denoted by N_b, N_g and N_d, respectively; P_{gk}^c and P_{dj}^c are initial scheduled transactions of real power generated and real power consumed at bus k [22].

2.2 Inequality Constraints

$$P_{gk}^{\min} \leq P_{gk} \leq P_{gk}^{\max}, \quad \forall k \in N_g \tag{6}$$

$$Q_{gk}^{\min} \leq Q_{gk} \leq Q_{gk}^{\max}, \quad \forall k \in N_g \tag{7}$$

$$P_{gk} - P_{gk}^{\min} \leq \Delta P_{gk} \leq P_{gk}^{\max} - P_{gk}, \tag{8}$$

$$V_m^{\min} \leq V_m \leq V_m^{\max}, \quad \forall m \in N_l \tag{9}$$

$$P_{ij} \leq P_{ij}^{\max}, \tag{10}$$

Here, upper bounds and lower bounds of associated variables are denoted by superscripts min and max. P_{ij} states real power flow in line between ith and jth bus while $P_{ij}^{\max}$ is its real power rating. N_l defines number of load buses and V_m is the magnitude of voltage associated with it.

2.3 Fitness Function

Often, the objective function serves as fitness function. In the present research, penalty approach [16] is followed, in which, inequality constraints behave as penalty functions and are added to main cost function to build up complete fitness function.

Explicit fitness function for CM issue is presented as Eq. (11) [20].

$$FF = TC_c + \alpha_1 \times \sum_{i=1}^{\text{cong}} \left(P_{ij} - P_{ij}^{\max}\right)^2 + \alpha_2 \times \sum_{j=1}^{VB} \left(\Delta V_j\right)^2 + \alpha_3 \times \left(\Delta P_g\right)^2 \tag{11}$$

where

$$\Delta V_j = \begin{cases} \left(V_j^{\min} - V_j\right); & \text{if } V_j \leq V_j^{\min} \\ V_j - V_j^{\max}); & \text{if } V \geq V_j^{\max} \end{cases}, \quad \Delta P_g = \begin{cases} \left(P_g^{\min} - P_g\right); & \text{if } P_g \leq P_g^{\min} \\ P_g - P_g^{\max}); & \text{if } P_g \geq P_g^{\max} \end{cases}$$

Fitness function FF is supposed to be optimized to obtain minimum cost. cong and VB represent the total number of congested lines in the network and total number of buses with unacceptable bus voltage profile, respectively. $\alpha 1, \alpha 2, \alpha 3$ are penalty factors. Numerical value of all three penalty factors is chosen as 10,000 [20]. Possibility of violations of inequality constraints is curbed by adding penalty terms to actual cost function.

3 Development of Modified Flower Pollination Algorithm (MFPA)

Flower pollination algorithm, inspired by pollination process in flowering plants is simple, yet powerful metaheuristics algorithm proposed by Yang [19]. An effective modification in FPA is proposed by Dubey et al. [21].

3.1 Flower Pollination Algorithm

Based on behavior of pollination process, role of pollinators and flower constancy, four rules are framed for application of FPA [19].

Rule 1 Biotic/global pollination happens through pollinators and their movement follows Levy flight pattern.

Rule 2 Abiotic/local pollination happens through self pollination.

Rule 3 Pollinating agents restrict their visit to some species only: flower constancy. Pollination probability is proportional to the resemblance of participating flowers.

Rule 4 A probability switch $\rho \in [0, 1]$ is designed to pick between local and global pollination in such a way that it is biased toward local pollination.

In FPA, diversification of searching agents obeys global pollination, whereas intensification obeys local pollination.

Pollens travel long distances in global pollination, as they are carried by pollinators which are often flying organisms [19].

Modified solutions through global pollination are mathematically expressed as:

$$x_i^{It+1} = x_i^{It} + L(\lambda)\left(\text{gbest} - x_i^{It}\right) \tag{12}$$

Here, x_i^{It} is the solution vector x_i at iteration It and gbest is the current best solution. $L(\lambda)$, a parameter associated with pollinating strength, is actually a step size. $L > 0$ is derived from Levy distribution.

$$L \sim \frac{\lambda \, \Gamma(\lambda) \sin(\pi \, \lambda/2)}{\pi} \frac{1}{s^{1+\lambda}}, \quad (s \gg s_0 \gg 0) \tag{13}$$

where $\Gamma(\lambda)$ is standard gamma function. Levy distribution works for larger step size ($s > 0$) [19]. Chosen value for λ is 1.5.

Modified solutions through local pollination and flower constancy rule are mathematically expressed as:

$$x_i^{It+1} = x_i^{It} + \varepsilon\left(x_j^{It} - x_k^{It}\right) \tag{14}$$

Here, x_j^{It} and x_k^{It} are pollens from same flower/same morph. This mimics the flower constancy in vicinity. ε is randomly drawn from a uniform distribution [0, 1].

Normally, pollination of flowers takes place at both global and local scale, randomly. Mathematically, it is modeled by probability switch ρ. In real world, chances of local pollination are on higher side. Thus, better results are obtained when ρ is kept at 0.8.

3.2 Modified FPA

Random occurrence of exploration and exploitation depending upon switch probability, often makes the FPA to lose direction and fail to identify global optima. Modified FPA (MFPA) is proposed, where (i) mutation in solutions during exploitation phase is modified by introducing user-controlled scaling factor 'F' and (ii) formerly achieved solutions undergo an additional 'intensive exploitation' phase to further tune the optimum solution. Solution is modified by employing random perturbations [21]. Mathematically,

$$x_i^{It+1} = x_i^{It} + F\left(x_j^{It} - x_k^{It}\right) \tag{15}$$

$$x_i^{New} = \text{gbest} + H * \Delta x_i^{New}, \quad H = \begin{cases} 1 & \text{if rand}_2 < \rho \\ 0 & \text{otherwise} \end{cases} \tag{16}$$

$$\Delta x_i^{New} = \text{rand}_3 + \left[(\text{rand}_4 - \text{rand}_5) \times \text{gbest}_{It})\right] \tag{17}$$

rand$_2$, rand$_3$, rand$_4$, rand$_5$ are random numbers drawn from uniform distribution in the range [0,1].

Execution of MFPA may be understood by computational steps for implementing MFPA for the solution of CM problem, discussed later.

4 Implementation of Modified FPA for CM

For the proposed research, required system data has been extracted from [16] and the results are compared with those presented in [16] and [20].

4.1 Test System Description

A medium-sized and large-sized test system has been used for investigations.

(i) Medium-sized modified IEEE 30 bus system has 06 generators, 24 load buses and 41 transmission lines; initial load for system is 283.4 MW and 126.2 MVAR.

(ii) Large-sized modified IEEE 57 bus system has 07 generators, 50 load buses and 80 transmission lines; initial load for system is 1250.8 MW and 335.9 MVAR.

4.2 Impact of Line Outage on Network: Case I

The outage of transmission line between bus 1 and bus 2 of modified IEEE 30 bus test system is considered. Consequently, lines 1–7 and 7–8 experience overloading of 14% and 05%, respectively. Power flow in these is 147.251 MW and 135.979 MW, respectively, with a total power violation of 23.23 MW. System's real power losses in state of contingency are 15.843 MW.

4.3 Impact of Line Outage and Increased Demand on Network: Case II

Sudden increase in demand by a factor of 1.5 at all PQ buses as well as outage of line 1–7 of modified IEEE 30 bus test system is dealt. As a result, lines 1–2, 2–8 and 2–9 become congested by 139%, 50% and 59%, respectively. Power flow in these is 310.9 MW, 97.35 MW and 103.5 MW, respectively, with a total power violation of 251.75 MW. System's real power losses in contingency state are 37.34 MW.

4.4 Impact of Variation in Line Power Limits: Case III and IV

In case III, power flow limits of line 5–6 to and 6–12 of modified IEEE 30 bus test system change to 175 and 35 MW in place of 200 and 50 MW, initially. Power flow in these lines is 195.397 MW and 49.315 MW, respectively. Thus, lines get congested by 12% and 41%, respectively, with a total power violation of 34.792 MW. System's losses with congestion are 27.338 MW.

Case IV considers variation in power flow limit of line 2–3 from 85 to 20 MW. Power flow in this line is 38.19 MW. As an outcome, line experiences congestion of 98.9% with 18.19 MW of power violation. System's losses in contingency state are 27.338 MW.

Table 1 gives tabular representation of Sects. 4.2–4.4.

Table 1 Real power flow (MW) during congestion and its relief for case I and II

Study	OLL[a]	Real power flow	Power flow with congestion relief by					Line limit (MW)
			MFPA	FPA [20]	PSO [16]	RSM [16]	SA [16]	
Case I	1–7	147.251	**128.7**	130	129.97	129.78	129.51	130
	7–8	135.979	**119.6**	120.78	120.78	120.60	120.35	130
Case II	1–2	310.9	**129.7**	130	129.97	129.91	129.78	130
	2–8	97.35	**63.99**	60.78	61.1	52.36	51.47	65
	2–9	103.5	**64.88**	65	64.67	55.43	54.04	65

[a]OLL: Overloaded lines

4.5 Alleviation of Congestion by Generation Rescheduling

MFPA is executed for the problem for 120 independent runs on MATLAB 2009b platform and the best outcome is exhibited. The population size of 20 and iteration count of 500 fits for all contingencies. Best results are achieved when value of scaling factor 'F' for case I, case II and case IV is kept at 0.8 and for case III at 0.7.

4.6 Computational Steps for Implementing MFPA for Solution of CM Problem

Step 1 Input busdata, branchdata, incremental and decremental bids for generating buses and generator operating limits.

Step 2 Initiate contingency that is to be simulated.

Step 3 Run power flow using Newton–Raphson method [23] and satisfying Eqs. (2–5). Thus, find the line limit violation and voltage violation at load buses.

Step 4 Randomly generate the initial population of solutions which is pollen/flower. The elements of these solution sets are the amount of rescheduled generation at each generating bus for complete removal of congestion.

Step 5 For each solution set, run load flow and evaluate fitness function using Eq. (17).

Step 6 Modify the solutions using Levy distribution/global pollination or local pollination depending on probability switch.

Step 7 Check feasibility limits and repair.

Step 8 Evaluate the fitness for the new solution sets and identify the best feasible solution.

Step 9 Modify the solutions through 'intensive exploitation' depending on switching probability.

Step 10   Evaluate the fitness and find the best feasible solution.
Step 11   If predefined iteration count has reached then stop, else go back to step 6.

5   Results and Discussions

Case I: Congestion is completely removed from the system. Cost incurred is 511.33 $/h. Power flow in formerly congested lines 1–7 and 7–8 is 128.7 MW and 119.6 MW, respectively. Loading factor for these lines is 99% and 92%, respectively. Total rescheduled generation is 25.714 MW. System's total real power losses decrease to 12.98 MW. When compared with FPA, PSO, RSM and SA, MFPA incurs lowest cost. Tables 1, 2 and 5 give detailed view on this. Figure 1 shows the comparative rescheduling of individual units.

Case II: Alleviation of congestion is successfully done with a total cost of 5315.17 $/h. Power flow in all congested lines, viz. 1–2, 2–8 and 2–9 is 129.7 MW, 63.99 MW and 64.88 MW, respectively, with loading factor of 99.7%, 98.4% and 64.88%, respectively. 168.409 MW of rescheduling is required to mitigate congestion. System's real power losses drop down to 26.075 MW. The cost achieved is lesser than those obtained with FPA, PSO, RSM and SA. Above findings are tabulated in Tables 1, 2 and 5 and comparison of rescheduling of individual units is shown in Fig. 2.

Case III: Line flows are within their limits with CM. Previously overloaded lines 5–6 and 6–12 now have a loading factor of 89.6 and 92.6% and power flow of 156.79 and 32.403 MW. Involved cost is 6279.11 $/h and total rescheduled generation is 156.89 MW. System's total real power losses decrease to 26.075 MW. When compared with FPA, PSO, RSM and SA, MFPA incurs lowest cost. Tables 3, 4 and 5 give detailed view on this. Figure 3 shows the comparative rescheduling of individual units.

Case IV: Mitigation of congestion is achieved with a total cost of 2882.01 $/h. Power flow in overloaded line, 2–3 is 19.78 MW, with loading factor of 98.9%. 70.62 MW of rescheduling is required to mitigate congestion. System's real power losses fall to 26.78 MW. The cost achieved is lesser than those obtained with FPA, PSO, RSM and SA. Above findings can be viewed in Tables 3, 4 and 5 and comparative rescheduling of individual units is shown in Fig. 4.

5.1   Convergence Behavior of MFPA

It is observed that MFPA produces a stable convergence in the tested cases of contingencies. Convergence characteristics of dealt four cases are shown in Figs. 5, 6, 7 and 8. Figures show that convergence is achieved for all operational variation cases.

Table 2 Generation rescheduling of generating units in MW for case I and II

Study	Method	Unit wise generation rescheduling						Gross ΔP_g
		ΔP_{g1}	ΔP_{g2}	ΔP_{g3}	ΔP_{g4}	ΔP_{g5}	ΔP_{g6}	
Case I	**MFPA**	**−9.914**	**15.740**	**−0.006**	**−0.022**	**0.012**	**0.020**	**25.714**
	FPA [20]	−9.1278	14.14	−0.206	−0.0188	0.189	1.013	24.703
	PSO [20]	−8.6123	10.4059	3.0344	0.0170	0.8547	−0.0122	22.936
	RSM [20]	−8.8086	2.6473	2.9537	3.0632	2.9136	2.9522	23.339
	SA [20]	−9.0763	3.1332	3.2345	2.9681	2.9540	2.4437	23.809
Case II	**MFPA**	**−8.849**	**76.849**	**0.001**	**64.97**	**10.14**	**7.6**	**168.409**
	FPA [20]	−8.589	74.024	0.000	13.5174	43.865	27.890	167.885
	PSO [20]	NA[a]	NA[a]	NA[a]	NA[a]	NA[a]	NA[a]	168.03
	RSM [20]	NA[a]	NA[a]	NA[a]	NA[a]	NA[a]	NA[a]	164.55
	SA [20]	NA[a]	NA[a]	NA[a]	NA[a]	NA[a]	NA[a]	164.53

[a]NA: Not available in referred literature

Fig. 1 Real power rescheduling for case I

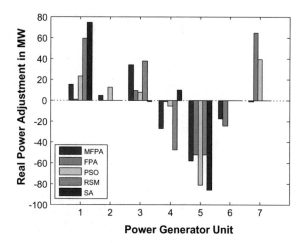

Fig. 2 Real power rescheduling for case II

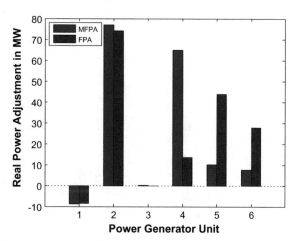

Table 3 Real power flow (MW) during congestion and its relief for case III and IV

Study	OLL[a]	Real power flow	Power flow with congestion relief by					Line limit (MW)
			MFPA	FPA [20]	PSO [20]	RSM [20]	SA [20]	
Case III	5–6	195.971	**156.79**	163.676	141	148.4	146.6	175
	6–12	49.351	**32.403**	35	34.67	35	34.84	35
Case IV	2–3	38.19	**19.78**	20	19.88	20	18.43	20

[a]OLL: Overloaded lines

Table 4 Generation rescheduling of generating units in MW for case III and IV

Study	Method	Unit wise generation rescheduling							Gross ΔP_g
		ΔP_{g1}	ΔP_{g2}	ΔP_{g3}	ΔP_{g4}	ΔP_{g5}	ΔP_{g6}	ΔP_{g7}	
Case III	**MFPA**	**15.26**	**4.32**	**33.73**	**−26.97**	**−57.99**	**−17.47**	**−1.15**	**156.89**
	FPA [20]	0.88	0.01	9.25	−1.36	−52.48	−24.55	64.33	152.86
	PSO [20]	23.14	12.45	7.49	−5.39	−81.22	0.00	39.03	168.72
	RSM [20]	59.27	0.00	37.45	−47.39	−52.12	0.00	0.00	196.23
	SA [20]	74.50	0.00	−1.52	9.95	85.92	0.00	0.00	171.89
Case IV	**MFPA**	**5.02**	**−28.51**	**33.03**	**0.33**	**−2.26**	**−0.55**	**−0.92**	**70.62**
	FPA [20]	0.01	−35.62	20.10	0.03	1.43	−0.03	13.97	71.19
	PSO [20]	NA[a]	NA[a]	NA[a]	NA[a]	NA[a]	NA[a]	NA[a]	76.31
	RSM [20]	NA[a]	NA[a]	NA[a]	NA[a]	NA[a]	NA[a]	NA[a]	89.32
	SA [20]	NA[a]	NA[a]	NA[a]	NA[a]	NA[a]	NA[a]	NA[a]	97.89

[a]NA: Not available in referred literature

Table 5 Comparison of cost for alleviating congestion for all cases of contingency

Study	Cost for alleviating congestion				
	MFPA	FPA [20]	PSO [20]	RSM [20]	SA [20]
CaseI	**511.33**	519.62	538.95	716.25	719.861
CaseII	**5315.17**	5320.8	5335.5	5988.05	6068.7
CaseIII	**6279.11**	6340.8	6951.9	7967.1	7114.3
Case IV	**2882.01**	2912.6	3117.6	3717.9	4072.9

Fig. 3 Real power rescheduling for case III

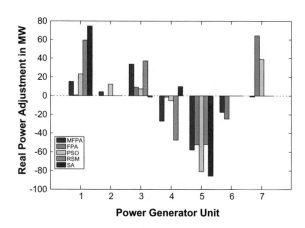

Fig. 4 Real power rescheduling for case IV

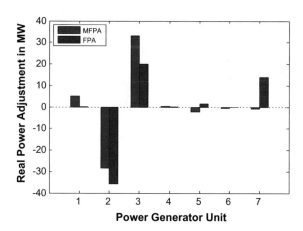

6 Conclusions

This paper establishes the superiority of recently proposed modified flower pollination algorithm (MFPA) over FPA, PSO, RSM and SA for the tested cases of line

Fig. 5 Convergence curve
for cost in case I

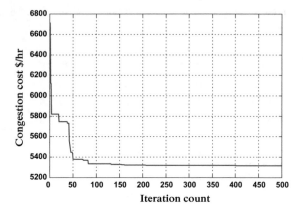

Fig. 6 Convergence curve
for cost in case II

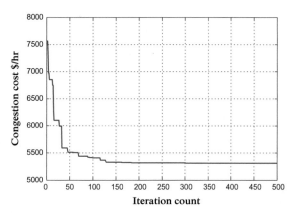

Fig. 7 Convergence curve
for cost in case III

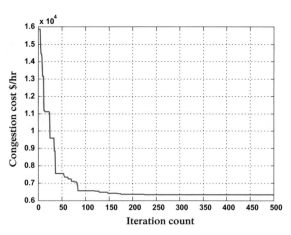

Fig. 8 Convergence curve for cost in case IV

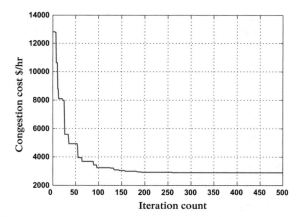

outage, increased demand and variation in line power limit contingencies. The performance of the proposed algorithm is tested for problem of congestion management on modified IEEE 30 bus system and modified IEEE 57 bus test system in a pool-based competitive electricity market. The proposed method based on modified FPA is found to further reduce the congestion cost per hour in tested four cases by 1.6%, 0.1%, 0.9% and 01%, respectively. The proposed algorithm proves to be promising in terms of cost, convergence and computational time.

Acknowledgements The authors acknowledge the financial support provided by AICTE New Delhi, India under the RPS research grant entitled "Addressing Power System Operational Challenges with Renewable Energy Resources Using Nature Inspired Optimization Techniques" sanctioned vide File No. 8-36/RIFD/RPS/POLICY-1/2016-17 dated August 2, 2017. The facilities and support provided by the Director and management of M.I.T.S Gwalior, India for carrying out this work are also sincerely acknowledged.

References

1. Lai, L.L.: Power System Restructuring and Deregulation, 1st edn. Wiley, New York (2001)
2. Christie, R.D., Wollenberg, B.F., Wangensteen, I.: Transmission management in the deregulated environment. P. IEEE **88**(2), 170–194 (2000)
3. Shahidehpour, M., Yamin, H., Li, Z.: Market Operations in Electric Power Systems. Wiley & Sons, Chichester (2002)
4. Bompard, E., Correia, P., Gross, G., Amelin, M.: Congestion-management schemes: a comparative analysis under a unified framework. IEEE Trans. Power Syst. **18**(1), 346–352 (2003)
5. Faliana, F.D., Ilic, M.: A mathematical framework for the analysis and management of power transactions under open access. IEEE Trans. Power Syst. **13**(2), 681–687 (1998)
6. Glavitsch, H., Alvarado, F.: Management of multiple congested conditions in unbundled operation of a power system. IEEE Trans. Power Syst. **13**(3), 1013–1019 (1998)
7. Pillay, A., Karthikeyan, S.P., Kothari, D.P.: Congestion management in power systems—a review. Electr. Power Energy Syst. **70**, 83–90 (2015)

8. Kumar, A., Srivastava, S.C., Singh, S.N.: Congestion management in competitive power market: a bibliographical survey. Electric Power Syst. Res. **76**, 153–164 (2005)

9. Overbye, T.J., Sauer, P.W., Marzinzik, C.M., Gross, G.: A user-friendly simulation program for teaching power system operations. IEEE Trans. Power Syst. **10**(4) (1995)

10. Talukdar, B.K., Sinha, A.K., Mukhopadhyay, S., Bose, A.: A computationally simple method for cost-efficient generation rescheduling and load shedding for congestion management. Int. J. Electr. Power Energy Syst. **27**(5), 379–388 (2005)

11. Balaraman, S., Kamaraj, N.: Application of differential evolution for congestion management in power system", Modern Applied Science **4**(8) (2010)

12. Kumar A., Srivastava, S.C.: AC power transfer distribution factors for allocating power transactions in a deregulated market. IEEE Power Eng. Rev., pp 42–43 (2002)

13. Gope, S., Dawn, S., Mitra, R., Goswami, A.K., Tiwari, P.K.: Transmission congestion relief with integration of photovoltaic power using lion optimization algorithm. Soft Computing For Problem Solving. Advances in Intelligent Systems and Computing, vol 816. Springer, Singapore (2019)

14. Zaeim-Kohan, F., Razmi, H., Doagou-Mojarrad, H.: Multi-objective transmission congestion management considering demand response programs and generation rescheduling. Appl. Soft Comput. **70**, 169–181 (2018)

15. Yadav, N.K.: Soft Comput. https://doi.org/10.1007/s00500-017-2792-3. 25 Apr 2019

16. Balaraman, S., Kamaraj, N.: Transmission congestion management using particle swarm optimization. J. Electr. Syst. **7**(1), 54–70 (2011)

17. Siddiqui, Anwar Shahzad, Sarwar, Md: An efficient particle swarm optimizer for congestion management in deregulated electricity market. J. Electr. Syst. Inf. Technol. **2**(3), 269–282 (2015)

18. Hemmati, R., Saboori, H., Jirdehi, M.A.: Stochastic planning and scheduling of energy storage systems for congestion management in electric power systems including renewable energy resources. Energy **133**(C), 380–387 (2017)

19. Yang, X.S.: Flower pollination algorithm for global optimization. In: Unconventional Computation and Natural Computation. Lecture Notes in Computer Science, vol. 7445, pp. 240–249 (2013)

20. Verma, S., Mukherjee, V.: A novel flower pollination algorithm for congestion management in electricity market. In: 3rd International Conference on Recent Advances in Information Technology, IEEE (2016)

21. Dubey, H.M., Pandit, M., Panigrahi, B.K.: A biologically inspired modified flower pollination algorithm for solving economic dispatch problems in modern power systems. Cogn. Comput. **7**(5), 594–608 (2015)

22. Kothari, D.P., Dhillon, J.S.: Power system optimization. PHI, New Delhi (2011)

23. Saadat, H.: Power System Analysis. Tata McGraw Hill Ltd, New Delhi (2002)

Intelligent Condition Monitoring of a CI Engine Using Machine Learning and Artificial Neural Networks

P. Naveen Kumar, S. R. Jeeva Karunya, G. Sakthivel and R. Jegadeeshwaran

Abstract The evolving automotive industry with the rapid growth of using fuel resources causes a shortfall of fuel availability. The engine maintenance has great importance and it is essential to develop a fault detection system and condition monitoring is done to reduce the damage-causing circumstances to improve engine safety. The engine tested at the rated speed of 1800 rpm by varying the loads. Captured vibration data is analyzed by using feature extraction and classification algorithms to obtain the best performance and minimum vibrations of blends. The best classification accuracy blend results are selected and used to develop a neural network model and to predict the vibration signatures. The neural network model is developed with a feedforward backpropagation algorithm and using Levenberg–Marquardt as training function. The developed ANN model is trained and vibration signatures are predicted for the selected blend B25 with the classification accuracy of 97%.

Keywords CI engine · Vibrations · Bio-diesel · Machine learning techniques · ANN

1 Introduction

The compression ignition engines (CI) are usually operated on account of its promising conditions over SI engines and a better economy. Because of more operation of vehicles, the non-renewable energy sources are draining faster and demand is increasing. To minimize the use of diesel, suitable alternate fuel is to be found and it should be sustainable. Here, in this study, fish oil is combined with diesel and biogas under different proportions which are used as a pilot fuel [1–3]. Mostly, low-frequency vibrations are formed due to its moving parts and high-frequency vibrations are caused due to piston slap, abnormal combustion, structural failure, in-cylinder pressure rise, wear of the piston, crankshaft speed fluctuations, knocking, and injection timing [4]. Condition monitoring is an approach of identifying engine health by

P. Naveen Kumar (✉) · S. R. Jeeva Karunya · G. Sakthivel · R. Jegadeeshwaran
School of Mechanical and Building Sciences, VIT, Chennai, India
e-mail: pallicheruvunaveen@gmail.com

© Springer Nature Singapore Pte Ltd. 2020
S. Agarwal et al. (eds.), *Machine Intelligence and Signal Processing*,
Advances in Intelligent Systems and Computing 1085,
https://doi.org/10.1007/978-981-15-1366-4_16

analyzing vibration data to recognize and classify the conditions by implementing appropriate condition monitoring (CM) techniques. The identification of the fault is known as fault detection and the type of fault is quantified by fault diagnosis. Here, one of the main objectives is to observe the continuous vibration data generated from the engine at different conditions. Another objective is to predict the vibration signatures of the engine from the observed signal features using a sensor mounted on the engine cylinder head [5] and data acquisition [6] systems. From the observed data, it is processed using machine learning techniques like feature extraction, feature selection, and feature classification [7, 8]. Supervised learning algorithms like random forest and Bayes net are used. From the obtained results, comparison study [8] of classification accuracy of classifiers is done for the selection of the best classifier and the best blend combination. The neural network is one of the artificial intelligence tools that are used for the prediction. The results obtained through machine learning have emerged with the neural networks and the ANN model is developed by using the algorithms and training function. A trained network model can be believed of an expert in the sort of data it has been specified to analyze. This skill can be used to provide predictions for new situations in order to solve the problem. Neural network models are advantages in engine optimization when compared to other optimizations. The root mean square (RMS) methodology was used to correlate the outputs.

2 Experimental Study

A single-cylinder, four-stroke compression ignition (CI) engine is connected to an eddy current kind dynamometer for loading and used to carry out experimental investigations with a dual fuel arrangement. A schematic figure and clear image of the examination setup are given in Figs. 2 and 3. The specifications of the engine used for this study are specified in Table 1. The testing was initially carried out with pure diesel along with the addition of methane through inlet manifold with three different flow rates. In sequent to that two different proportions of CH_4 and CO_2 intimating the simulated form of biogas were passed to the inlet manifold. Based on the methane flow rate and CO_2 adjustment is made to diesel injection through the governor to maintain the engine at a constant speed. For monitoring engine vibrations, accelerometer is mounted on the engine cylinder head. The vibration data for 100 cycles is employed and the average is considered, to remove fault at numerous cycles. The same experiment is repetitive for various loads from no load (0 Nm) to full load (20 Nm). The vibration data is collected from the sensor using the DAQ device and analyzed. The photographic view of the sensor mounted is shown in Fig. 1.

In methane enhancement, the test fuel initially used is diesel. The methane is added with it through inlet manifold under the flow rate mixture of methane and CO_2 as detailed in Table 2. The same experiment is done with ethyl ester of fish oil of various fuel blends (B20, B25, B30, B35, and B40) as the test fuel. Then the corresponding vibration signals are captured.

Table 1 Engine specifications

Parameter	Value
Bore	87.52 mm
Stroke	80 mm
Cubic capacity	0.481 L
Number and arrangement of cylinders	One-vertical
Working cycle	Four-stroke diesel
Combustion principle	Compression ignition
Compression ratio	17:1
Peak pressure	73.6 bar

Fig. 1 Photographic view of the accelerometer mounted on engine cylinder head

3 Methodology

The vibration signals are captured for all the specified conditions as stated through the accelerometer. The collected raw vibration signal data is further analyzed using two phases.

1. Machine learning techniques
2. Neural network model.

The first phase is machine learning techniques. Supervised learning algorithms like decision trees and Bayes net are used. They form the tree structure starting from the training data and built a predictive model. The goal is to achieve classification accuracy by a minimum number of features. Machine learning has three significant

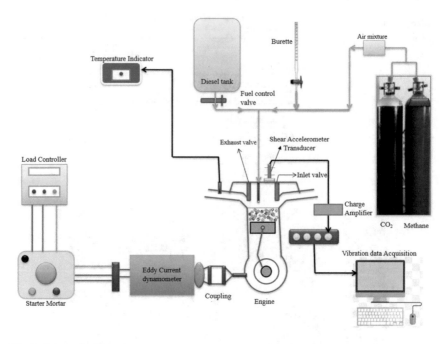

Fig. 2 Schematic diagram

Table 2 Biogas proportion for blend B25

Fuel blend name	Methane (CH$_4$) flow rate in LPM (liter per minute)	Carbon dioxide (CO$_2$) flow rate in LPM (liter per minute)
B25	4	0
B25	8	0
B25	12	0
B25	4.8	3.2
B25	7.2	4.8

stages; (i) feature extraction, (ii) feature selection, and (iii) feature classification as shown in Fig. 3.

The second phase is neural network modeling, from the results of best-selected blend data is used to develop the ANN model and ANN model is trained by varying the neurons in the hidden layer, and the network is tested and if it satisfies the conditions, then predicts the vibration signatures, if it is not satisfied, the neurons in the hidden layer are varied and go for testing (Fig. 4).

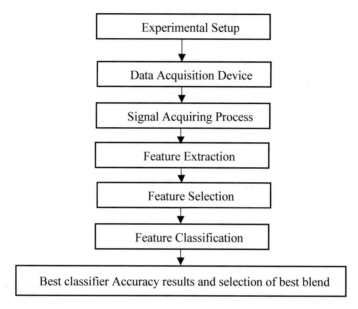

Fig. 3 Machine learning methodology

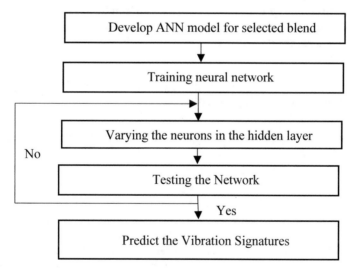

Fig. 4 Neural network methodology

4   Machine Learning Techniques

From the captured data for dissimilar running conditions of the engine with a change of fuels and load conditions, vibration signals are obtained and data is used for the

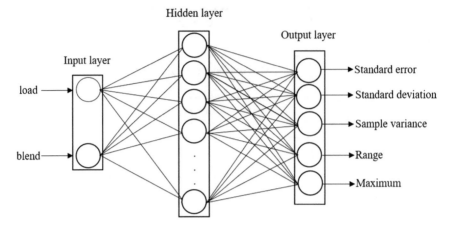

Fig. 5 Architecture of the ANN model

machine learning process and we observed the disparity in the signals for each load condition. So, it is clearly showing we can classify by using the classifiers. The two different classifiers used are random forest and Bayes net.

Decision trees are a standard and efficient way of representing the information form an algorithm. The decision tree uses a tree graph model for generating the classification rules through the nodes, where each branch is connected to the nodes which are the classification attributes.

4.1 Random Forest Algorithm

It is a supervised learning algorithm that makes the forest with several trees. The more robust and high classification accuracy is obtained by increasing the number of trees in the forest. The decision tree grows using the classification and regression trees. The tree building starts with passing the root node to the binary nodes and the splitting of data is done accordingly [9]. The most promising class random forest will increase the accuracy of rising the tree. Classification and regression tree splits possible points based on the first variable. This applies the criteria for splitting and evaluation by plummeting the impurity. The ranking order of the tree starts according to the best split of nodes. The selected variable will mostly reduce the impurity of the nodes. Here, creating pseudocode consists of splitting of data and increasing the number of trees in the forest which is training the algorithm. This algorithm creates pseudocode and split into two stages such as creating pseudocode and pseudocode to perform the prediction. The random forest uses the trained pseudocode for prediction. The over-fitting of data problems will not be found in the algorithm. It is used for identifying the most important features among the available features in a training set data. The growth of the tree will be terminated when there is no possibility of splitting data nodes and when the extreme limit is reached, it will stop the further tree growth by pruning.

4.2 Bayes Net Algorithm

The Bayesian network is a probabilistic graphical model or also known as a type of statistical model that the set of variables and their conditional dependencies on a directed acyclic graph (DAG). It takes an event E and starts predicting for several possible roots. Bayesian networks can represent and solve decision problems with the help of influence diagrams. The variables are dependent on each other which one variable will be influencing the other variable. In the graphical model of DAG, the variables are represented as nodes. Each node is associated with a probability function.

The Bayesian network performs three main tasks:

(i)   Inferring unobserved variables
(ii)   Parameter learning
(iii) Structure learning.

In the first task, the nominated variable relation forms the complete model and reducing the expected loss proportion by choosing the values for the variable. For complex difficulties, decision error probability is reduced and the common inference method is used for the elimination of variables. The joint probability represents the full network and learning starts with single distribution and increases from single to multi-distribution. Direct maximization of the distribution will create the complication for the given variables and leads to increasing the size of the model dimensionally. In the third task, significant defining of the network will be complex for humans and network structure learns from the data. The basic learning idea drives back to improving the established algorithm.

5   Neural Network Model

Artificial neural networks are computing structures which are stimulated from a biological neural network that reflects the human nervous system. The ANN model is based on the collection of associated nodes known as the neurons. Each neuron is associated with additional neurons to mount the network. The neural network itself does not an algorithm in addition to frame the network and numerous neural network algorithms are used to work with the complex data. The architecture of the ANN model is shown in Fig. 6. Through hidden layers, the information passes from the input layer to the output layer. The number of hidden layers and neurons in each hidden layer can be ranging according to the difficulty of the data set [10] (Fig. 5).

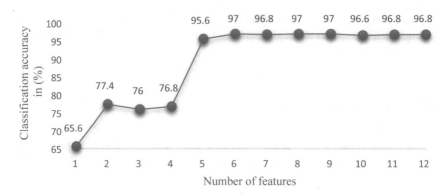

Fig. 6 Classification accuracy versus number of features for random forest algorithm

In ANN, 70% of data is taken for training and 30% of data is taken for testing. The 2-1-1-5 network configuration is formulated for the present experiment, where 2 indicates the input layer with 2 neurons and 5 indicates the output layer with 5 neurons and the 1-1 represents the number of hidden layers.

From the results obtained by machine learning techniques for selection of best classifier and selection of best blend combinations. It is known B25 blend is the best and further by applying the neural network approach, it is processed for predicting the statistical features for the selected blend combination. Here, load and blend are the input parameters and various parameters like standard error, standard deviation, sample variance, range, and maximum are the statistical features which are considered as output parameters for the neural model as detailed in Fig. 5.

5.1 Feedforward Backpropagation Network (FFBP)

It is a static function in which the data passes through only one direction where the training process of the network is done initially data is passed through the input layer to hidden layer with adding weights and layers is moved forward direction to the output layer for predicting the outputs. The activation function which is both nonlinear sigmoid (or) logistic function for the hidden layers and linear function for the hidden layer making contact to the output layer. In this feedforward network, the mapping between the input layer, hidden layer, and output layer is made by the neurons. The variance between the output value and trained value gives the error occurred [11]. The error is backpropagated and minimizing the error by changing the weights and bias. Then the trained dataset is used to predict the new set of values.

6 Results and Discussions

In this study fault diagnosis of a CI engine detection using machine learning techniques and neural network, the model has been discussed. The various statistical features are been extracted from the rough vibration signal for the blend B25; out of the different features, the most relevant features selected are standard deviation, sample variance, standard error, minimum, mean, kurtosis, skewness, and range. The decision tree classifier and Bayes net classifier are used to find out the classification accuracy of the systems. Each classifier algorithm has its advantages in providing classification accuracy. The classification accuracy and the confusion matrix of the two different tree classifiers are compared to give the best classification accuracy and selection of the best classifier.

6.1 Effects of Features

From a specified vibration signal, the number of features extracted like standard deviation, sample variance, standard error, minimum, mean, kurtosis, skewness, and range. Certain features will help to notice the fault condition and some features may not provide required classification accuracy and hence, those features can be neglected or removed to obtain the high classification accuracy rate as shown in Table 3.

From the below table, it is witnessed that the maximum classification accuracy is 97% seen for six features and the disparity in classification accuracy is observed due to adding some features.

Table 3 Classification accuracy of the selected features for random forest algorithm

No. of features	Classification accuracy in (%)
1	65.6
2	77.4
3	76
4	76.8
5	95.6
6	97
7	96.8
8	97
9	97
10	96.6
11	96.8
12	96.8

Table 4 Confusion matrix for random forest algorithm

Testing conditions	*a	*b	*c	*d	*e
*a	98	1	0	1	0
*b	0	98	2	0	0
*c	0	5	93	2	0
*d	0	1	3	96	0
*e	0	0	0	0	100

where *a = L0M7.2C4.8, *b = L5M7.2C4.8, * c = L10M7.2C4.8, *d = L15M7.2C4.8, and *e = L20M7.2C4.8

6.2   Classification Using Random Forest Algorithm

The best classification results for the B25 blend are discussed in detail by using the obtained confusion matrix for the random forest algorithm is shown in Table 4. Here, random forest algorithm is capable of recognizing the fault signal condition and the good signal condition. At very few instincts, the signal is misclassified with an overall classification accuracy of 97%. It is noticed that over 100 signals, 98 signals are correctly classified for condition L0M7.2C4.8, 98 signals for condition L5M7.2C4.8, 93 signals for condition L10M7.2C4.8, 96 signals for condition L15M7.2C4.8, 100 signals for condition L20M7.2C4.8. It also observed that from the confusion matrix that for condition L0M7.2C4.8, 1 instinct is classified in L10M7.2C4.8 and 1 is classified in L15M7.2C4.8.

The random forest algorithm shows the maximum classification accuracy results of 97% by utilizing eight features only and it is selected as the best classifier.

6.3   Classification Using Bayes Net Algorithm

From a specified vibration signal, the number of features is extracted like standard deviation, sample variance, standard error, minimum, mean, kurtosis, skewness, and range

These features will help to notice the fault condition and some features may not give the required classification accuracy and hence, those features can be removed to obtain the high classification accuracy rate as shown in Table 5. From the below table, it is recognized that the maximum classification accuracy is 95% seen for three features.

Figure 7 shows the variation's accuracy. Hence, it is observed that the Bayes net algorithm is capable of distinguishing with good accuracy and detection of fault state conditions is achieved with an accuracy of 95% for only three features.

The best classification results for the B25 blend are discussed in detail by using the obtained confusion matrix for the Bayes net algorithm as shown in Table 6. Here, Bayes net algorithm is able to identify the fault condition signal and the good

Table 5 Classification accuracy of the selected features for Bayes net algorithm

No. of Features	Classification accuracy in (%)
1	68.4
2	92
3	95
4	95
5	94.6
6	95
7	94
8	92.2
9	92.6
10	92.2
11	92.2
12	92.2

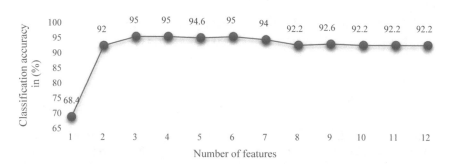

Fig. 7 Classification accuracy versus number of features for Bayes net algorithm

Table 6 Confusion matrix for Bayes net algorithm

Testing conditions	*a	*b	*c	*d	*e
*a	95	1	3	0	0
*b	4	86	10	0	0
*c	1	4	94	1	0
*d	0	0	0	100	0
*e	0	0	0	0	100

Where *a = L0M4.8C3.2, *b = L5M4.8C3.2, * c = L10M4.8C3.2, *d = L15M4.8C3.2, and *e = L20M4.8C3.2

condition signal. At very few instincts, the signals are misclassified with an overall classification accuracy of 97%. It is noticed that over 100 signals, 95 signals are correctly classified for condition L0M4.8C3.2, 86 signals for condition L5M4.8C3.2, 94 signals for condition L10M4.8C3.2, 100 signals for condition L15M4.8C3.2, 100

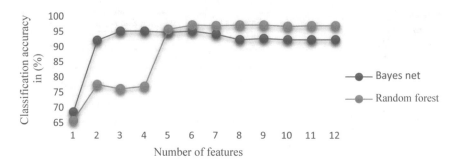

Fig. 8 Classification accuracy versus number of features for random forest algorithm and Bayes net algorithm

signals for condition L20M4.8C3.2. It also observed that from the confusion matrix for condition L0M4.8C3.2, 1 instinct is classified in L10M4.8C3.2 and 3 are classified in L15M4.8C3.2.

When comparing both classifiers random forest shows the best classification accuracy over Bayes net algorithm as shown in the above Fig. 8.

6.4 Neural Network Model

From the results gained by machine learning techniques for selection of best classifier and selection of best blend combinations, it is clearly identified B25 blend is the best and further applying the neural network approach, it is processed for predicting the statistical features for the selected blend combination.

From the above Table 7, the average correlation coefficient values are displayed and compared; from this, the best average correlation coefficient observed is 0.97922 for B25 blend with M7.2C4.8 combination by satisfying the conditions of overall

Table 7 Feed forward backpropagation algorithm for B25 blend with combinations

Combinations	M4	M8	M12	M4.8C3.2	M7.2C4.8
Neurons and algorithm	FF-4,4 N	FF-12 N	FF-15,15 N	FF-4 N	FF-4 N
Standard error	0.9348	0.994	0.882	0.9868	0.9772
Standard deviation	0.9471	0.994	0.871	0.99	0.9727
Sample variance	0.9093	0.993	0.887	0.9891	0.9811
Range	0.9281	0.934	0.931	0.9213	0.9702
Maximum	0.9034	0.954	0.976	0.9637	0.9949
Average correlation coefficients	0.92454	0.9738	0.9094	0.97018	**0.97922**

individual R values is greater than 0.9 and for the individual R values for standard error is 0.9772, standard deviation is 0.9727, sample variance is 0.9811, range is 0.9702, and maximum is 0.9949.

7 Conclusion

In this study, the health condition monitoring and prediction of vibration signatures of a CI engine were carried out using two techniques such as machine learning and neural network. The selection of the best blend combination with high classification accuracy using classifiers is done in machine learning part. Further, prediction of vibration signatures is carried out through the artificial neural network in the next part of the study. The results show that the classification accuracy of the random forest algorithm for blend B25 with combination M7.2C4.8 gives 97% of classification and it is selected as the best blend and further processed with the neural network by developing ANN model and for blend B25 prediction results obtained by using feedforward backpropagation network algorithm. M4 combination gives an average correlation coefficient of 0.92454, M8 combination gives an average correlation coefficient of 0.9738, M12 combination gives an average correlation coefficient of 0.9094, M4.8C3.2 combination gives an average correlation coefficient of 0.97018, M7.2C4.8 combination gives an average correlation coefficient of 0.97922.

References

1. Sakthivel, G., Nagarajan, G., Illangakumaran, M., Gaikwad, A.B.: Comparative analysis of performance, emission and combustion parameters of diesel engine fuelled with ethyl ester of fish oil and its diesel blends. Fuel **132**, 116–124 (2014)
2. Ahmad, et al.: Vibration analysis of a diesel engine using biodiesel and petro diesel fuel blends (2012)
3. Feroskhan, M., Ismail, S.: Investigation of the effects of biogas composition on the performance of a biogas–diesel dual fuel CI engine. Biofuels **8**, 593–601 (2016)
4. Jindal, S.: Vibration signatures of a biodiesel fueled CI engine and effect of engine parameters. **3**(1), 151–160 (2012)
5. Somashekar, V., Satish, K., Jamuna, A.B., Ranjitha, P.: Vibration signatures analysis of IC engine. **2**(13) (2013)
6. Nandi, A.K., Liu, C., Wong, M.L.D.: Intelligent vibration signal processing for condition monitoring. In: Proceedings of International Conference Surveillance, pp. 1–15 (2013)
7. Sharma, A., Sugumaran, V., Devasenapati, S.B.: Misfire detection in an IC engine using vibration signal and decision tree algorithms. Measurement 370–380 (2013)
8. Jegadeeshwaran, R., Sugumaran, V.: Comparative study of decision tree classifier and best first tree classifier for fault diagnosis of automobile hydraulic brake system using statistical features. Measurement **46**, 3247–3260 (2013)
9. Yang, B.S., Di, X., Han T.: Random forest classifier for machine fault diagnosis. J. Mech. Sci. Technol. 1716–1725 (2008)

10. Ilangakumaran, M., Sakthivel, G., Nagarajan, G.: Artificial neural network approach to predict the engine performance of fish oil bio diesel with diethyl ether using back propagation algorithm. Int. J. Ambient. Energy **37**(5), 446–455 (2016)
11. Bhowmik, S., Panua, R., et al.: Artificial Neural network prediction of diesel engine performance and emissions fueled with diesel, kerosene, and ethanol blends: a fuzzy based optimization. J. Energy Resour. Technol. **139** (2017)

Bacterial Foraging Optimization in Non-identical Parallel Batch Processing Machines

Arindam Majumder and Dipak Laha

Abstract The paper presents a modified bacterial foraging optimization (BFO) algorithm to solve non-identical parallel batch processing machine scheduling problems with the objective of minimizing total weighted tardiness. The proposed algorithm combines the shortest position value-based heuristic with the basic BFO algorithm to enhance the solution quality. The computational experimental results based on ninety randomly generated problem instances reveal that the proposed algorithm performs better than the existing particle swarm optimization algorithm.

Keywords Parallel batch processing machine · Total weighted tardiness · Bacterial foraging optimization · BFOA

1 Introduction

The machines having capability to process multiple jobs at a time are known as batch processing machines. These machines are used in integrated circuit (IC) chip manufacturing industries [1], metalworking industries [2], aircraft industry [3], furniture manufacturing industries [4], shoe making industries [5] and semiconductor wafer manufacturing industries [6]. However, it is seen in the scheduling literature that the research on scheduling of batch processing machines primarily focusses in the semiconductor wafer manufacturing industries [6–11]. In these industries, the batch processing machines are mainly used for assembly and testing purposes.

Thus, a numerous effort has been given by the previous researchers for increasing the productivity and reducing the cost penalty related to the lateness of delivery. Initially, researchers developed and solved mixed integer linear programming (MILP) models for scheduling of jobs in parallel batch processing machines [7, 12–15]. The two commercial solvers, namely IBM ILOG CPLEX [12–14] and XPRESS MP® [15] were mostly used for this purpose. However, the MILP models are solvable if

A. Majumder (✉)
Mechanical Engineering Department, National Institute of Technology, Agartala 799046, India

D. Laha
Mechanical Engineering Department, Jadavpur University, Kolkata 700032, India

© Springer Nature Singapore Pte Ltd. 2020   215
S. Agarwal et al. (eds.), *Machine Intelligence and Signal Processing*,
Advances in Intelligent Systems and Computing 1085,
https://doi.org/10.1007/978-981-15-1366-4_17

the problem size is small or less complex [15]. This motivates the researchers to propose and implement heuristics and metaheuristics for solving parallel batch processing machine scheduling problems. The most commonly used metaheuristics to solve parallel batch processing machine scheduling problems are simulated annealing [15], genetic algorithm [7], particle swarm optimization algorithm [13], greedy randomized adaptive search procedure [16], ant colony optimization algorithm [17] and cuckoo search algorithm [18]. In the recent times, bacterial foraging optimization algorithm plays a vital role in the field of process and parametric optimization because of its capability for solving real-life problems in several fields of engineering and management [19–23].

Therefore, in this present research work, an attempt has been taken for finding out the applicability of BFO algorithm in batch processing machine scheduling problems. For implementing this algorithm in parallel batch processing machines scheduling, a SPV-based heuristic is combined with BFO algorithm. The objective of the problem is considered as total weighted tardiness (TWT), which is to be minimized. Later, the results obtained by the modified BFO algorithm are compared with those of the existing PSO algorithm.

2 Problem Definition

Suppose, a set of F job families are processed on M number of parallel BPMs having different capacities (S_m). Each of the job family (f) contains a set of J_f jobs. The jobs included in each family have same processing time and priority. The job size for each family is considered as less than the machine's capacity. The jobs of one or more families are assigned to a machine in batches. All the job families allowed to process in any of the batch processing machines provided the machine is not occupied by the other job family. Each batch may contain more than one job families. However, splitting of job families is considered to be unacceptable. The known parameters for each job family are processing time (p_f), job size (s_f), job due date (d_f) and job priority (w_f). The time to process each batch (b) is the processing time required for the job having longest processing time among all the jobs of the batch. We consider the two decisions, i.e., (i) batching of jobs and (ii) batch sequencing independently.

If the same job size of all the part families is considered and only one machine is provided having capability to process each part family individually, then, the considered problem is reduced to a scheduling problem having one machine and the corresponding objective as TWT. Since Lawler (1977) proves the single machine scheduling problem as NP-hard, therefore, the considered problem can also be treated as NP-hard.

3 Discrete Bacteria Foraging Algorithm

Bacteria foraging optimization algorithm is a nature-inspired optimization algorithm based on the group foraging strategy of e-coli bacteria. In real world, a set of flagella helps the bacterium to move from one to another place. When the bacterium spins the flagella clockwise, the bacterium tumbles for slow moving toward nutrient environment. However, the anticlockwise spinning of flagella helps the bacterium to swim and move with faster rate. If the bacterium gets food sufficiently then it grows and starts reproduction by splitting them into replicas at appropriate temperature. On the other hand, due to slow or drastic change in environment either the bacteria group situated in the region is destroyed or the group disperses into a new region. Based on this phenomenon of e-coli bacteria, Passino [24] developed BFOA with four important steps, discussed below:

3.1 Chemotaxis

In this step of BFOA, each of the bacterium moves in a new location using swimming and tumbling operation. The equation used for swimming and tumbling is given below:

$$\theta^j(a+1,b,c) = \theta^j(a,b,c) + C(j)\frac{\Delta(j)}{\sqrt{\Delta^t(j)\Delta(j)}} \tag{1}$$

where, $C(j)$ denotes the step size considered in the arbitrary direction specified by the tumble. Δ denotes a vector in the arbitrary direction having each element in between $[-1$ and $1]$.

3.2 Swarming

The step enables the bacteria to group in a nutrient-rich region by staying away from toxic locations. For grouping, each of the bacterium cells gives signal to other cells by the release of attractants and repellents. In the algorithm, the signaling between the cells is simulated by the equation given below:

$$\mathbf{J_{CC}}(\boldsymbol{\theta}, \mathbf{P}(a,b,c)) = \sum_{i=1}^{n} \mathbf{J_{CC}}(\boldsymbol{\theta}, \boldsymbol{\theta}^i(a,b,c))$$

$$= \sum_{i=1}^{n}\left[-\mathbf{d_{at}}e^{\left(-\mathbf{w_{at}}\sum_{z=1}^{P}(\theta_z - \theta_z^i)^2\right)}\right] + \sum_{i=1}^{n}\left[-\mathbf{h_{rp}}e^{\left(-\mathbf{w_{rp}}\sum_{z=1}^{P}(\theta_z - \theta_z^i)^2\right)}\right] \tag{2}$$

where, $J_{cc}(\theta, P(a, b, c))$ is the part which is to be combined the actual objective function to make the objective function as time varying, n is the population size of bacteria, p is the number of dimensions of the search space, and $\theta = [\theta_1, \theta_2, ..., \theta_p]_t$ denotes the point location in the search space. d_{at}, w_{at}, h_{rp} and w_{rp} are different coefficients employed for signaling.

3.3 Reproduction

The process helps to simulate the killing of bacteria S_r $(S/2)$ having least health and splitting of healthier bacteria into replicas.

3.4 Elimination-Dispersal

The step includes the elimination and dispersal stage of bacteria's life cycle through elimination of each bacterium with a probability of P_{ed} and replaced by a newly generated bacterium within the search space.

The detailed flowchart describing the different steps of bacteria foraging algorithm is shown in Fig. 1. However, it is not possible to use this algorithm directly due to its inability to search in a discrete search space. Therefore in this present study, a HAD heuristic [25] inspired SPV rule-based heuristic to assign the jobs in the batches of the machines is integrated with BFOA. The different steps of this heuristic given below:

1. Sequence the jobs in increasing order of their weight values in order to build an unassigned job set.
2. Initially when all the machines are free, allocate the first job in the batch of the machine having highest capacity. Discard the job from unassigned job set.
3. When the first job is acceptable in more than one batch of different machines, then, assign the job in the batch, which gives minimum TWT. Discard the job from the unassigned job set.
4. When the first job is acceptable in multiple batch of a same machine, then, assign the job in the first available batch and discard it from the unassigned job set.
5. When the first job is not fitting in any of the batches of the machines then open a batch in each machine and go to step 3.

4 Computational Results

In order to test the capability of the modified BFO algorithm, ninety randomly generated instances of non-identical parallel batch processing machines scheduling are

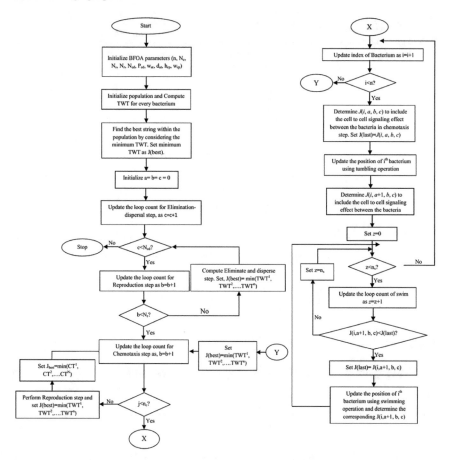

Fig. 1 Flow chart of bacteria foraging optimization algorithm

solved. These test instances are generated by considering the two machines for each instance and number of jobs as 5, 7, 9 11, 13, 15, 50, 100 and 200. The other input parameters associated with the generation of these test instances are shown in Table 1. The results obtained by the modified BFO algorithm are then compared with the solutions achieved by the PSO algorithm proposed in [25]. The relative difference between the TWT obtained by the modified BFO and the existing PSO for each instance is determined in terms of percentage. Equation 6 represents the mathematical expression for calculating relative percentage difference (RPD).

Where,

$$z_f = \text{Discrete Uniform}\left[\alpha.\left(1 - \frac{D_t}{2} \right), \alpha.\left(1 + \frac{D_t}{2} \right) \right] \tag{3}$$

$$\alpha = (1 - D_s).C_{\max}^\# \tag{4}$$

Table 1 Process parameters used to generate random instances [25]

Parameter	Notation	Value
Machine capacity	S_m	[40, 45]
Job size	s_f	Discrete random [2, 30]
Processing time of job family	p_f	Discrete random [0, 48]
Priority of job family	w_f	Discrete random [8, 48]
Due date of job family	d_f	$\beta.(P_f + z_f)$
Due date adjustment factor	β	0.2
Due data spread out	D_s	0.3
Due date tightness	D_t	0.5

$$C_{max}^{\#} = \text{makespan determined through Full batch LPT rule} \qquad (5)$$

$$RPD_1(\%) = \left(\frac{TWT_1^{Algorithm} - TWT_1^{Best}}{TWT_1^{Best}} \right) \times 100 \qquad (6)$$

where

$$TWT_1^{Algorithm} = \text{TWT value for lth instance obtained by the algorithm} \qquad (7)$$

$$TWT_1^{Best} = \text{Best TWT value for lth instance obtained by any of two algorithms} \qquad (8)$$

In this study, the process parameters of BFO algorithm [26] are as follows: $d_{at} = h_{rp} = 0.5$, $w_{at} = 0.2$, $w_{rp} = 5.05$, Number of swimming steps (N_s) = 10, Number of chemotaxis steps (N_c) = 2, Probability of elimination-dispersal (P_{ed}) = 0.95, Number of elimination-dispersal steps (N_{ed}) = 4, Number of reproduction steps (N_{re}) = 50, Number of bacteria (n) = 200. The process parameters considered for the existing PSO algorithm is taken from [25]. Both modified BFO and existing PSO algorithms are coded in MATLAB 2017, a programming environment. The execution of these algorithms is carried out on a notepad having Intel i5-2450M CPU with 4 GB RAM running at 2.50 GHz.

Figure 2 shows the average RPD for the BFOA and PSO algorithms. From the results, it has clearly seen that the modified BFO algorithm performs significantly better than the existing PSO algorithm for 50, 100 and 200 job size problems. This is evident due to the consistent convergence rate of the modified BFOA as compared to the existing PSO algorithm shown in Fig. 3. However, the proposed algorithm shows a moderately better performance while solving problems with 13 and 15 job sizes. The results also indicate both the algorithms are comparable for 5, 7, 9 and 11 job size problems.

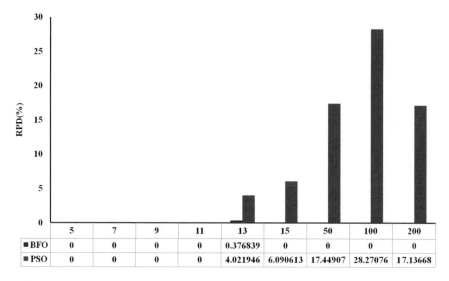

	5	7	9	11	13	15	50	100	200
■ BFO	0	0	0	0	0.376839	0	0	0	0
■ PSO	0	0	0	0	4.021946	6.090613	17.44907	28.27076	17.13668

Fig. 2 Box plot for particle swarm optimization algorithm and bacteria foraging optimization algorithm

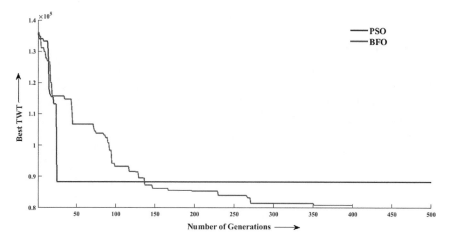

Fig. 3 Convergence plot of particle swarm optimization algorithm and bacteria foraging optimization algorithm for the problem instance having 50 job size

In order to compare the performance consistency, Fig. 4 presents the box plot of RPDs obtained by both the algorithms. From the results, it appears that for job sizes of 5, 7, 9 and 11 both the algorithms are consistent while solving batch processing machine scheduling problems. However, the PSO algorithm shows significantly inconsistent to obtain optimal solution for 13, 15, 50, 100 and 200 job size problems.

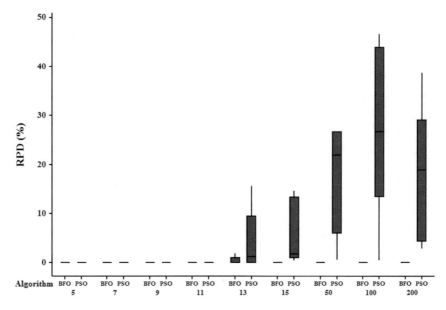

Fig. 4 Box plot for particle swarm optimization algorithm and bacteria foraging optimization algorithm

5 Conclusion

The study considers the scheduling of job families in parallel batch processing machines with non-identical machine capacity to minimize the total weighted tardiness. In the proposed algorithm, a SPV-based heuristic is combined with the basic BFO algorithm for grouping of job families in the batch and allocating the batches in the machines. During initialization of population, each of the individual is represented by assigning weight value to each job family. A number of randomly generated instances with two machines and different jobs are solved in order to find the effectiveness of the proposed BFO. The computational results reveal the superiority of the proposed algorithm over the exiting PSO algorithm.

References

1. Ahmadi, J.H., Ahmadi, R.H., Dasu, S., Tang, C.S.: Batching and scheduling jobs on batch and discrete processors. Oper. Res. **40**(4), 750–763 (1992)
2. Ram, B., Patel, G.: Modeling furnace operations using simulation and heuristics. In: Proceedings of the 30th Conference on Winter Simulation, pp. 957–964. IEEE Computer Society Press, Washington, D.C., USA (1998, Dec)
3. Van De Rzee, D.J., Van Harten, A., Schuur, P.C.: Dynamic job assignment heuristics for multi-server batch operations-a cost based approach. Int. J. Prod. Res. **35**(11), 3063–3094 (1997)

4. Yuan, J.J., Liu, Z.H., Ng, C.T., Cheng, T.E.: The unbounded single machine parallel batch scheduling problem with family jobs and release dates to minimize makespan. Theoret. Comput. Sci. **320**(2–3), 199–212 (2004)
5. Fanti, M.P., Maione, B., Piscitelli, G., Turchiano, B.: Heuristic scheduling of jobs on a multi-product batch processing machine. Int. J. Prod. Res. **34**(8), 2163–2186 (1996)
6. Gallego, M.C.V.: Algorithms for scheduling parallel batch processing machines with non-identical job ready times. Doctoral dissertation, Florida International University (2009)
7. Damodaran, P., Hirani, N.S., Velez-Gallego, M.C.: Scheduling identical parallel batch processing machines to minimise make span using genetic algorithms. Eur. J. Ind. Eng. **3**(2), 187–206 (2009)
8. Damodaran, P., Velez-Gallego, M.C.: Heuristics for makespan minimization on parallel batch processing machines with unequal job ready times. Int. J. Adv. Manuf. Technol. **49**(9–12), 1119–1128 (2010)
9. Hulett, M., Damodaran, P.: A particle swarm optimization algorithm for minimizing total weighted tardiness of non-identical parallel batch processing machines. In: IIE Annual Conference. Proceedings, p. 901. Institute of Industrial and Systems Engineers (IISE) (2015)
10. Suhaimi, N., Nguyen, C., Damodaran, P.: Lagrangian approach to minimize makespan of non-identical parallel batch processing machines. Comput. Ind. Eng. **101**, 295–302 (2016)
11. Hulett, M., Damodaran, P.: An Artificial Bee Colony Algorithm for Minimizing Total Weighted Tardiness of Non-identical Parallel Batch Processing Machines. In IIE Annual Conference Proceedings, pp. 848–853. Institute of Industrial and Systems Engineers (IISE) (2017)
12. Chang, P.Y., Damodaran, P., Melouk, S.: Minimizing makespan on parallel batch processing machines. Int. J. Prod. Res. **42**(19), 4211–4220 (2004)
13. Damodaran, P., Diyadawagamage, D.A., Ghrayeb, O., Vélez-Gallego, M.C.: A particle swarm optimization algorithm for minimizing makespan of non-identical parallel batch processing machines. Int. J. Adv. Manuf. Technol. **58**(9–12), 1131–1140 (2012)
14. Chung, S.H., Tai, Y.T., Pearn, W.L.: Minimising makespan on parallel batch processing machines with non-identical ready time and arbitrary job sizes. Int. J. Prod. Res. **47**(18), 5109–5128 (2009)
15. Damodaran, P., Vélez-Gallego, M.C.: A simulated annealing algorithm to minimize makespan of parallel batch processing machines with unequal job ready times. Expert Syst. Appl. **39**(1), 1451–1458 (2012)
16. Damodaran, P., Vélez-Gallego, M.C., Maya, J.: A GRASP approach for makespan minimization on parallel batch processing machines. J. Intell. Manuf. **22**(5), 767–777 (2011)
17. Cheng, B., Wang, Q., Yang, S., Hu, X.: An improved ant colony optimization for scheduling identical parallel batching machines with arbitrary job sizes. Appl. Soft Comput. **13**(2), 765–772 (2013)
18. Majumder, A., Laha, D., Suganthan, P.N.: A hybrid cuckoo search algorithm in parallel batch processing machines with unequal job ready times. Comput. Ind. Eng. **124**, 65–76 (2018)
19. Panda, S.K., Padhee, S., Anoop Kumar, S.O.O.D., Mahapatra, S.S.: Optimization of fused deposition modelling (FDM) process parameters using bacterial foraging technique. Intell. Inf. Manag. **1**(02), 89 (2009)
20. Nouri, H., Hong, T.S.: A bacteria foraging algorithm based cell formation considering operation time. J. Manuf. Syst. **31**(3), 326–336 (2012)
21. Liu, C., Wang, J., Leung, J. Y. T., Li, K.: Solving cell formation and task scheduling in cellular manufacturing system by discrete bacteria foraging algorithm. Int. J. Prod. Res. 1–22 (2015)
22. Dasgupta, S., Das, S., Biswas, A., Abraham, A.: Automatic circle detection on digital images with an adaptive bacterial foraging algorithm. Soft. Comput. **14**(11), 1151–1164 (2010)
23. Bermejo, E., Cordón, O., Damas, S., Santamaría, J.: A comparative study on the application of advanced bacterial foraging models to image registration. Inf. Sci. **295**, 160–181 (2015)
24. Passino, K.M.: Biomimicry of bacterial foraging for distributed optimization and control. IEEE Control Syst. Mag. **22**(3), 52–67 (2002)

25. Hulett, M., Damodaran, P., Amouie, M.: Scheduling non-identical parallel batch processing machines to minimize total weighted tardiness using particle swarm optimization. Comput. Ind. Eng. **113**, 425–436 (2017)
26. Majumder, A., Laha, D., Suganthan, P.N.: Bacterial foraging optimization algorithm in robotic cells with sequence-dependent setup times. Knowl. Based Syst. **172**, 104–122 (2019)

Healthcare Information Retrieval Based on Neutrosophic Logic

Sanjeev Kumar Sinha and Chiranjeev Kumar

Abstract In order to deal with enormous unstructured healthcare data, it becomes paramount to develop efficient tools for retrieval of useful data subject to a query posed by entities. The entity can be a doctor, nurse, patient, or relative. Uncertainty, imprecision, and incompleteness arise from various forms and are present when a query is posed to the unstructured database. A popular technique for information retrieval is Fuzzy Logic-based information retrieval. Primarily, the Mamdani Fuzzy Inference Engine is used for ranking the documents based on the query. Recently, Neutrosophic Logic was proposed for efficient handling of data which is uncertain, incomplete, and imprecise. This research aims to validate the effectiveness of Neutrosophic Logic for healthcare document retrieval using Neutrosophic Logic and analyze its merit or demerit over the traditional Fuzzy Logic-based retrieval strategy. The experimental results confirm efficiency measured in terms of MSE scores. The proposed work has been validated for cancer-related healthcare documents.

Keywords Healthcare · Neutrosophic Logic · Information retrieval (IR) · Fuzzy Logic · Electronic healthcare reports · Healthcare information retrieval · HIS

1 Introduction

A lot of healthcare data is created, presented, and retrieved in computerized format. This data is generated in various processes which represent the flow of task in particular healthcare procedures. Healthcare procedure can be related to admission and discharge, health monitoring and advice, operation procedure, diagnostics procedure, preventive health, home health monitor, healthcare reporting and documentation, hospitalization, nursing, healthcare data management, etc. Few examples

S. K. Sinha (✉) · C. Kumar
Department of Computer Science and Engineering, Indian Institute of Technology (ISM), Dhanbad, India
e-mail: sanjeev.2013dr1029@cse.iitism.ac.in

C. Kumar
e-mail: chiranjeev@iitism.ac.in

© Springer Nature Singapore Pte Ltd. 2020
S. Agarwal et al. (eds.), *Machine Intelligence and Signal Processing*,
Advances in Intelligent Systems and Computing 1085,
https://doi.org/10.1007/978-981-15-1366-4_18

of healthcare workflow are monitoring a remote patient and advising the treatment for him/her, standardization of patient healthcare data (e.g., diagnostics report, admission, and discharge summary), management of appointment calendar and health data for clinicians and patient over cloud, operation theater management, and e-health services. This gives a clear idea to envision the complexity of the process as well as the amount of data that would be generated via the processes, and specifically, most of them are unstructured in nature. Now to refer a particular set of data, physicians, doctors, and healthcare staff have to dig down into each data/document and have to check for the relevancy.

Information retrieval (IR) [1] deals with extraction and ranking of required information from a corpus which is most significant to a specific client need and a specified query from a particular information set. Output ranking on the document helps in decision system. In this paper, we are focusing on healthcare information retrieval. In this case, the corpus can be an accumulation of clinical reports, labeled MRI images, test records, the Intranet/web archives of clinical information. Ranking of the document will help the clinician to seek out most relevant result which he/she is expecting. It is clinician's choice how he/she decides to consider the document; he may be interested in top 5% references to help him in decision making. Content information retrieval is very essential in today's scenario of data explosion. This field has increased enormous fame keeping in view the accessibility of content as reports, sites, tweets, item surveys, news articles, official records in an association and sites to make reference to few.

The motivation behind the present research is the development of efficient healthcare information retrieval (HIR) system. Working on the same lines, we firstly aim to develop a model for healthcare retrieval based on Neutrosophic Logic which is an improvement over exiting model of Fuzzy Logic-based system HIR. Due to its ability to capture indeterministic characteristics of the data result, application of Neutrosophic Logic has recently gained immense importance in other areas of applications. Also, Neutrosophic Logic is being used in healthcare decision systems like medical diagnostics [2]. One of the healthcare application areas of Neutrosophic Logic has been in medical imaging decision system [3], where image detection is performed for blood vessels detection in retinal images. However, no such work to the best of our knowledge has been documented in the literature on the topic of health information retrieval using Neutrosophic Logic. The aim of our work is to develop a Neutrosophic Logic-based model to help the clinicians retrieve correct set of medical record for their reference based on a particular need.

The paper is organized as follows. Section 2 presents a brief review of existing techniques. Section 3 introduces the Neutrosophic Logic scheme for HIR and presents the proposed technique. Results and conclusion are given in Sect. 4.

2 Healthcare Information Retrieval (HIR)

IR has seen phenomenon development in past several decades. The first IR system was proposed by Luhn [1] in which an exact match of a query and document was achieved. Now, there has been a number of retrieval engines present at our disposal that suits our complex requirements. Not only has man's need for information changed but also the processing techniques to retrieve is changing which is proportional to the changing needs both in requirements and domains. The query and document are two key factors in designing an information retrieval system. The first important factor is database. Healthcare database is huge and is ever increasing with time. The query posed may not be sufficient to produce the appropriate documents in an exact match technique of retrieval.

Fuzzy Rough set-based techniques [4] have been used for generic query-based information retrieval which uses Wikipedia as a knowledge data source. The authors have primarily worked on "terror"-based information retrieval. The novel method utilizes the Fuzzy Rough set-based similarities which use Wikipedia to compute Fuzzy equivalence relations. Fuzzy Logic has been traditionally used for IR. Lately, Fuzzy Logic-based technique for information retrieval was modified by Gupta et al. [5], in which two inference engines were used. The inputs used in [5] are as follows: (a) TF: term frequency of document, (b) IDF: inverse document frequency of document, (c) the inverse document length, (d) term frequency of query, and (e) inverse query frequency. Alhabashneh [6] have proposed Fuzzy rule approach for large-scale information retrieval.

A healthcare information retrieval Hersh, 2015 [7], is gaining popularity recently. Availability of electronic health record system and clinical healthcare software in the leading hospitals in many of the developed and developing country clearly indicates the result of research and advancement in this area. Some of the hospital-centric commercial services related to clinical side are being offered on Web and mobile by the hospitals and diagnostic laboratories. Considering the initial stage of HIR application in health care, we have documented the following reasons for its research and development.

1. Presence of healthcare software's generated data in forms of electronic healthcare reports (EHR).
2. Large repositories of information about disease present on Web.
3. Regular news updated on healthcare front.
4. Breaking scientific discoveries.
5. Patients' self-document records.
6. Medical historic textual data present in various forms.

The following section discusses the concept of Neutrosophic Logic.

3 Fuzzy Logic and Neutrosophic Logic-Based HIR

Fuzzy sets [8] have since its inception been used in large number of applications which vary from electronics, mechanical engineering, natural language processing, image processing, to mention few. The conception behind Fuzzy Logic is that in a Fuzzy set, each member of universe has a membership with which it belongs to the Fuzzy set. The membership takes value between 0 and 1. Neutrosophic Logic [9] describes that each set is consisting of memberships of triplet (truth, indeterminacy, and falsity). The philosophy behind the development of Neutrosophic Logic is that every element of universe may belong to it with a varying degree of membership, and some elements may belong to more than one set which causes a high indeterminacy, and hence the concept. Ansari et al. [10] proposed the concept of Neutrosophic Classifier. A fuzzy based ontology model has been proposed by Zeinab et al. [11]. We have used the concept of HIR using multilevel Neutrosophic logic to minimize interdependency in result set.

The input variables used in FIS follow from a part of work by [10], viz:

1. TF: term frequency
2. IDF: inverse document frequency
3. N: inverse noun count.

The output variables used in FIS is Rank, and the FIS system is Mamdani Fuzzy Inference Engine with centroid-based defuzzifier.

To develop a model which ensures better results in terms of correctness of the document set retrieved, we have proposed six input variables in our model which help in handling the complexity of information retrieval in a better way.

The input variables used in NIS as proposed are as follows:

1. Truth-TF: term frequency
2. Truth-IDF: inverse document frequency
3. Truth-N: inverse noun count
4. Indeterminacy-TF: term frequency
5. Indeterminacy-IDF: inverse document frequency
6. Indeterminacy-N: inverse noun count.

The output variables used in NIS are:

1. Truth-Rank
2. Indeterminacy-Rank.

The two NIS systems are based on Mamdani Inference with centroid-based defuzzifier [12].

$$x^* = \frac{\sum_{i=1}^{n} x_i . \mu_c(x_i)}{\sum_{i=1}^{n} \mu_c(x_i)} \tag{1}$$

Figure 1 describes the Mamdani Fuzzy Inference Engine (FIS) [8] for IR, and the second part of Fig. 1 shows the proposed Neutrosophic Inference Engine (NIS),

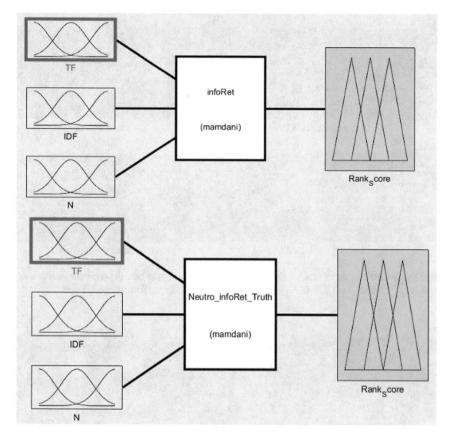

Fig. 1 Fuzzy Logic and Neutrosophic Logic-based healthcare information retrieval system

for IR. Figure 2 depicts the memberships of term frequency and inverse document frequency of the FIS and NIS respectively. Figure 3 describes the memberships used in FIS and NIS, respectively, by the Inverse Noun Frequency. Figure 4 (i) to (iii) describes the Neutrosophic Indeterminacy of the three inputs chosen for IR, while Fig. 4a describes the membership used by the output of FIS engine, via the retrieval rank. Figure 4b depicts the rank in terms of Neutrosophic Truth components, and Fig. 4c shows the Neutrosophic Indeterminacy component for the NIS. Figures 5 and 6 show the evaluation of rank computations using Fuzzy Inference Engine and proposed Neutrosophic Inference Engine, respectively. The inference and ranking is performed in MATLAB.

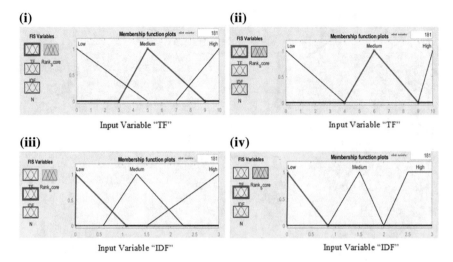

Fig. 2 Membership functions for (i) term frequency using Fuzzy set, (ii) term frequency using Neutrosophic Truth, (iii) IDF using Fuzzy sets, and (iv) IDF using Neutrosophic Truth

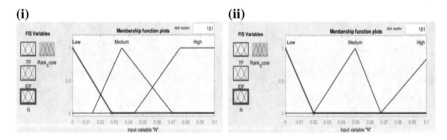

Fig. 3 Membership functions for inverse Noun frequency using (i) Fuzzy sets and using (ii) Neutrosophic Truth

4 Results and Conclusion

The query considered was "cancer," and the first 50 documents retrieved from large corpus based on ranking were considered for re-ranking. The dataset used for this is TREC Clinical data [13] and also BMC cancer open access article set [14]. The dataset consisted of top 50 documents related to cancer. Each document set itself contain around 500–1000 words. We have developed and implemented two models, viz:

1. Model 1 is the Fuzzy Logic-based IR system.
2. Model 2 is the proposed Neutrosophic Logic-based IR system.

The composite rule used in our proposed Neutrosophic Logic-based model for application in data set is:

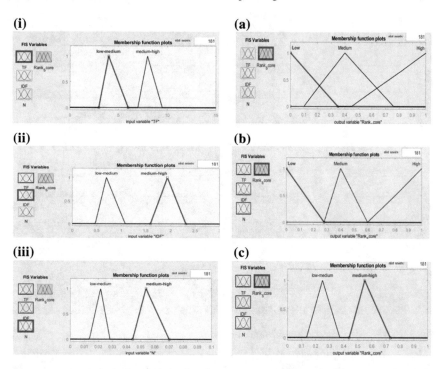

Fig. 4 (i)–(iii) Fuzzy set-based memberships for TF, IDF, Noun count, **a** Fuzzy set-based memberships of HIR Rank and **b–c** Neutrosophic memberships for truth and indeterminacy of HIR Rank Score

```
NumMFs=3
MF1='low':'trimf',[-0.4 0 0.4]
MF2='medium':'trimf',[0.1 0.5 0.9]
MF3='high':'trimf',[0.6 1 1.4]
[Input2]
Name='Neutro-Inderminancy'
Range=[0 1]
NumMFs=3
MF1='low':'trimf',[-0.4 0 0.4]
MF2='medium':'trimf',[0.1 0.5 0.9]
MF3='high':'trimf',[0.6 1 1.4]
[Output1]
Name='output1'
Range=[0 1]
NumMFs=3
MF1='low':'trimf',[-0.4 0 0.4]
MF2='medium':'trimf',[0.1 0.5 .9]
MF3='high':'trimf',[0.6 1 1.4]
```

```
[Rules]
1 1, 1 (1) : 1
2 1, 2 (1) : 1
2 3, 1 (1) : 1
3 1, 3 (1) : 1
3 2, 3 (1) : 1
3 1, 2 (1) : 1
2 2, 2 (1) : 1
2 3, 1 (1) : 1
1 3, 1 (1) : 1
1 3, 1 (1) : 1
```

Table 1 shows the FIS-based rankings and the proposed NIS-based rankings produced on the datasets of cancer documents. For evaluations, the mean square error is computed with the actual rankings of the documents as marked by the Google search

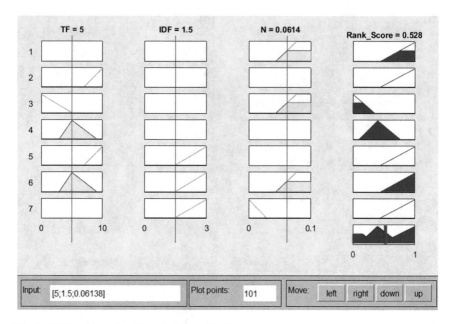

Fig. 5 HIR Ranking using Fuzzy Inference Engine

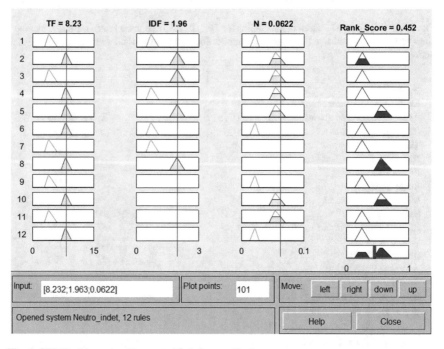

Fig. 6 HIR Ranking using Neutrosophic Inference Engine

Table 1 Ranks generated by first 50 documents by FIS and proposed NIS systems

Sr.No.	FIS Rank	NIS Rank	Sr.No.	FIS Rank	NIS Rank
1	14	14	26	21	20
2	18	18	27	25	15
3	49	38	28	28	24
4	17	49	29	36	29
5	44	19	30	2	50
6	22	12	31	20	43
7	23	22	32	41	45
8	46	35	33	43	41
9	32	46	34	45	1
10	19	32	35	8	6
11	38	37	36	9	8
12	12	23	37	35	10
13	37	33	38	39	13
14	31	42	39	7	40
15	15	31	40	1	47
16	24	16	41	40	39
17	29	21	42	4	5
18	33	25	43	26	7
19	42	28	44	34	9
20	3	36	45	48	4
21	11	27	46	5	17
22	27	30	47	10	26
23	30	11	48	6	34
24	50	3	49	13	44
25	16	2	50	47	48

engine. The mean square errors (MSE) between the FIS and the proposed technique in comparison with actual ranking value has been observed and we have seen that there is a significant reduction in MSE of the proposed technique in comparison to the FIS technique. The paper presents a novel technique of healthcare information retrieval. The work proposed improves the mean square error considerably. Also, the correlation between output FIS and NIS was found to be 0.6093 guidelines. Further, on manual verification of collected set, there was an improvement seen by clinician as per his interpretation of subject knowledge.

5   Future Work

This model can be applied to test other dataset such as arthritis or diabetics and also on a larger data set to test the correctness of result for making a generalized conclusion about advantages of using Neutrosophic-based model for healthcare information retrieval. In this work, textual medical record has been used, and system can be further enhanced for structured medical records like using clinical document architecture (CDA)-based medical records as input to the system. System can be used in social network information retrieval for convergence of textual data to analyze a specific trend [15].

References

1. Luhn, H.P.: A new method of recording and searching information. Am. Documentation (pre-1986) **4**(1), 14 (1953)
2. Edward Samuel, A., Narmadhagnanam, R.: Neutrosophic Refined Sets in Medical Diagnosis Fuzzy Mathematical Archive **14**(1), 117–123. ISSN: 2320 –3242 (P), 2320 –3250 (2017)
3. Sharma, A., Rani, S.: An automatic segmentation & detection of blood vessels and optic disc in retinal images. In: ICCSP (2016)
4. Yadav, N., Chatterjee, N.: Fuzzy rough set based technique for user specific information retrieval: a case study on wikipedia data. Int. J. Rough Sets Data Anal. (IJRSDA) **5**(4), 32–47 (2018)
5. Gupta, Y., Saini, A., Saxena, A.K.: A new fuzzy logic based ranking function for efficient information retrieval system. Expert Syst. Appl. **42**(3), 1223–1234 (2015)
6. Alhabashneh, O., Iqbal, R., Doctor, F., James, A.: Fuzzy rule based profiling approach for enterprise information seeking and retrieval. Inf. Sci. **394**, 18–37 (2017)
7. Hersh, W.R.: Information Retrieval for Healthcare, pp. 467–505 (2015)
8. Zadeh, L.A.: Fuzzy sets. Inf. Control **8**(3), 338–353 (1965)
9. Smarandache, F.: A unifying field in logics: neutrosophic logic. In: Philosophy. American Research Press, pp. 1–141 (1999)
10. Gupta, A.Q., Biswas, R., Aggarwal, S.: Neutrosophic classifier: an extension of fuzzy classifier. Elsevier-Appl. Soft Comput. **13**, 563–573 (2013)
11. Attia, Z.E., Gadallah, A.M., Hefny, H.M.: An enhanced multi-view fuzzy information retrieval model based on linguistics. IERI Proc. **7**, 90–95 (2014)
12. Bezdek, J.C., Ehrlich, R., Full, W.: FCM: the fuzzy c-means clustering algorithm. Comput. Geosci. **10**(2–3), 191–203 (1984)
13. Medical Data Set. http://www.trec-cds.org/. 2014 through 2018
14. BMC Cancer Data Article Set. https://bmccancer.biomedcentral.com/
15. Bouadjenek, M.R., Hacid, H., Bouzeghoub, M.: Social networks and information retrieval, how are they converging? A survey, a taxonomy and an analysis of social information retrieval approaches and platforms. Inf. Syst. **56**, 1–18 (2016)

Convolutional Neural Network Long Short-Term Memory (CNN + LSTM) for Histopathology Cancer Image Classification

Zanariah Zainudin, Siti Mariyam Shamsuddin and Shafaatunnur Hasan

Abstract Deep learning algorithm such as Convolutional Neural Networks (CNN) is popular in image recognition, object recognition, scene recognition and face recognition. Compared to traditional method in machine learning, Convolutional Neural Network (CNN) will give more efficient results. This is due to Convolutional Neural Network (CNN) capabilities in finding the strong feature while training the image. In this experiment, we compared the Convolutional Neural Network (CNN) algorithm with the popular machine learning algorithm basic Artificial Neural Network (ANN). The result showed some improvement when using Convolutional Neural Network Long Short-Term Memory (CNN + LSTM) compared to the multi-layer perceptron (MLP). The performance of the algorithm has been evaluated based on the quality metric known as loss rate and classification accuracy.

Keywords Deep Learning · Image Recognition · Convolutional Neural Network · Long Short-Term Memory · Histopathology Cancer Image · Artificial Neural Network

1 Introduction

Image processing is basic operations where images can be enhanced for better quality, as well as classified or detected, and many other types of operations can be performed. Image classification is one main part of image processing where it can allow automatic allocation of the same image into one class [1]. In image classification, there are two types of classifications, which are supervised and unsupervised classification. The

Z. Zainudin (✉) · S. M. Shamsuddin · S. Hasan
School of Computing, Faculty of Engineering, Universiti Teknologi Malaysia, 81310 Skudai, Johor, Malaysia
e-mail: zanariah86@gmail.com

S. M. Shamsuddin
e-mail: sitimariyams@gmail.com

S. Hasan
e-mail: shafaatunnur@utm.my

© Springer Nature Singapore Pte Ltd. 2020
S. Agarwal et al. (eds.), *Machine Intelligence and Signal Processing*,
Advances in Intelligent Systems and Computing 1085,
https://doi.org/10.1007/978-981-15-1366-4_19

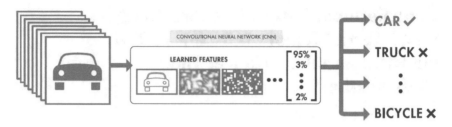

Fig. 1 Convolutional Neural Network [2]

process of image classification involves two steps, training the model and followed by testing/validation. Training process will take all the features and characteristics from all trained images and build a strong/robust model where it can classify the new data from testing dataset/validation dataset. Testing dataset will categorize the test images under various classes in which the model was trained. Result from testing model will be measured when the model assigning of predicted class versus the real class inside testing dataset. This is where the performance measure will be done where its means squared error (MSE) and accuracy have been recorded to see if the trained models are good enough.

Since the 1990s, there has been some research on deep structured learning, which has also been called deep learning emerged as a new area of machine learning research. Deep learning can be defined as a class of machine learning techniques that exploits many layers of nonlinear information processing for supervised or unsupervised feature extraction and transformation and for pattern analysis and classification. Deep learning application in Convolutional Neural Network (CNN) is widely used for image classification as shown in Fig. 1.

Convolutional Neural Networks (CNNs) are the most popular neural network model being used for image classification problem. CNNs were introduced in the 1990s by LeCun's [3], but then, it has been outperformed by another algorithm such as Support Vector Machine (SVM) and others in classification problem. In 2012, Krizhevsky et al. [4] did some improvements in CNNs by showing higher image classification accuracy on the ImageNet Large Scale Visual Recognition Challenge (ILSVRC) [5]. They succeeded on training a large CNN network using a larger dataset which included 1.2 million labelled images, with some modification on LeCun's CNN architecture (for example, introducing the $\max(x, 0)$ rectifying nonlinearity and 'dropout' regularization layer). Based on previous research, we know that some popular dataset benchmarks, the MNIST dataset by Lecun [3] and CIFAR-10 databases by Krizhevsky [5] were frequently used as benchmarks to see the performance of a classification algorithm. Even though the best-known result has achieved above 90% of accuracy which has been discovered [6], it was still popular among researchers, and it is still a hot topic these days. The performance measures for previous researchers on image recognition have typically been measured using loss rate and classification accuracy.

Long Short-Term Memory (LSTM) networks were found by Hochreiter and Schmidhuber in 1997 who have also discovered many high accuracy in many application domains. Recurrent neural networks (RNNs) are well known for Long Short-Term Memory (LSTM) which network uses not only the current input but their own internal hidden state when Long Short-Term Memory (LSTM) is used in the output activation. Long short-term memory (LSTM) is another deep learning algorithm that can avoid gradient vanishing problem during optimization problems. LSTM has been visualized as recurrent gates called as 'forget' gates. This 'forget' gates are where LSTM can prevent gradient vanishing problem, and it will bring all the error backwards with many numbers of virtual layers. That is why LSTM memory can be used for the millions of events of previous steps before. LSTM also functions on long delays between some delays in some events. Many applications use stacks of LSTM and train their models to find suitable weight for the matrix where it can maximize the probability of the label sequences in a training set and give the best result in output layer. In this study, we focus more on Convolutional Neural Network (CNN) and Long Short-Term Memory (LSTM). The next subsection will explain more on histopathology image and mitosis/non-mitosis.

2 Introduction to Mitosis and Non-Mitosis

Cancer is referred to as a dangerous disease that causes abnormal changes in tissue cells. Usually, a normal person's cells will replace the dying cells through a normal process, but if the person is detected to have cancer cells, the dying cells will be replaced with abnormal and cancerous cells and will turn into a tumour. A tumour can be detected as benign or malignant [7]. Breast cancerous cells can be detected on three locations, which are:

- Lobules, which produce milk
- Ducts, the area that is connected from lobules to the nipple
- Stromal tissues, which keep fat and breast tissues.

In order to detect breast cancer, various clinical tests have been used in hospitals such as ultrasound scan, mammogram and biopsy. Normally, many patients use mammogram to detect if they have cancer cells such as benign or malignant. But this test has some disadvantages where a younger woman may get non-accurate results because a younger woman's breast will give a clearer picture compared to an older woman's. As an alternative test, mostly younger women are recommended to use biopsy test to get an accurate result. Usually, biopsy test is a supportive test conducted after the first test. The biopsy process is the most efficient test to detect cancerous cell [8].

The cells, which are not detected as cancer cells but have been misclassified as cancer cells because of their similarity to cancer cells, called benign. On the other hand, the malignant, the cancerous cell, has a potential to grow and become dangerous to the patients. This research will focus more on breast cancer. Visual appearance of

Fig. 2 Visual appearance of
mitosis in dataset AMIDA13

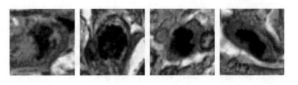

mitosis and non-mitosis in histopathology images is one of the components required
for developing a classifier to classify the grading of cancer histopathology image.
Most commonly, the images of mitotic cells as shown in Fig. 2 are dark, jagged and
have non-similarity texture and structure [7–9].

Due to this complexity of the mitosis, some of them cannot be detected by only
using bare eyes. In terms of shape, texture and structure that may be found to look-a-
like mitosis cell throughout the tissues of breast cancer slide, will make the detection
mitosis cell extremely difficult. Additionally, the standard stain normalization has
been used to improve the histopathology image. Most of the previous researchers
used some approaches to detect mitosis cell by first identifying possible objects, area
or region that will be accepted or rejected as mitosis cells based on some similarity
criterion [10–12]. This research does not employ candidate detection phase; it trains
a pixel-level deep convolution network, which takes all the training pixels along with
their immediate surrounding pixels as input and builds a deep neural network model
for mitosis cells which are directly used on test data to perform classification.

3 Literature Review

In this subsection, literature review for related works is presented in Table 1 [13–22].

4 Research Framework

In this subsection, the research framework for this study is presented in Fig. 4. The
main goal of this study is to achieve the objective where it classifies all cancer cells
correctly. In this paper, the research framework is carried out into five phases as
below:

- **Phase 1**: Preliminary Investigation
- **Phase 2**: Dataset Collection
- **Phase 3**: Image Classification Using Deep Learning
- **Phase 4**: Performance Measurement
- **Phase 5**: Results and Discussion.

In phase one, we need to achieve the main objective in this research work to detect
cancer cell correctly using the selected datasets. The first step in this research is the

Table 1 Related works about deep learning

Authors	Results	Problem with previous methods
Yu, Lin, & Lafferty, 2011	The loss rate is 0.77 using the Hierarchical Sparse Coding	Encode local patches independently. The spatial neighbouring of the image was been ignored
Ciresan et al., 2013	Result shows that Deep Neural Network has accuracy from 0.751–0.782	Not mentioned
Chan et al., 2014	Deep learning PCANet shows accuracy result from 99.58% and 86.49% using 60% train 40% for testing (Extended Yale B Dataset)	Not mentioned
Donahue et al., 2015	Results show that their proposed method LSTM Decoder has achieved 28.8% compared to previous work statistical machine translation (SMT) 26.9%	Using Statistical Machine Translation approach has been widely used among previous research which gives results from 24.9% to 26.9%.
Wahlstr, 2016	The result for joint training error is 0.0011 and separate training error is 0.0011	Not mentioned
Wahab et al., 2017	They proposed Balanced-Convolutional Neural Network has been used for better training the dataset which shows result 0.79 compared to other techniques such as CNN generated features based	Mentioned that mitoses are not balanced with non-mitotic nuclei crated skewness in the dataset for the classifiers and affect the performance measures
Shen et al., 2017	The author's use Multi-crop Convolutional Neural Network (MC-CNN) and obtained accuracy result 87.16%	Not mentioned
Feng et al., 2018	Accuracy results obtained 98.27% for benign subset and 90.54%	Not mentioned

(continued)

Table 1 (continued)

Authors	Results	Problem with previous methods
Xu et al., 2018	Results from 40.7 to 67.9% using CaffeNet Model	High memory will be used and it affects execution of model training
Zainudin et al., 2018	Using Convolutional Neural Network. Accuracy results are obtained from 81.24% to 91.67%	They also stated that traditional machine learning algorithm is good, but the result depends on the complexity of the image itself. For this experiments, they used BP-ANN, Random Forest, K-Means and Naïve Bayes
Zainudin et al., 2019	Using different layer architecture Convolutional Neural Network. Accuracy results are obtained from 84.49% to 98.99%	Using different dataset to find good accuracy and loss rate

Fig. 3 Visual appearance of non-mitosis in dataset AMIDA13

preliminary investigation, which is also called as literature review, gives clear understanding on how to set up the deep learning and also deep learning problem. This is important to set the real situation and get a proper solution that can solve the problem of this study. After that, phase two will begin. The datasets applied in this study were collected from many free public datasets available in previous studies based on image processing. In this study, we use datasets collected from free database such as AMIDA13 as shown in Figs. 2 and 3. In phase three, the objective is to create the classifier for each class in each dataset and also to build the good classifier to detect all of the classes using AMIDA13 image dataset based on Fig. 4. In phase four, performance measurement takes all experimental results and reports them from algorithm created by all selected algorithm such as multi-layer perceptron, Convolutional

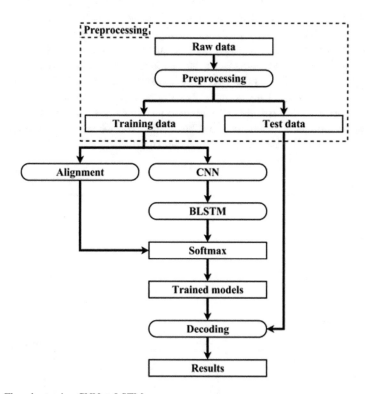

Fig. 4 Flowchart using CNN + LSTM

Neural Network (LENET), Convolutional Neural Network (ALEXNET) and Convolutional Neural Network Long Short-Term Memory (CNN + LSTM). This step also includes taking all the hyper parameter tuning when setting up the experiments and see whether the results are good with the proposed algorithms.

Finally, the last phase of this study will be discussed in the next subsection, and all results will be analysed to get better understanding on the performance measure on cancer cells detection.

5 Experimental Results

In this subsection, experimental results for this study are presented in Tables 2 and 3. The main objective of this study is to detect the cancer cells which are also known as mitosis from histopathology image and also reduce false negative rate (FPR). The dataset is split into two types of datasets which included training dataset containing 80% and testing dataset containing 10% and another 10% for validation process. In this study, we also used two types of architecture of CNN to see the improvement with the results. This is important to ensure that all images have been fed to the models and to see the difference between them. The objective is to feed the classifiers to new datasets. Tables 2 and 3 describe the results using different types of CNN architecture.

Based on Tables 2 and 3 above, we can see that other classifiers had good results but using CNN + LSTM as a classifier gave a significant result. The accuracy rate from the CNN was significant compared to other classifiers. The lowest result shows that multi-layer perceptron (MLP) had average accuracy from 38.28% to 42.97% from three trainings. The accuracy has been improved when Convolutional Neural Network (LENET), Convolutional Neural Network (ALEXNET) and Convolutional

Table 2 Accuracy rate

	Training 1 (%)	Training 2 (%)	Training 3 (%)
Multi-layer Perceptron (MLP)	38.28	41.41	42.97
CNN (LENET)	63.28	56.25	54.69
CNN (ALEXNET)	54.06	53.75	55.00
CNN + LSTM	**88.43**	**86.50**	**85.83**

Table 3 Loss rate

	Training 1	Training 2	Training 3
Multi-layer Perceptron (MLP)	9.57	9.19	8.94
CNN (LENET)	5.19	7.05	7.30
CNN (ALEXNET)	7.40	7.45	7.25
CNN + LSTM	**3.07**	**3.75**	**3.762**

Table 4 Classification on cancer cell

Image	Classification
Image A	Mitosis
Image B	Mitosis
Image C	Mitosis
Image D	Non-Mitosis
Image E	Non-Mitosis
Image F	Non-Mitosis

Neural Network Long Short-Term Memory (CNN + LSTM) were used. This is because CNN has more layers to train the dataset compared to other classifiers. The time performance of CNN + LSTM was a bit slow compared to normal CNN, but improvement could be observed if GPU was used. As we can see, some of the results on mitosis and mitosis detection are quite promising for the Convolutional Neural Network Long Short-Term Memory (CNN + LSTM) as visualized in Table 4.

6 Conclusion

The experimental results show all the results using multi-layer perceptron, Convolutional Neural Network (LENET), Convolutional Neural Network (ALEXNET) and Convolutional Neural Network Long Short-Term Memory (CNN + LSTM). The result from Convolutional Neural Network Long Short-Term (CNN + LSTM) outperformed other algorithms. Even though there were many classifier algorithms especially in medical image, Convolutional Neural Network Long Short-Term Memory (CNN + LSTM) had a great accuracy compared to others. This is because Convolutional Neural Network Long Short-Term Memory (CNN + LSTM) has the advantages of CNN and LSTM algorithms. For these reasons, the researcher proposed on using this algorithm called Convolutional Neural Network Long Short-Term (CNN + LSTM) after considering both smaller regions mitosis and the image quality of histopathology image. Convolutional Neural Network Long Short-Term Memory (CNN + LSTM) method was used to detect the mitosis and non-mitosis inside breast cancer histopathology image.

Acknowledgements This work is supported by Universiti Teknologi Malaysia (UTM), Malaysia under Research University Grant No. 17H62, 03G91and 04G48, and ASEAN-Indian Research Grant. The authors would like to express their deepest gratitude to Cyber Physical System Research Group, as well as School of Computing, Faculty of Engineering, Universiti Teknologi Malaysia for their continuous support in making this research a success.

References

1. Gurcan, M.N., Boucheron, L.E., Can, A., Madabhushi, A., Rajpoot, N.M., Yener, B.: Histopathological image analysis: a review. IEEE Rev. Biomed. Eng. **2**, 147–171 (2009)
2. Phung, S.L., Bouzerdoum, A.: Matlab library for convolutional neural networks. Vis. Audio Signal Process. Lab Univ. Wollongong, no. November, pp. 1–18 (2009)
3. LeCun, Y., Cortes, C., Burges, C.: THE MNIST DATABASE of handwritten digits. Courant Inst. Math. Sci., 1–10 (1998)
4. Krizhevsky, A., Sutskever, I., Hinton, G.E.: Alexnet. Adv. Neural Inf. Process. Syst., 1–9 (2012)
5. Krizhevsky, A., Sutskever, I., Hinton, G.E.: ImageNet classification with deep convolutional neural networks. Adv. Neural Inf. Process. Syst., 1–9 (2012)
6. Hsieh, P.C., Chen, C.P.: Multi-Task Learning on Mnist Image Datasets. In: ICLR2018, no. 1998 (2018), ICLR 2018 Conference, pp. 1–7
7. Hao Chen, P.A.H., Dou, Q., Wang, X., Qin, J.: Mitosis detection in breast cancer histology images via deep cascaded networks. Thirtieth AAAI Conf. Artif. Intell., 1160–1166 (2016)
8. Paul, A., Mukherjee, D.P.: Mitosis detection for invasive breast cancer grading in histopathological images. IEEE Trans. Image Process. **24**(11), 4041–4054 (2015)
9. Saha, M., Chakraborty, C., Racoceanu, D.: Efficient deep learning model for mitosis detection using breast histopathology images. Comput. Med. Imaging Graph. **64**, 29–40 (2018)
10. Veta, M., van Diest, P.J., Kornegoor, R., Huisman, A., Viergever, M.A., Pluim, J.P.W.: Automatic nuclei segmentation in H&E stained breast cancer histopathology images. PLoS One **8** (2013)
11. Veta, M., et al.: Assessment of algorithms for mitosis detection in breast cancer histopathology images. Med. Image Anal. **20**(1), 237–248 (2015)
12. Roux, L., et al.: Mitosis detection in breast cancer histological images An ICPR 2012 contest," *J. Pathol. Inform.*, vol. 4, no. 1, p. 8, 2013
13. Xu, H., Chen, Y., Lin, R., Kuo, C.-C.J.: Understanding convolutional neural networks via discriminant feature analysis. APSIPA Trans. Signal Inf. Process. **7**, 1–15 (2018)

14. Donahue, J., et al.: Long-term Recurrent Convolutional Networks for Visual Recognition and Description, vol. 38, no. 3, pp. 170–172 (2013)
15. Chan, T., Jia, K., Gao, S., Lu, J., Zeng, Z., Ma, Y.: PCANet: A Simple Deep Learning Baseline for Image Classification? arXiv Prepr, pp. 1–15 (2014)
16. Yu, K., Lin, Y., Lafferty, J.: Learning image representations from the pixel level via hierarchical sparse coding. In: Proceedings of the IEEE Comput. Soc. Conf. Comput. Vis. Pattern Recognit., pp. 1713–1720 (2011)
17. Wahlstr, N.: Learning Deep Dynamical Models from Image Pixels (2016)
18. Wahab, N., Khan, A., Lee, Y.S.: Two-phase deep convolutional neural network for reducing class skewness in histopathological images based breast cancer detection. Comput. Biol. Med. **85**, 86–97 (2017)
19. Feng, Y., Zhang, L., Yi, Z.: Breast cancer cell nuclei classification in histopathology images using deep neural networks. Int. J. Comput. Assist. Radiol. Surg. **13**(2), 179–191 (2018)
20. Zainudin, Z., Shamsuddin, S.M., Hasan, S., Ali, A.: Convolution neural network for detecting histopathological cancer detection. Adv. Sci. Lett. **24**(10), 7494–7500 (2018)
21. Shen, W., et al.: Multi-crop convolutional neural networks for lung nodule malignancy suspiciousness classification. Pattern Recognit. **61**, 663–673 (2017)
22. Zainudin, Z., Shamsuddin, S.M., Hasan, S.: Deep layer CNN architecture for breast cancer histopathology image detection. In: Advances in Intelligent Systems and Computing (2020)

Forecasting with Multivariate Fuzzy Time Series: A Statistical Approach

Mahua Bose and Kalyani Mali

Abstract In recent times, forecasting using fuzzy time series (FTS) models is a very popular research topic. In multivariate fuzzy time series forecasting models, for a given sequence, similar sequences are extracted by explicit matching of rules. In these models, index value of fuzzy set with the maximum membership value is used for rule matching. In this case, two situations can arise: (1) more than one similar sequence can be found in the training set and (2) no matching rule is found. In order to get an accurate forecast, an efficient defuzzification technique should be adopted. To address this problem, a new multivariate forecasting model is presented based on the Bayesian approach. The proposed model is applied to standard datasets: (1) Taiwan Stock Exchange (1997–2003) and (2) US Civilian Unemployment. Forecast accuracy is measured using root mean square error (RMSE). Rank test is also performed. Proposed idea shows better result than some of the popular fuzzy time series forecasting models.

Keywords Bayesian · Feature · Interval · Probability

1 Introduction

In fuzzy time series forecasting models [1–3], observations are represented by a sequence of fuzzy sets [4]. These models are capable of processing linguistic variables (i.e., good, bad, medium). This is a limitation of classical time series model. Fuzzy time series models do not require large training samples as in statistical models. In fuzzy time series models, there exists a relationship between the current observation with one or more its previous observation. If there exists a single state $f(t-1)$ on which next state $f(t)$ depends, then the relationship is called first order and the model

M. Bose (✉) · K. Mali
Department of Computer Science & Engineering, University of Kalyani, Nadia, West Bengal, India
e-mail: e_cithi@yahoo.com; mahua.bose.wb@gmail.com

K. Mali
e-mail: kalyanimali1992@gmail.com

© Springer Nature Singapore Pte Ltd. 2020   247
S. Agarwal et al. (eds.), *Machine Intelligence and Signal Processing*,
Advances in Intelligent Systems and Computing 1085,
https://doi.org/10.1007/978-981-15-1366-4_20

is first-order forecasting model. In case of dependency between a current state $f(t)$ with more than one previous state, i.e., $f(t - 1), f(t - 2) \ldots f(t - n)$, it called high-order model [1–3, 5, 6].

Efficiency of these models is dependent on two factors: (1) data partitioning technique and (2) forecasting technique. Forecasting is done following two basic steps: defining fuzzy logical relationship (FLR) and defuzzification.

Different data partitioning techniques have already been reported in the literature. Similarly, forecasting can also be done using explicit matching of rules and artificial neural network. Application of fuzzy time series models is done in many forecasting areas [7], namely student enrollment, rainfall, stock market, agricultural production, and temperature.

Models that rely on rule-based technique, problem occurs when rules with same left-hand side, have different right-hand side. Another problem arises when there is no matching rule present in the dataset. Former can be solved using various defuzzification techniques. In the defuzzification techniques presented so far, generally centroid method or weighted average method has been applied. Average or weighted average of midpoints of intervals (corresponding to the fuzzy sets) is taken as predicted value. In addition to that, different schemes of assigning weights have been presented [8–20]. In case of missing patterns, the previous value is projected as forecast. But this idea does not produce good results. Models using ANN [21–27] for training and forecasting overcome above two issues. But out-of-sample forecast is not satisfactory and computation time is high for large datasets.

Therefore, the aim of this study is to develop a model that can overcome these issues. Forecasting model presented in this paper is based on Bayesian approach [28, 29].

Advantage of the proposed work is that its training time is less than that of ANN and explicit matching of rule is not required. This idea outperforms some of the fuzzy time series forecasting models developed in the past.

This paper is organized as follows: In Sect. 2, basic concept Bayesian approach is discussed. Section 3 contains the proposed technique for forecasting with illustration. In Sect. 4, comparative study of the experimental results is given. Finally, conclusion is given in Sect. 5.

2   Bayesian Approach

Bayesian classifier estimates the probabilities of occurrence of different attribute values for different classes in the training data. On the basis of that, it predicts the class for out-of-sample pattern.

Let $X = [X_1, X_2, \ldots, X_m]$ be the set of features for a training set. Let $P(X)$ be the probability that a training pattern has that feature set X, regardless of the class to which it belongs to. Let us consider $n(n > 1)$ classes in the training set.

$P(D_k)$ is the prior probability of a training pattern belongs to a class D_k ($1 \leq k \leq n$).

Now the posterior probability $P(D_k|X)$ that a training pattern with feature set X belongs to a class D_k is defined as follows:

$$P(D_k|X) = P(D_k)P(X|D_k)/P(X) \tag{1}$$

where $P(X|D_k)$ is the conditional probability that a training pattern of class D_k has feature set X. In order to estimate $P(X|D_k)$, we need to consider all possible values of features. It is usually assumed that the attributes are independent.

$$P(X|D_k) = \prod_{i=1}^{m} P(X_i|D_k) \tag{2}$$

It is known as Naïve Bayes classifier. To classify a pattern with feature set X, following expression is to be maximized.

$$P(D_k) \prod_{i=1}^{m} P(X_i|D_k) \tag{3}$$

Let us consider there are $|D_k|$ patterns in class k. Then

$$P(D_k) = (|D_k| + 1)/\left(n + \sum_{j=1}^{n} |D_j| \right) \tag{4}$$

Let us consider "d" possible values of a feature X_i are val_{i1}, val_{i2}, …, val_{id}.

$$(1 \leq i \leq m).$$

Let $\left(\left| D_k^{ij} \right| \right)$ be the number of training pattern of class D_k for which the value of attribute X_i is val_j ($i1 \leq j \leq id$).
Now, according to Bayesian estimation,

$$P\{(X_i = \text{val}_j)|D_k\} = \left(\left| D_k^{ij} \right| + 1 \right)/(id + |D_k|) \tag{5}$$

where $(1 \leq k \leq n)$, $(1 \leq i \leq m)$ and $(i1 \leq j \leq id)$.
In the Bayesian estimation, one is added to the number of training pattern of class D_k for which the value of feature X_i is val_j. Due to this fact, in Bayesian estimation, probability is never equal to zero.

3 Proposed Work

Detained description of the proposed work (Fig. 1) with illustration is given below.

Step 1. For every two consecutive observations at t and t-1, compute the difference between them. In order to get all positive values, add a positive value (Z) to the differences. Now, universe of discourse U is defined as follows:

$$U = \left[\left(\text{Data}_{\min} - W_1\right) - \left(\text{Data}_{\max} + W_2\right)\right]$$

where W_1 and W_2 are two positive values.

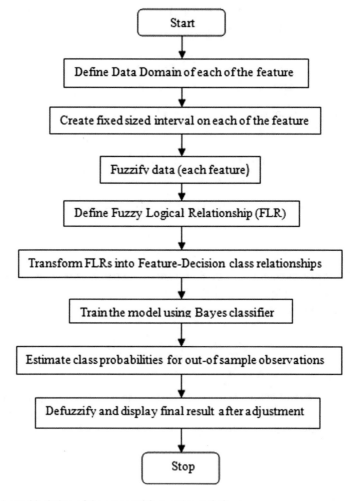

Fig. 1 A graphical view of the proposed forecasting technique

Table 1 A fragment of dataset (1999)

Date	TAIEX	DJI	NASDAQ
7/3/1999	–	–	–
8/3/1999	6431.96	9727.61	2397.62
8/4/1999	6493.43	9693.76	2392.94
8/5/1999	6486.61	9772.84	2406.00
8/6/1999	6436.80	9897.44	2412.25
8/7/1999	6462.73	9876.35	2381.53
8/8/1999	–	–	–

Lower bound (LB) $= \text{Data}_{min} - W_1$ and Upper bound(UB) $= \text{Data}_{max} + W_2$

Step 2. Each feature is partitioned into n equal length intervals.

Example: Let us consider TAIEX dataset (1999). After taking consecutive differences, a positive constant value 515 is added to each difference value. Minimum and maximum values are 8.54 and 934.96, respectively. So, LB = 0, UB = 945, and U = [0-945]. Similar operations are performed on other features (Table 1).

In the TAIEX dataset, there are 9 intervals (interval length = 105). This is main factor. Secondary factors DJI and NASDAQ are divided into 5 and 3 intervals, respectively.

Step 3. Define Fuzzy Sets on each interval (for each feature):

Example: Let us consider feature 1 (main factor). For TAIEX dataset, $n = 9$.

Fuzzy sets $A_1, A_2, ..., A_n$ are defined on the universe of discourse U as follows. Each interval p_i belongs to a particular $A_j, j = 1, 2, ..., n$.

$$A_1 = 1/p_1 + 0.5/p_2 + \cdots + 0/p_{n-1} + 0/p_n$$
$$A_2 = 0.5/p_1 + 1/p_2 + \cdots + 0/p_{n-1} + 0/p_n$$
$$\cdots\cdots\cdots\cdots\cdots\cdots\cdots\cdots\cdots\cdots\cdots\cdots\cdots\cdots\cdots\cdots$$
$$A_{n-1} = 0/p_1 + 0/p_2 + \cdots + 1/p_{n-1} + 0.5/p_n$$
$$A_n = 0/p_1 + 0/p_2 + \cdots + 0.5/p_{n-1} + 1/p_n$$

Fuzzy sets are defined as:

A_1 poor
A_2 very low
A_3 low
A_4 below normal
A_5 normal
A_6 above normal
A_7 high
A_8 very high
A_9 abnormal.

Table 2 Fuzzification of dataset: fragment of dataset (1999)

Time period	Difference values after adjustment			Fuzzified values		
	Features			Features		
	1	2	3	1	2	3
1	576.47	241.15	105.32	A_6	B_3	C_2
2	508.18	354.08	123.06	A_5	B_4	C_2
3	465.19	399.60	116.25	A_5	B_4	C_2
4	540.93	253.91	79.28	A_6	B_3	C_1
–	–	–	–	–	–	–

Data values are fuzzified into the interval with highest membership (Table 2). Similarly, other two features are fuzzified. For feature 2 (DJI), five fuzzy sets, i.e., B_1, B_2, B_3, B_4, and B_5, are defined. Three fuzzy sets, i.e., C_1, C_2, C_3, are defined on feature 3 (NASDAQ).

Step 4. Establish fuzzy logical relationships: In this work, first-order FLR is established. Let us consider primary factor is M. There are h secondary factors S_1, S_2, ..., S_h. Then the first-order relationship is represented as,

$$M(t-1), S_1(t-1), \ldots, S_h(t-1) \rightarrow M(t)$$

Example: From Table 2, we can write following rules (FLR),

$A_6, B_3, C_2 \rightarrow A_5$
$A_5, B_4, C_2 \rightarrow A_5$
$A_5, B_4, C_2 \rightarrow A_6$

Relationship is established between previous fuzzified values of three features with current value of feature 1 (main factor).

Step 5. Transform FLRs into feature-decision class relationships (Table 3). L.H.S (precedent) of the above rules will act as features, and R.H.S (consequent) is the decision. From the rules defined in step 4, features and decision classes are shown in table. To show a decision class, index number of the corresponding fuzzy set is used in Table 3.

Step 6. Train the model with Bayes classifier

Table 3 Features and decision classes

Feature			Decision class
1	2	3	
A_6	B_3	C_2	5
A_5	B_4	C_2	5
A_5	B_4	C_2	6
–	–	–	–

For each decision class k ($k = 1, 2, ..., n$), calculate $P(D_k)$ using Eq. (4). In this example, number of decision class is 9. So in Eq. 4, $n = 9$. Number of training pattern in class k is represented by $|D_k|$. Total number of training patterns in each of the nine decision classes is represented by $\sum_{j=1}^{9} |D_j|$.

Then, for each possible values ($j = 1, 2,..., d$) of each feature ($i = 1, 2, ...,m$) calculate conditional probabilities applying Eq. 5.

In this study, number of decision classes is always equal to the number of possible values of feature 1 (main factor). In this particular example, number of possible values of feature 1, feature 2, and feature 3 are nine, five, and three, respectively.

Step 7. Estimate class probabilities for each pattern

For each new pattern from test set, estimate class probabilities using Eq. 3.

Let us consider a new pattern A_8, B_4, C_1 at time "$t - 1$." We have to predict the next value of feature 1 (main factor) at time t.

$$A_8, B_4, C_1 \rightarrow ?$$

Now, for each decision class k ($k = 1, 2, ..., 9$) compute Probability$_k$ as follows:

$$P(D_k) * P\{(X_1 = A_8)|D_k\} * P(D_k)$$
$$* P\{(X_2 = B_4)|D_k\} * P(D_k) * P\{(X_3 = C_1)|D_k\}$$

Step 8. Perform defuzzification

In Bayesian estimation, an unknown pattern is assigned to the class with highest probability. But in this paper, we are predicting a value not a class. So, in order to get a crisp value as forecast, we have applied a simple defuzzification strategy.

If there are n number of decision classes represented by fuzzy sets A_1, A_2, ..., A_n and mid_1, mid_2, ..., mid_n are the midpoints of the intervals corresponding to the fuzzy sets, then for each pattern, predicted value (difference) is obtained by calculating defuzzified value as follows:

$$\text{Defuzzified value} = \sum_{k=1}^{n} (mid_k * \text{Probability}_k) \Big/ \sum_{k=1}^{n} \text{Probability}_k \qquad (6)$$

Probability$_k$ ($k = 1, 2, ..., n$) is computed using Eq. 3.

Step 9. Calculate final forecast after adjustment: To get a forecast for time t, subtract the positive value (Z) from the predicted difference ($t, t - 1$) and then add it with the actual observation at $t - 1$.

Forecast (t) = actual value ($t - 1$) + defuzzified value ($t, t - 1$) − positive value (Z).

Example: Actual value of 27/12/1999 is 8415.07. Then, the forecast for the next day (28/12/1999) = $8415.07 + 523.9134 - 515 = 8423.98$

4 Experimental Results and Discussion

4.1 Datasets

(1) For the first four experiments, following three daily datasets are downloaded:

   (i) TAIEX (Taiwan Stock Exchange Capitalization Weighted Stock Index) dataset (1997–2000)
   (ii) DJI (Dow Jones Industrial Average) dataset (1997–2000)
   (iii) NASDAQ (National Association of Securities Dealers Automated Quotations) dataset (1997–2000).

Data from January to October are used for training and for testing purpose dataset covering the period Nov-Dec are used.

(2) For the fifth experiment, monthly civilian unemployment dataset (from 1/1/1948 to 1/12/2013) containing unemployment rate and unemployment duration average, and employment to population ratio are collected [30]. Data during the period 1948–2012 are used to train the model, and the rests are served as testing set.

4.2 Comparative Analysis

For the first four experiments, TAIEX is used as main factor and the secondary factors are DJI and NASDAQ. In the fifth experiment, unemployment rate is the main factor. Other two features are secondary features.

Proposed model generates forecast for the main factor from the past values (one period) of the main factor and secondary factors.

Experiment is carried out by varying number of intervals. Intervals with lowest error are reported in Tables 4 and 5 only. Lowest RMSE values are marked in bold. Error estimation is done using root mean square error (RMSE).

$$\text{RMSE} = \sqrt{\sum_{i=1}^{n} ((\text{Actual} - \text{predicted}) * (\text{Actual} - \text{predicted}))/n} \qquad (7)$$

Here, n represents the number of data sample.

For TAIEX dataset (1997–2000), comparative studies (Table 4) show that the proposed model produces lowest average RMSE value (118.23).

According to Table 4, none of the models outperforms the others [5, 10, 17–19, 23] in all four datasets. In order to prove superiority of the proposed model, rank test is performed. Rank is assigned to each model (in ascending order) starting from the

Table 4 TAIEX forecasting (1997–2000): comparative analysis of RMSE values

Models	1997	1998	1999	2000	Average
Univariate Model [5]	154	134	120	176	146
Model [18] (Using NASDAQ)	139.85	119.22	102.6	119.98	120.41
Model [18] (Using Dow Jones & NASDAQ)	138.51	117.87	101.33	121.27	119.75
Model [19] (Using NASDAQ)	138.41	113.88	102.11	131.3	121.43
Univariate Model [16]	136.96	115.85	102.24	123.33	119.59
Model [17] (Use NASDAQ)	133.45	113.37	104.99	124.52	119.08
Model [17] (Use DJI)	133.82	112.11	103.9	127.32	119.23
Model [10] (using NASDAQ)	130.02	115.26	101.82	128.95	119.01
Model [10] (using DJI)	130.64	118.4	99.28	129.23	119.39
Univariate Model [23]	129	116	103	126	118.5
Proposed (using NASDAQ & DJI)	136.07	109.68	98.07	129.08	118.23

Table 5 Unemployment rate forecasting (comparative analysis)

Models	RMSE
Model [5]	0.26
Model [18]	0.2
Model [19]	0.19
Model [20]	0.18
Model [30]	0.14
Model [31]	0.135
Proposed	0.115

lowest RMSE value for each dataset one by one. Then average rank of each of the model is calculated. Average rank of the proposed model is smallest (Fig. 2).

Using unemployment dataset (Table 5), proposed model is compared with other univariate models [5, 18–20] and multivariate model [30]. The proposed technique shows best result with lowest RMSE value (0.115).

5 Conclusion and Future Work

This study presents a multi-feature fuzzy time series forecasting model using the concept of Bayesian estimation. Five multi-feature time series data are considered for the experiment. Comparative analysis with other techniques shows a significant improvement in RMSE value for TAIEX and unemployment data. This is a first-order model. Presently, the proposed work deals with three features. In future, the model can be extended to high order model with more features.

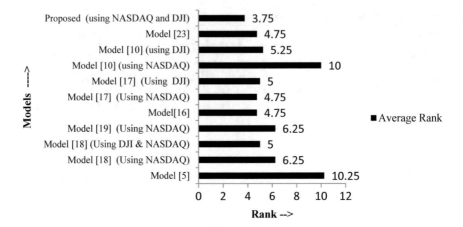

Fig. 2 Average rank of models

References

1. Song, Q., Chissom, B.: Fuzzy time series and its models. Fuzzy Sets Syst. **54**, 269–277 (1993)
2. Song, Q., Chissom, B.: Forecasting enrollments with fuzzy time series—Part I. Fuzzy Sets Syst. **54**, 1–9 (1993)
3. Song, Q., Chissom, B.: Forecasting enrollments with fuzzy time series—Part II. Fuzzy Sets Syst. **64**, 1–8 (1994)
4. Zadeh, L.A.: Fuzzy set. Inf. Control **8**, 338–353 (1965)
5. Chen, S.M.: Forecasting enrollments based on fuzzy time series. Fuzzy Sets Syst. **81**, 311–319 (1996)
6. Lee, L.W., Wang, L.H., Chen, S.M., Leu, Y.H.: Handling forecasting prob lems based on two-factors high-order fuzzy time series. IEEE Trans. Fuzzy Syst. **14**(3), 468–477 (2006)
7. Bose, M., Mali, K.: Designing fuzzy time series forecasting models: a survey. Int. J. Approximate Reasoning **111**, 78–99 (2019)
8. Chen, S.-M., Tanuwijaya, K.: Multivariate fuzzy forecasting based on fuzzytime series and automatic clustering techniques. Expert Syst. Appl. **38**, 10594–10605 (2011)
9. Chen, M.Y., Chen, B.T.: A hybrid fuzzy time series model based on granular computing for stock price forecasting. Inf. Sci. **294**, 227–241 (2015)
10. Chen, S.-M., Jian, W.-S.: Fuzzy forecasting based on two-factors second-order fuzzy-trend logical relationship groups, similarity measures and PSO techniques. Inf. Sci. **391–392**, 65–79 (2017)
11. Cheng, S.-H., Chen, S.-M., Jian, W.-S.: Fuzzy time series forecasting based on fuzzy logical relationships and similarity measures. Inf. Sci. **327**, 272–287 (2016)
12. Cai, Q., Zhang, D., Zheng, W., Leung, S.C.H.: A new fuzzy time series forecasting model combined with ant colony optimization and auto-regression. Knowl.-Based Syst. **74**, 61–68 (2015)
13. Bisht, K., Kumar, S.: Fuzzy time series forecasting method based on hesitant fuzzy sets. Expert Syst. Appl. **64**, 557–568 (2016)
14. Uslu, V.R., Bas, E., Yolcu, U., Egrioglu, E.: A fuzzy time series approach based on weights determined by the number of recurrences of fuzzy relations. Swarm Evol. Comput. **15**, 19–26 (2014)
15. Singh, P., Borah, B.: An efficient time series forecasting model based on fuzzy time series. Eng. Appl. Artif. Intell. **26**, 2443–2457 (2013)

16. Bose, M., Mali, K.: A novel data partitioning and rule selection technique for modeling high-order fuzzy time series. Appl. Soft Comput. **63**, 87–96 (2018)
17. Chen, S.M., Chen, S.W.: Fuzzy forecasting based on two-factors second-order fuzzy-trend logical relationship groups and the probabilities of trends of fuzzy logical relationships. IEEE Trans. Cybern. **45**, 405–417 (2015)
18. Wang, L., Liu, X., Pedrycz, W., Shao, Y.: Determination of temporal in formation granules to improve forecasting in fuzzy time series. Expert Syst. Appl. **41**(6), 3134–3142 (2014)
19. Chen, S.M., Manalu, G.M.T., Pan, J.S., et al.: Fuzzy forecasting based on two-factors second-order fuzzy-trend logical relationship groups and particle swarm optimization techniques. Cybern. IEEE Trans. **43**(3), 1102–1117 (2013)
20. Lu, W., Chen, X., Pedrycz, W., Liu, X., Yang, J.: Using interval information granules to improve forecasting in fuzzy time series. Int. J. Approximate Reasoning **57**, 1–18 (2015)
21. Yu, T.H.-K, Huarng, K.: A neural network-based fuzzy time series model to improve forecasting. Expert Syst. Appl. **37**, 3366–3372 (2010)
22. Singh, P., Borah, B.: High-order fuzzy-neuro expert system for daily temperature Forecasting. Knowl.-Based Syst. **46**, 12–21 (2013)
23. Wei, L.Y., Cheng, C.H., Wu, H.H.: A hybrid ANFIS based on n-period moving average model to forecast TAIEX stock. Appl. Soft Comput. **19**, 86–92 (2014)
24. Wei, L.Y.: A GA-weighted ANFIS model based on multiple stock market volatility causality for TAIEX forecasting. Appl. Soft Comput. **13**(2), 911–920 (2013)
25. Wei, L.-Y.: A hybrid ANFIS model based on empirical mode decomposition for stock time series forecasting. Appl. Soft Comput. **42**, 368–376 (2016)
26. Su, C.-H., Cheng, C.-H.: A hybrid fuzzy time series model based on ANFIS and integrated nonlinear feature selection method for forecasting stock. Neurocomputing **205**, 264–273 (2016)
27. Chen, M.Y.: A high-order fuzzy time series forecasting model for internet stock trading. Future Gener. Comput. Syst. **37**, 461–467 (2014)
28. Maron, M.E.: Automatic indexing: an experimental inquiry. J. ACM **8**(3), 404–417 (1961). https://doi.org/10.1145/321075.321084
29. Russell, S., Norvig, P.: Artificial Intelligence: A Modern Approach, 2nd edn. Prentice Hall (2003)
30. Deng, W., Wang, G., Zhang, X., Xu, J., Li, G.: A multi-granularity combined prediction model based on fuzzy trend forecasting and particle swarm techniques. Neurocomputing **173**, 1671–1682 (2016)
31. Bose, M., Mali, K.: An improved technique for modeling fuzzy time series, In: the proceedings of 2nd international conference on computational intelligence, communications, and business analytics, Kalyani Govt. Engg. College, West Bengal (Communications in Computer and Information Science, vol. 1030, Springer, 2019) (2018)

Nature-Inspired Algorithm-Based Feature Optimization for Epilepsy Detection

Anurag Singh, Suraj Sharma, Vivek Mukundan, Tapendra Kumar and Nalini Pusarla

Abstract Epilepsy or seizure is a prevalent neurological disorder present among all age of people. However, there is still a lack of precise automated detection methods for epilepsy. We present a computer-aided diagnosis system using nature-inspired algorithms for automatic detection of epilepsy from EEG data. Unlike the traditional approaches, we propose to employ nature-inspired algorithms such as Genetic Algorithm (GA) and Particle Swarm Optimization (PSO) for feature optimization before the detection process. Time–frequency domain features have been extracted using discrete wavelet transform from the time-series EEG data. Out of these features, discriminatory and optimized set of features are obtained using GA and PSO, which are used further to diagnose a person into an epileptic and non-epileptic class. As compared to the non-optimization-based approaches, the proposed method performs better in terms of improved detection accuracy.

Keywords Electroencephalogram (EEG) · Epilepsy · Genetic algorithm · KNN · Disease detection · Classification

A. Singh (✉) · S. Sharma · V. Mukundan · T. Kumar · N. Pusarla
International Institute of Information Technology, Naya Raipur, Atal Nagar, Chhattisgarh 493 661, India
e-mail: anurag@iiitnr.edu.in

S. Sharma
e-mail: suraj17101@iiitnr.edu.in

V. Mukundan
e-mail: vivekm17101@iiitnr.edu.in

T. Kumar
e-mail: tapendra17101@iiitnr.edu.in

N. Pusarla
e-mail: nalini@iiitnr.edu.in

© Springer Nature Singapore Pte Ltd. 2020
S. Agarwal et al. (eds.), *Machine Intelligence and Signal Processing*,
Advances in Intelligent Systems and Computing 1085,
https://doi.org/10.1007/978-981-15-1366-4_21

1 Introduction

Epilepsy is a transient neurological disorder caused by sudden brain's abnormality and unrestricted electrical activity of thousands of nerve cells [1]. More than 50 million populations are suffering from epilepsy all around the world [2]. Long-term medications are necessary for curing epilepsy and if unsuccessful brain abscission is suggested [3]. In most of the cases, seizure inception is uncertain and occurs during sleep so a person cannot live independently and may lead to death [4, 5]. Electroencephalogram (EEG) signal non-invasively measures the electrical activity of the human brain and proved to be quite important diagnostic tool for detecting many neurological disorders [5]. In contrast to imaging techniques, EEG is significantly used in detecting and diagnosing epileptic seizure from patient's brain waves [6]. As shown in Fig. 1, a person may have sudden attack of epilepsy which is reflected as the sudden transients with high amplitude in the normal EEG signals. EEG is tolerant to movements like eye blinks (eyes open and eyes closed), ECG, contraction of muscles and external disturbances like power line noise and electrode placement positions which is intractable with imaging techniques [7]. In general, electrodes are placed on special positions on the scalp which is 10–20 international system to record the brain waves [8]. EEG signals recorded for a long time provide detailed information of amplitude and frequency of epileptic seizures, which helps in clinical diagnosis and treatment while it is difficult with short-term EEG recordings in most of the epileptic subjects [9]. These recordings are from the surface of the cerebral cortex, whereas deeper neuronal activity is not reflected in EEG [10]. Visual inspection of long-term EEG by a group of neurologists for seizure detection is an enfeeble task. It is a quite time-consuming process for even experienced neurologists to manually inspect the epileptic seizure in the EEG signals [11]. Also, the ever-increasing number of mental patients has spurred the need of an efficient automatic identification system for accurate prediction and detection system for seizure activities using non-invasive EEG signals [12]. Diagnosing seizure at an early stage is beneficial for patients for better therapies and also reduces provocation on clinicians [1]. EEG is preprocessed for artifacts removal and then further processed to extract the clinically relevant information called features, embedded in the EEG signals. Feature extraction plays a very important role in the automated diagnosis of epilepsy. So far, several techniques

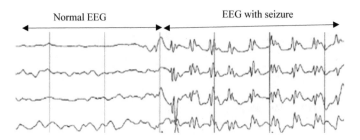

Fig. 1 Normal and epileptic seizure present in four different EEG signals

have been proposed in the literature for detecting seizures but still, it is a foremost challenging issue in the absence of a precise and low-complex system.

Murro et al. developed an automated epilepsy detection algorithm with high sensitivity and low false alarm rate using time domain-based features [13]. Efficient epileptogenic localization and seizure inception was addressed using Fourier transformed directed transfer function method to measure the flow of information in different regions of the cerebral cortex. This algorithm worked on canine epilepsy based on dog's EEG in the frequency domain and the performance was benchmarked [14]. Altunay et al. recommended a seizure automation algorithm which observes the spiky nature of seizures in EEG waves. They attained 94.5 and 90.5% for non-seizure and seizure intervals [15]. Zhang experimented on long-term EEG to improve the efficiency for automatic epileptic detection using KNN as a classifier. They evaluated 21 patient's EEG data with sensitivity 91.01 and specificity 95.77% [16]. Hesse and James performed signal subspace correlation on 25-channel EEG for tracing and diagnosing epileptic intervals. Both worked on implementing blind source separation methods on seizure dataset for achieving better results [17]. Kiranyaz et al. developed a universal technique which addressed the limitations of long-term monitoring of EEG for seizure classification. An ensemble of binary classifiers along with particle swarm optimization not only detected the seizure but seizure onset too which resulted in average sensitivity greater than 89%, specificity of above 93% [18]. Ovchinnikov et al. proposed a new framework on spikes discharge in brain waves using wavelet analysis. The continuous wavelet transform (CWT) method is suitable for real-time diagnosis of spikes in ECoG rhythms of rodents [19]. Conradsen et al. came up with multimodal seizure intelligent system with electromyogram (EMG) as input. The authors have obtained high latency using wavelet packet transformation (WPT) subbands of EMG for SVM classification [20]. Kannathal et al. differentiated normal and seizure subjects using nonlinear entropy algorithms [21]. Faust et al. used parametric methods like ARMA, Burg for power spectrum calculations of brain wave which automatically classify normal and epileptic subjects [22]. Shoeb and Guttag demonstrated a technique on CHB-MIT Scalp EEG dataset by measuring the energy of the two-second segment and filtered components as feature vectors. SVM classifier along with kernel basis functions is used to identify seizure [23]. Xiang et al. implemented a method for inspection of epileptic seizures based on fuzzy entropy [24]. A multivariate autoregressive model was entrenched along with partial directed coherence technique for locating seizures. The reported performance was 98.3% classification rate, sensitivity 67.88%, and specificity of 99.34% but not pointing on confinement of seizure zones [25]. A classification algorithm was further proposed for seizure and seizure-free intervals by Gaussian filtered binary patterns of short-term EEG with reduced number of computations [26].

Most of the aforementioned methods do not consider optimizing features before performing classification task, thus taking more processing time as well as low detection rate. Motivating from this fact, in this paper, we have explored nature-inspired optimization algorithms for best feature selection in epilepsy detection problem. Here, we are presenting an optimization-based epilepsy detection approach which requires very less number of features with better diagnosis performance. The redundancy in the feature space has been removed using nature-inspired optimization algorithms like PSO and GA, and the reduced dimensional feature vector is employed further for classification purpose. Thus, the proposed approach inhibits very low complexity with improved detection accuracy as compared to the existing methods without feature optimization.

2 Methodology

A flow chart of the proposed methodology is given in Fig. 2. The broad steps involved are preprocessing, feature extraction, feature optimization followed by classification. A preprocessing is done to the signals in order to remove noise. For this, the signal is passed through a bandpass filter of fourth order. The range of the BPF used in this model is from 0.35 to 40 Hz. Many methods such as the Fourier Transform (FT) and Short-Time Fourier Transform (STFT) have already been proposed and tested for analyzing the signal but they suffer from certain shortcomings. The FT is incapable of efficiently handling non-stationary signals such as EEG signals as it provides

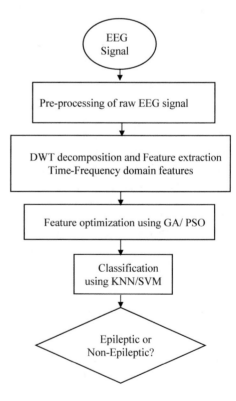

Fig. 2 Flowchart of the proposed approach

no time resolution. STFT uses a fixed time–frequency resolution and hence suffers from appropriate window selection. Increasing the resolution in time decreases the resolution in frequency and vice versa. Hence, we use discrete wavelet transform for feature extraction, which is excellent time–frequency analysis tool and quite suitable for non-stationery EEG signals. Because of its high energy compaction property, it helps efficiently localize the characteristic waveforms of EEG signal into delta (<4 Hz), theta (4–8 Hz), alpha (8–13 Hz), and beta (13–30 Hz) waves. A batch-wise processing model is followed for the processing of the input EEG signal. Signals are fed batch-wise in a 2 s window with a 1.5 s overlapping window [27]. Each batch has a batch frequency of 256 Hz/Sample. The discrete wavelet transform (DWT) technique is applied to extract time–frequency domain features. Selection of suitable mother wavelet and the number of levels of decomposition is very important in the analysis of signals using DWT. The wavelet can be chosen depending on how smooth the signal is and also on the basis of the amount of computation involved. The number of levels of decomposition is chosen based on the dominant frequency components of the signal. The levels are chosen such that those parts of the signal that correlate well with the frequencies required for classification of the signal are retained in the wavelet coefficients. We have employed Daubechies wavelet (dB2) with five levels of wavelet decomposition [27]. A total of nine features are extracted from the signal. These features include mean, mean absolute deviation, root mean square, standard deviation, average power, logarithmic power, minimum and maximum, and range. Each feature is calculated on each of the six wavelet subbands. Hence, a 54-feature vector is obtained for further processing.

2.1 Feature Optimization

The optimizing module of the proposed algorithm focuses on reducing the feature space in order to infuse more discriminability among features by discarding redundancy. The low dimensionality of the feature space would also help to reduce the computational complexity involved in the detection/classification process. Since, nature-inspired optimization algorithms are widely known for their optimization capability, we employed two such algorithms, genetic algorithm (GA) [28] and particle swarm optimization (PSO) [29] for the feature optimization purpose. GA is based on the mechanics of natural genetics and biological evolution. The first step is to initialize the population. Here, the features are the number of genes for the algorithm. Recall that we had the feature vector as a 54×1 vector. Now, this vector is initialized with a population of 0's and 1's. Now, the population is initialized in a 10×54 vector matrix. Now, two parents are selected randomly from the population and then they are subjected to crossover. Crossover refers to the generation of an offspring produced by the mutual interaction between the parents. In this model, one-point as well as multi-point crossover techniques are used to produce offspring. The probability of crossover is taken as Pc = 0.85.

After performing the crossover, we obtain two children namely child1 and child2. The next step is mutation wherein the genes of the children are varied with a certain probability (taken as Pm = 0.01). Mutation is performed so that the population is more diverse and aims at exploring the search space. Mutation is more likely to result in convergence than crossover. The feature vector matrix consisting of 54 features is now bit-flipped with a low probability. Now, the features corresponding to bit 1 are selected and the features corresponding to bit 0 are omitted. This process of selection of features continues until it finds an optimized set of features giving maximum accuracy. The process terminates when maximum accuracy achieved or it reaches to maximum number of iterations.

PSO is another nature-inspired algorithm, which is based on the natural flocking mechanism of birds and fishes. It is widely used in optimization problems where a dataset is reduced to a subset which contains the best features. Firstly, the particles are initialized with a zero velocity as well as position vectors. Now, a randomly generated position and velocity are applied to move the particle in the search space. Now, these particles are subjected to move in the search space. The algorithm returns two values namely the global best position and global best feature. These features are the best that are selected from the population and they carry significant characteristics of the disease. The run-time complexity of GA/PSO is marginal, 0.766 s as observed on a core i3, 4 GB RAM CPU; hence, there is no significant increase in the overall computational complexity of the proposed detection algorithm.

Once the feature optimization is done, we divided the feature set into 70:30 ratios, i.e., 70% for training and 30% for testing. The data used for training is not used for testing. We created two classes, epileptic and non-epileptic. All the seizure-free datasets kept in the non-epileptic case and the EEG data with seizure activity is placed in the epileptic class. As it becomes a binary classification, we have used K-NN and SVM for the classification purpose, as they are suitable for two-class classification problems. SVM with RBF kernel is used keeping all other tuning parameters as their default values. Detection accuracy is calculated in terms of different state-of-the-art performance measure [27].

3   Results and Discussion

The performance evaluation of the proposed methodology has been carried out using the publically available epilepsy database [30]. The database contains five sets of EEG signal denoted by alphabets A–E. Each set consists of 100 single-channel EEG signal of 23.6.sec duration. Sets A and B contain EEG data from healthy person with eyes open (A) and eyes closed (B), respectively. Sets C and D contains the pre-ictal and post-ictal EEG signals from a patient who had suffered from epilepsy earlier but now recovered it, i.e., they contain seizure-free EEG activities. The set E contains the EEG segments with seizure activities. The data has been processed batch-wise with overlapping window. First, we calculate the detection accuracy using traditional approach, i.e., extracting different features and then feeding it to classifiers for detection purpose. However, we intended to reduce the feature space in order to boost the classification accuracy and reduce the computational complexity, thus making the detection process faster. To achieve this objective, we propose to use the Evolutionary Computation (EC), algorithms, PSO, and GA for feature optimization, i.e., to select best features.

The quantitative results obtained are given in Tables 1 and 2 in terms of different performance measures, accuracy, precision, specificity, and sensitivity [27]. We have considered a total of 54 time–frequency domain features and GA and PSO are used to choose the best set of features. Table 1 depicts the numerical results when different combinations of features are chosen. It can be seen clearly that as we choose the best set of features using GA, the detection performance of the algorithm improves. The best accuracy of 97.52% using KNN classifier is obtained when three features were selected out of 54. This is because GA optimizes the feature space and helps to choose only relevant and discriminative features rejecting the redundant features. GA also helps in a reduction of feature space by around 94.44%, reducing computational complexity significantly. Using PSO, there is no significant gain in accuracy is obtained as shown in Fig. 3; however, computational efficiency has been surely increased by reducing the number of features from 54 to 6 with almost same accuracy level. So, as compared to PSO, GA may be more suitable for feature optimization in epilepsy detection problem.

We also analyzed the performance of the proposed algorithm with respect to tuning parameters of GA, like number of generations and population size. The number of generations decides the iterations for search of the optimized solution according to the fitness value in each generation. The population size may help in getting better Pareto

Table 1 Detection performance parameters using GA for feature optimization

No. of features	Accuracy (%)	Precision (%)	Specificity (%)	Sensitivity (%)
3	97.52	92.50	99.74	99.87
6	95.97	89.96	99.45	99.75
12	95.68	90.57	99.45	100
15	95.51	92.87	100	100
18	95.91	93.29	100	100
21	95.79	92.02	100	100
24	94.88	92.87	99.77	100
54	93.90	92.21	100	100

Table 2 Variation of detection accuracy with respect to population size in GA-based feature optimization

Population size	Accuracy (%)
10	97.53
20	96.38
30	95.23
40	95.46

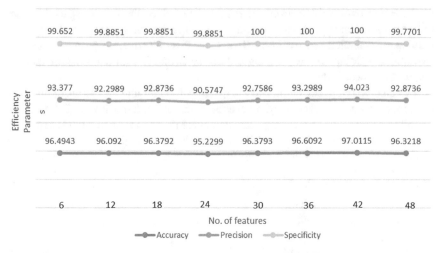

Fig. 3 Variation of different quality measures (accuracy, precision, and specificity) with respect to number of features when PSO is employed for feature optimization

curve, i.e., optimized set of solutions. These two factors may have different impact on the feature optimization task and hence, final accuracy results may change. We have simulated our results for varying number of generations and population size. Results are given in Table 2. It is observed that the optimized feature set, and consequently best result, is obtained at number of generations equal to 1 at population size of 10. As we increase the number of population size beyond 10, the accuracy gradually decreases. Also, the experiments have been carried out with different number of generations. It was found that the optimized sets of features are obtained at single generation only and the final detection accuracy decreases at higher generations.

Comparative analysis of the proposed approach using GA with similar existing methods is also done and results have been shown in Table 3. It is clear from this analysis that optimizing the feature space before the classification step not only decreases the computational complexity but also improves the detection accuracy

Table 3 Comparison of the proposed method with the existing methods

Methods	Accuracy (%)
Permutation Entropy + SVM [31]	93.55
DWT-based approximate entropy (ApEn) + Artificial neural network and SVM [32]	92.5
DWT-based fuzzy approximate entropy + SVM [33]	95.85
DWT-based features + ANN [27]	89.02
Time–frequency features + Relief F ranking + Bayesian Network [33]	95.0
Time–frequency features + hypothetical testing + fuzzy classifier [35]	96.48
Proposed method	97.53

as compared to the methods which do not employ feature optimization. This is because of the capability of the optimization algorithm (GA and PSO) used in the proposed method, which maintains the discriminative nature of features even in lower-dimensional space, thus helping improve the accuracy further. Few methods [34, 35] which also employ feature optimization like our method have also been compared. Methods [34, 35] use Relief F ranking and hypothetical testing, respectively, for selecting best features. As evident from the comparison table, the proposed method with nature-inspired feature optimization outperforms the above methods too.

4 Conclusion

An efficient approach for feature optimization is proposed for epilepsy detection application using EEG signals. Evolutionary algorithms like genetic algorithm (GA) and particle swarm optimization (PSO) have been explored to select optimized set of features out of several time–frequency domain features. The best set of features selected by the GA/PSO is then employed for diagnosis purpose whether a person is epileptic or non-epileptic. Posing it as a binary classification problem, KNN and SVM were used for the classification task. It was found that DWT-based time–frequency features optimized with GA give better detection accuracy using KNN classifier as compared to the existing methods which do not optimize features.

References

1. Witte, H., Iasemidis, L.D., Litt, B.: Special issue on epileptic seizure prediction. IEEE Trans. Biomed. Eng. **50**(5), 537–539 (2003)
2. Nashef, L., Ryvlin, P.: Sudden unexpected death in epilepsy (SUDEP): update and reflections. Neurol. Clin. **27**(4), 1063–1074 (2009)
3. Tellez-Zenteno, J., Hernandez Ronquillo, L., Moien-Afshari, F., Samuel Wiebe, S.: Surgical outcomes in lesional and non-lesional epilepsy: a systematic review and meta-analysis, Epilepsy Res. **89** (2010)
4. Hughes, J.R.: A review of sudden unexpected death in epilepsy: prediction of patients at risk. Epilepsy Behav. **14**(2), 280–287 (2009)
5. Hassan, A.R., Bhuiyan, M.I.H.: An automated method for sleep staging from EEG signals using normal inverse Gaussian parameters and adaptive boosting. Neurocomputing **219**(5), 76–87 (2017)
6. Kuzneicky, R.I.: MRI in cerebral developmental malformations and epilepsy. Magn Reson. Imaging **13**, 1137–1145 (1995)
7. Hassan, A.R., Bhuiyan, M.I.H.: A decision support system for automatic sleep staging from EEG signals using tunable Q-factor wavelet transform and spectral features. J. Neurosci. Methods **271**(15), 107–118 (2016)
8. Shoeb, A.: Application of Machine Learning to Epileptic Seizure Onset Detection and Treatment. Ph.D thesis, Massachusetts Institute of Technology (2009)
9. Hassan, A.R., Subasi, A.: A decision support system for automated identification of sleep stages from single-channel EEG signals. Knowl. Based Syst. **128**(15), 115–124 (2017)

10. Wang, G., Worrell, G., Yang, L., Wilke, C., He, B.: Interictal spike analysis of high-density EEG in patients with partial epilepsy. Clin. Neurophysiol. **122**(6), 1098–1105 (2011)
11. Sanei, S., Chambers, J.A.: EEG Signal Processing. Wiley, New York (2007)
12. Lehnertz, K.: Non-linear time series analysis of intracranial EEG recordings in patients with epilepsy–an overview. Int. J. Psychophysiol. **34**(1), 45–52 (1999)
13. Murro, A.M., King, D.W., Smith, J.R., Gallagher, B.B., Flanigin, H.F., Meador, K.: Computerized seizure detection of complex partial seizures. Electroencephalogr. Clin. Neurophysiol. **79**(4), 330–333 (1991)
14. Franaszczuk, P.J., Bergey, G.K.: Application of the directed transfer function method to mesial and lateral onset temporal lobe seizures. Brain Topogr. **11**(1), 13–21 (1998)
15. Altunay, S., Telatar, Z., Erogul, O.: Epileptic EEG detection using the linear prediction error energy. Expert Syst. Appl. **37**(8), 5661–5665 (2010)
16. Zhang, Y., Zhou, W., Yuan, S., Yuan, Q.: Seizure detection method based on fractal dimension and gradient boosting. Epilepsy Behav. **43C**(27), 30–38 (2014)
17. Hesse, C.W., James, C.J.: Tracking and detection of epileptic form activity in multichannel ictal EEG using signal subspace correlation of seizure source scalp topographies. Med. Biol. Eng. Compu. **45**(10), 909–916 (2007)
18. Kiranyaz, S., et al.: Automated patient-specific classification of long-term electroencephalography. J. Biomed. Inform. **49**, 16–31 (2014)
19. Ovchinnikov, A., et al.: An algorithm for real-time detection of spike-wave discharges in rodents. J. Neurosci. Methods **194**(1), 172–178 (2010)
20. Conradsen, I., Beniczky, S., Wolf, P., Kjaer, T.W., Sams, T., Sorensen, H.B.: Automatic multimodal intelligent seizure acquisition (MISA) system for detection of motor seizures from electromyographic data and motion data. Comput. Methods Programs Biomed. **107**(2), 97–110 (2012)
21. Kannathal, N., Choo, M.L., Acharya, U.R., Sadasivan, P.K.: Entropies for detection of epilepsy in EEG. Comput. Methods Programs Biomed. **80**(3), 187–194 (2005)
22. Faust, O., Acharya, U.R., Min, L.C., Sputh, B.H.: Automatic identification of epileptic and background EEG signals using frequency domain parameters. Int. J. Neural Syst. **20**(2), 159–176 (2010)
23. Shoeb, A., Guttag, J.: Application of machine learning to epileptic seizure onset detection. In: 27th International Conference on Machine Learning (ICML), Haifa (2010)
24. Xiang, J., Li, C., Li, H., Cao, R., Wang, B., Han, X., Chen, J.: The detection of epileptic seizure signals based on fuzzy entropy. J. Neurosci. Methods. **243**, 18–25 (2015)
25. Wang, G., Sun, Z., Tao, R., Li, K., Bao, G., Yan, X.: Epileptic seizure detection based on partial directed coherence analysis. IEEE J. Biomed. Health Inform. **20**(3), 873–879 (2016)
26. Tiwari, A.K., Pachori, R.B., Kanhangad, V., Panigrahi, B.K.: Automated Diagnosis of Epilepsy Using Key-Point-Based Local Binary Pattern of EEG Signals. IEEE J. Biomed. Health. Inf. **21**(4), 888–896 (2017)
27. Patnaik, L.M., Manyam, O.K.: Epileptic EEG detection using neural networks and postclassification. Comput. Methods Programs Biomed. **91**, 100–108 (2008)
28. Goldberg, D.E., Holland, J.H.: Genetic algorithms and machine learning. Mach. Learn. **3**(2), 95–99 (1988)
29. Eberhart, R., Kennedy, J.: A new optimizer using particle swarm theory. In: MHS'95, Proceedings of the Sixth International Symposium on Micro Machine and Human Science, pp. 39–43 (1995)
30. Andrzejak, R.G. et al.: Indications of nonlinear deterministic and finite-dimensional structures in time series of brain electrical activity: dependence on recording region and brain state. Phys. Rev. E. **64**: 061907 (2001) [Online]. Available: http://www.meb.unibonn.de/epileptologie/science/physik/eegdata.html
31. Nicolaou, N., Georgiou, J.: Detection of epileptic electroencephalogram based on permutation entropy and support vector machines. Expert Syst. Appl. **39**(1), 202–209 (2012)
32. Kumar, Y., Dewal, M.L., Anand, R.S.: Epileptic seizures detection in EEG using DWT-based ApEn and artificial neural network. Sign. Image Video Proc. **8**(7), 1323–1334 (2014)

33. Kumar, Y., Dewal, M.L., Anand, R.S.: Epileptic seizure detection using DWT based fuzzy approximate entropy and support vector machine. Neurocomputing **133**, 271–279 (2014)
34. Pippa, E., Zacharaki, E.I., Mporas, I., Tsirka, V., Richardson, M.P., Koutroumanidis, M., Megalooikonomou, V.: Improving classification of epileptic and non-epileptic EEG events by feature selection. Neurocomputing **171**, 576–585 (2016)
35. Harpale, V., Bairagi, V.: An adaptive method for feature selection and extraction for classification of epileptic EEG signal in significant states, J. King Saud University Comput. Inform. Sci. (2018)

A Combined Machine-Learning Approach for Accurate Screening and Early Detection of Chronic Kidney Disease

Klinsega Jeberson, Manish Kumar, Lordwin Jeyakumar and Raghav Yadav

Abstract Chronic kidney disease (CKD) largely affects people worldwide and is now seen as a common health threat attributable to its escalating prevalence and huge costs associated with dialysis and transplantation, which are therapies for advanced CKD. Due to various factors, the detection of the condition in patients is being delayed. Also, the markers like serum creatinine and albumin which are commonly used for CKD screening in clinical practice seem to be inadequate. In this study, an approach combining J48 and LADTree has been proposed. The outcomes indicate that the proposed approach performed exceptionally well with perfect accuracy in cross validation. The proposed model exhibited better performance than doctor 1 and doctor 2 on a test set containing 100 records in an additional validation process with a performance difference of 10% and 20%, respectively. The model may be adopted to construct decision support system that enables inexperienced physicians to accurately detect CKD using few laboratory values. Low-cost gadgets based on the proposed model may be developed which may extend the reach of health-care professionals outside the clinic and therefore aid timely diagnosis of the condition in large proportion of population at risk.

Keywords Chronic kidney disease · LADTree · J48 · ADTree · Decision support system

K. Jeberson (✉) · R. Yadav
Department of Computer Science & Information Technology, Sam Higginbottom University of Agriculture, Technology and Sciences, Prayagraj, India
e-mail: klinsega.jeberson@shiats.edu.in

M. Kumar
Indian Institute of Information Technology, Allahabad, Prayagraj, India

L. Jeyakumar
St. John's Research and Development Centre, Agriculture and Agri-Food, Newfoundland and Labrador, Canada

© Springer Nature Singapore Pte Ltd. 2020
S. Agarwal et al. (eds.), *Machine Intelligence and Signal Processing*,
Advances in Intelligent Systems and Computing 1085,
https://doi.org/10.1007/978-981-15-1366-4_22

1 Introduction

Chronic kidney disease (CKD) is a disorder marked by steady decline of renal function over time, leading to end-stage renal disease (ESRD) and later death of the individual if untreated [1]. National Kidney Foundation statistics reveal that 10% of the population worldwide is affected by CKD [2], and researchers have estimated that rate of kidney failure will rise excessively in developing countries like China and India, where the number of elderly individuals is on the rise [3]. Researchers have also conveyed that no less than 2284 million individuals would have died prematurely for the reason that dialysis or transplantation which is the treatments for ESRD were exorbitant [4]. Due to rising incidence rates and huge costs associated with therapies, CKD is recognized as a health threat globally.

In this study, combined classifiers were constructed out of the individual classifiers: alternating decision tree (ADTree), Decision table, J48 and Logit Boost alternating decision tree (LADTree). The reason for directing the study in this route is the successful performance of hybrid methods having been reported in various studies. Some of them include the use of such models in the prediction of coronary artery disease [5], classification of diabetes [6] and in intrusion detection system [7]. Also, the four individual classifiers ADTree, Decision table, J48 and LADTree have shown promising results in various classification problems [8–14].

2 Related Work

Jena and Kami [15] applied different methods on the same CKD dataset. The models were validated using tenfold cross-validation technique. Multilayer perception was reported to demonstrate the highest accuracy of 99.75%. Akben [16] proposed a new approach based on predetermined attribute limits to classify CKD. Basar et al. [17] conducted a study on the performance of adaptive boosting (AdaBoost) ensemble learning algorithm in CKD diagnosis and compared the results with that of individual classifiers. Chen et al. [18] investigated the working of a pair of fuzzy learning algorithms on the CKD data and compared their performance with that of partial least squares discriminant analysis classifier. Quinlan [19] has analyzed the performance of logistic regression, SVM with a linear kernel, SVM with RBF kernel, decision tree, random forest and AdaBooston the CKD data. Mean imputation method was employed to fill the missing values prevailing in the dataset. Salekin and Stankovic [20] have employed LASSO for feature selection and random forest for classification.

Table 1 Attributes and their description

Attribute	Description	Attribute	Description
age (A_1)	Age	sod (A_{13})	Sodium
bp (A2)	Blood pressure	pot (A_{14})	Potassium
sg(A_3)	Specific gravity	hemo (A_{15})	Hemoglobin
al (A_4)	Albumin	pcv (A_{16})	Packed cell volume
su (A_5)	Sugar	wbcc (A_{17})	White blood cell count
rbc (A_6)	Red blood cells	rbcc (A_{18})	Red blood cell count
pc (A_7)	Pus cells	htn (A_{19})	Hypertension
pcc(A_8)	Pus cell clumps	dm (A_{20})	Diabetes mellitus
ba (A_9)	Bacteria	cad (A_{21})	Coronary artery disease
bgr (A_{10})	Blood glucose random	appet (A_{22})	Appetite
bu (A_{11})	Blood urea	pe (A_{23})	Pedal edema
sc (A_{12})	Serum creatinine	ane (A_{24})	Anemia

2.1 Dataset Used

The dataset used in this study was taken from UCI repository, and the same has been donated by Apollo Hospitals in a city called Madurai in South India [21]. The dataset consists of 400 instances out of which 250 instances belong to the class 'ckd,' and the remaining 150 instances belong to the class 'notckd.' It includes 25 attributes comprising 14 numeric, ten nominal and one class attribute. The description of the attributes is shown in Table 1.

2.2 Classification Techniques

2.2.1 Decision-Tree and Rule-Based Algorithms

Decision-tree and rule-based classification algorithms were adopted in this study as they are 'white-box models.'

2.2.2   Combined Classification Models

In this study, six combined classifiers were constructed by combining four individual learning algorithms: ADTree (ADT) [13], Decision table (DT) [8], J48 and LADTree (LT) in pairs. The six classifiers constructed are ADT–DT, ADT–LT, ADT–J48, DT–LT, DT–J48 and LT–J48.

2.2.3   Proposed Model

The proposed model is a combination of LT and J48 classification algorithms. They are combined using the average of probabilities rule as shown in pseudocode (Sect. 2.2.2.3.)

LADTree

In a J class problem, for an instance i, there are J responses each taking values in $\{-1, 1\}$. The vector $F_j(x)$ represents the predictions which are the sum of the predictions of the classifier group on the instance x over the J classes.

$$p_j(x) = \frac{e^{F_j(x)}}{\sum_{k=1}^{J} e^{F_k(x)}}, \quad \sum_{k=1}^{J} F_k(x) = 0 \tag{1}$$

A detailed explanation of LADTree is presented by [22].

J48

J48 is the Java version of the classification algorithm C4.5, which is an enhancement of Iterative Dichotomiser 3 (ID3) algorithm [19].

$$\text{GainRatio} = \frac{\text{Gain(A)}}{\text{Split Info}_A(D)} \tag{2}$$

Split information value produced by partitioning training dataset D into k subsets corresponding to k outcomes on attribute A can be computed as

$$\text{SplitInfo}_A(D) = -\sum_{i=1}^{k} \frac{|D_i|}{|D|} \times \log_2\left(\frac{|D_i|}{|D|}\right) \tag{3}$$

Information gained by branching on attribute A is defined as

$$\text{Gain(A)} = \text{Inf}(D) - \text{Inf}_A(D) \tag{4}$$

Let J represents the number of classes, and P_j denotes the probability of cases in D that belongs to the jth class. The information gained before partitioning D can be expressed as

$$\text{Inf}(D) = -\sum_{j=1}^{J} P_j \log_2(P_j)$$

(5)

and the corresponding information gained after partitioning D with respect to attribute A with k possible values is

$$\text{Inf}_A(D) = \sum_{i=1}^{k} \frac{|D_i|}{D} \times \text{Inf}(D_i)$$

(6)

Gain ratio reduces the bias demonstrated by IG toward attributes with large number of values, by taking into account the size and number of braches during the selection procedure. The detailed theory behind J48 and its feature selection technique is presented by [23].

Pseudocode of the Proposed Approach

Training phase
Input: D: Set of examples
Output: Classifier
Apply LADTree learning algorithm on the training data to generate the LADTree tree.
Apply J48 learning algorithm on the training data to generate the J48 tree.
Classification phase
Input: X: instance to be classified
Output: class label
For the given instance X, let Pl_{ckd}, Pl_{nckd}, Pj_{ckd}, Pj_{nckd} be the probabilities returned by the classifiers LADTree and J48 that X belongs to the class 'ckd' and 'notckd,' respectively.
Calculate the average probabilities AP_{ckd} and AP_{nckd} of class 'ckd' and 'notckd,' respectively, using the formula

$$AP_{ckd} = \frac{Pl_{ckd} + Pj_{ckd}}{2}$$

(7)

$$AP_{nckd} = \frac{Pl_{nckd} + Pj_{nckd}}{2}$$

(8)

Assign X to the class with largest average probability.

2.3   Validation and Evaluation

2.3.1   Cross Validation

Tenfold cross-validation method was employed to validate the models generated by the classification algorithms. In tenfold cross-validation technique, the dataset D is randomly subdivided into ten mutually exclusive, approximately equal-sized subsets, D1, D2,…, D10. Training and testing are iterated ten times.

2.3.2   Performance Metrics

The performance metrics used for evaluating the classifier are accuracy, sensitivity, specificity, precision, recall, area under receiver operating characteristic (AUROC) and number of features.

$$\text{Accuracy} = \frac{(TN + TP)}{(TN + TP + FN + FP)} \tag{9}$$

$$\text{Sensitivity} = \frac{TP}{(TP + FN)} \tag{10}$$

$$\text{Specificity} = \frac{TN}{(TN + FP)} \tag{11}$$

$$\text{Precision} = \frac{TP}{(TP + FP)} \tag{12}$$

Where TP (True positive) is the total number of positive instances that have been rightly classified as positive, TN (True negative) is the total number of negative instances that have been rightly classified as negative, FN (False negative) is the total number of positive instances that have been wrongly classified as negative. FP (False positive) is the number of negative instances that have been wrongly classified as positive.

2.3.3   Validation Against Decision making of Doctors

As a means of comparing the rendering of the proposed model with that of doctors, a test set containing 100 instances with no missing values was prepared from the CKD dataset. The test dataset represented both the classes 'ckd' and 'notckd' in the ratio 3:2. Two doctors, one from Madurai Medical College, Madurai, Tamil Nadu and another from Dr. Somervell Memorial CSI Medical College, Thiruvananthapuram, Kerala, both having more than 20 years' experience were provided with test records containing only the laboratory values which according to them were commonly used for CKD screening in clinical practice and got the records classified. The diagnostic performance of the proposed model was compared with that of the doctors.

3 Results

The ten classifiers (four individual + six combined) were trained on the CKD dataset, and the resultant models were validated using tenfold cross-validation technique. All the models performed well producing accuracies ranging from 99 to 100% (Table 2). Of all the ten classifiers, the combined classifiers performed better than the individual ones with their predictive abilities ranging from 99.25 to 100%. Of the six combined classifiers, ADT–DT and LT–J48 demonstrated the highest accuracy of 100% and showed perfect sensitivity and specificity which indicates an accurate and unbiased classification. They also exhibited 100% precision, recall and AUC (Table 3). The

Table 2 Performance of four individual and six combined classifiers

Classifier	Accuracy (%)	Sensitivity (%)	Specificity (%)
ADT	99.75	99.60	100.00
DT	99.00	99.60	98.00
LT	99.00	99.60	98.00
J48	99.00	98.40	100.00
ADT–DT	100.00	100.00	100.00
ADT–LT	99.50	99.20	100.00
ADT–J48	99.75	99.60	100.00
DT–LT	99.25	98.80	100.00
DT–J48	99.50	99.60	99.30
LT–J48	100.00	100.00	100.00

Table 3 Precision, recall and AUC of the classifiers on the class 'CKD'

Classifier	Precision (%)	Recall (%)	AUROC (%)
ADT	100.00	99.60	100.00
DT	98.80	99.60	99.21
LT	100.00	98.40	99.95
J48	98.80	99.60	99.94
ADT–DT	100.00	100.00	100.00
ADT–LT	100.00	100.00	100.00
ADT–J48	100.00	99.60	99.99
DT–LT	100.00	98.80	99.99
DT–J48	99.60	99.60	99.99
LT–J48	100.00	100.00	100.00

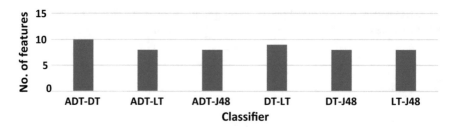

Fig. 1 Number of features used by the combined classifiers for classifying CKD

Classifier	Features used
ADT–DT	A_3, A_4, A_5, A_6, A_{10}, A_{12}, A_{15}, A_{20}, A_{22}, A_{23}
ADT–LT	A_3, A_4, A_6, A_{10}, A_{12}, A_{15}, A_{19}, A_{20}
ADT–J48	A_3, A_4, A_6, A_{10}, A_{12}, A_{15}, A_{20}, A_{23}
DT–LT	A_3, A_4, A_5, A_6, A_{12}, A_{15}, A_{22}, A_{23}
DT–J48	A_3, A_4, A_5, A_{12}, A_{15}, A_{20}, A_{22}, A_{23}
LT–J48	A_3, A_4, A_6, A_{12}, A_{15}, A_{19}, A_{20}, A_{23}

Table 4 Features used by the six combined classifiers to classify CKD

perfect behavior of the two models may be credited to the inherent, excellent feature selection techniques in the classifiers that constituted the two combined models. Though both the classifiers performed equally well, LT–J48 was considered the optimal classifier owing to the lesser number of attributes it used for classification. LT–J48 also yielded compact trees indicating no symptoms of overfitting.

The number of parameters used by the combined models to classify the instances falls in the range 8–10 (Fig. 1) The predictor attributes used by the best classifier are A_3, A_4, A_6, A_{12}, A_{15}, A_{19}, A_{20} and A_{23} (Table 4). It was observed that few features like A_3, A_4 and A_5 appeared in almost all the classifiers representing their relative importance over other features in CKD classification. It was also observed that the attributes A_3, A_4 and A_{15} occurred in all the combined models ascertaining to be the most crucial parameters used in differentiating CKD cases from not CKD ones. Most of the models constructed in this study showed slightly deteriorated sensitivity which may be attributed to the presence of missing values or noisy data.

The model LT–J48 was then validated against the decision making of two clinicians who participated in the validation process. The test dataset contained 100 records out of which 60 had positive class and 40 had negative class. The clinicians were provided with only those laboratory values which they commonly used for CKD detection. The attributes they used for classifying records were A_4, A_{11}, A_{12}, A_{19} and A_{20}. The results proved superior performance of the proposed model over doctor1 and doctor 2 with an accuracy difference of 10% and 20%, respectively (Fig. 2). Clinician 1 and clinician 2 classified the records with (83.3%, 100%) and (70%, 95%) sensitivity, specificity, respectively. The relatively poor sensitivity raises concern about the possibility of CKD patients going undiagnosed. It was observed

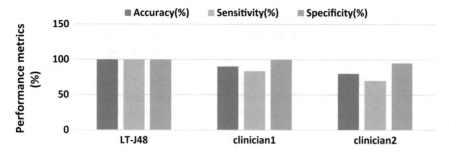

Fig. 2 Comparison of classification performance of LT–J48 and clinicians

that features like A_3, A_5, A_6 which have high occurrence among the models were not considered for CKD screening in clinical practice [24].

4 Discussion

The best predictive accuracies obtained in this and prior studies on the same CKD dataset have been presented in Table 4. The values presented show that all the approaches proposed in the literature yielded promising results. The slightly degraded performance of other models is attributable to various aspects. One of the factors associated is the absence of feature selection process in some of the related studies [16, 19]. Prior work demonstrates the superior performance of learning algorithms such as MLP and SVM after feature selection [15, 25]. Selection of appropriate feature selection technique and manually deciding on the number of features are also crucial in improving the model performance. For example, choosing the top five

Table 5 Accuracy achieved by LT–J48 and other models proposed in prior studies

Author (Year)	Learning algorithm	Accuracy
Jena and Kamila (2015)	Multilayer perceptron	99.75
Akben (2016)	New approach	98.00
Başar et al. (2016)	AdaBoost J48	99.50
Chen et al. (2016)	FuRES	98.10
Reddy and Cho (2016)	SVM	98.00
Salekin and Stankovic (2016)	LASSO + random forest	99.80
Serpen (2016)	C4.5	98.25
Polat et al. (2017)	Best first + SVM	98.50
Subasi et al. (2017)	Random forest	100.00
This study	Combined LADTree–J48	100.00

LASSO attributes in one of the studies [20] reduced the accuracy of the random forest classifier slightly. Had the researchers chosen the top six attributes, there is greater probability that the accuracy of the model improves as the sixth LASSO attribute was sc (serum creatinine) which was established as one of the crucial attributes used for discriminating CKD and not CKD cases in this study. In addition to generating accurate models, feature selection reduces the overall diagnostic cost, and this is undoubtedly a desired characteristic for diagnostic models to be used in medical field.

Cautious selection of the value of k in k-fold cross validation is imperative as it can affect the performance of the model. The inappropriate k value leads to deteriorated performance of the classifier. This is further established from the results reported in prior study where k is set as 3 [26]. When k is set as low as 3, lesser training data (as low as 267 instances in CKD dataset) is supplied to the learning algorithm in each fold resulting in suboptimal classifiers. This can especially be a problem when the dataset at hand is not large enough implying insufficient data for model construction. The value of k is data dependent, and in most of the related studies and also the proposed method, k was configured as 10 where 90% of the instances will be utilized for training constructing better classifiers (Table 5).

One of the factors assumed to have degraded the performance of the classifiers proposed in the literature is the application of missing value imputation on the dataset. Reddy and Cho [27] have implemented missing value imputation by mean method. SVM was reported showing the highest accuracy of 98%, the second-best random forest with 96% and AdaBoost demonstrating 95% accuracy. More sophisticated approaches for handling missing values like regression and decision trees generate more reasonable values and may contribute toward better classification accuracy. Learning algorithms like J48, random forest and MLP have inherent ability to handle missing values and have shown exceptional results on the CKD dataset which has lot amount of missing values. Subasi et al. [28] have reported 100% accuracy using random forest classifier on the same CKD data without performing missing value imputation.

Though the performance difference between the proposed model and those reported in the literature is marginal, the proposed model proves better than its counterparts. In medical decision making, white-box models are preferred over black-box models as the professionals would like to understand the logic behind the decision making, and hence multilayer perceptron [15] and SVM [25, 27] would be poor choices. And for the same reason, the statistical model published in the literature [16] would take a back seat. Moreover, these models do not select core features and hence do not contribute to cost reduction. AdaBoost classification proposed by the researchers [17] may not be considered a better technique as it is sensitive to noise and outliers. Medical data is prone to noise and problem might be encountered in future when increasing data to generate more generalized models.

Though the fuzzy logic-based classification technique FuRES [18] produces a set of interpretable fuzzy rules, it is not recommended for CKD classification as the membership functions associated with the rules are not interpretable for the end users.

And studies have reported degraded predictive accuracy of fuzzy classifiers over non-fuzzy classifiers [29] and thus may not be appropriate for practical applications. The proposed model is better than C4.5 model [26] owing to its more accurate predictions. Even though random forest [28, 20] performs at par with the proposed approach, it is not considered the classifier of choice as it uses more classification rules and thereby more features for classification and hence do not contribute to cost reduction.

In spite of the highlights, the study also suffers few limitations. The dataset used in this study consists of merely 400 records and has been donated by Apollo Hospitals, Madurai, India. The assumption is that the data contained may be of people from a particular locality. This triggers an uncertainty in the generalization ability of the model. With huge datasets, covering individuals from different age groups, races, gender, etc., it will be possible to develop better generalized models. Another shortcoming of this study is that no nephrologist has participated in the classification process which might have weakened the score in the clinician's end. But the purpose of the study was to construct a competent model that can aid both nephrologist and non-nephrologist clinicians in CKD screening, and it is presumed that the proposed classifier can certainly make a significant difference if adopted.

5 Conclusion

A combined model based on LADTree and J48 for CKD screening was proposed in this study. The proposed model yielded classification accuracy of 100% in cross validation. It also demonstrated 100% accuracy on a test set of 100 records, which was substantially higher than the diagnostic accuracy shown by the two doctors who participated in the study. The proposed model used eight attributes for classifying the records. The model also exhibited 100% sensitivity and specificity in both validations. It is believed that the proposed model can be really beneficial to the clinicians and health-care providers, enabling them to accurately diagnose their patients for CKD. Low-cost gadgets outfitted with the classifier may be developed that can possibly spread out the reach of health-care professionals in and outside clinical settings, thereby enabling prompt diagnosis of the condition in greater percentage of population at risk.

References

1. Lee, M.C., Wu, S.F.V., Hsieh, N.C., Tsai, J.M.: Self-management programs on eGFR, depression, and quality of life among patients with chronic kidney disease: ameta-analysis. Asian Nurs. Res. **10**(4), 255–262 (2016)
2. Damien, P., Lanham, H.J., Parthasarathy, M., Shah, N.L.: Assessing key cost drivers associated with caring for chronic kidney disease patients. BMC Health Serv. Res. **16**(1), 690 (2016)
3. Jha, V., Garcia-Garcia, G., Iseki, K., et al.: Chronic kidney disease: global dimension and perspectives. Lancet **382**(9888), 260–272 (2013)

4. Liyanage, T., Ninomiya, T., Jha, V., Neal, B., Patrice, H.M., Okpechi, I., Rodgers, A.: Worldwide access to treatment for end-stage kidney disease: a systematic review. Lancet **385**(9981), 1975–1982 (2015)

5. Verma, L., Srivastava, S., Negi, P.C.: A hybrid data mining model to predict coronary artery disease cases using non-invasive clinical data. J. Med. Syst. **40**(7), 1–7 (2016)

6. Kahramanli, H., Allahverdi, N.: Design of a hybrid system for the diabetes and heart diseases. Expert Syst. Appl. **35**(1), 82–89 (2008)

7. Xiang, C., Yong, P.C., Meng, L.S.: Design of multiple-level hybrid classifier for intrusion detection system using Bayesian clustering and decision trees. Pattern Recogn. Lett. **29**(7), 918–924 (2008)

8. Amasyali, M.F., Demirhan, A., Bal, M.: Analysis of changes in market shares of commercial banks operating in Turkey using computational intelligence algorithms. Adv. Artif. Intell. **2014**, 1 (2014)

9. Gandhi, N., Armstrong, L.: Applying data mining techniques to predict yield of rice in humid subtropical climatic zone of India. In: Proceedings of the International Conference on Computing for Sustainable Global Development, pp. 1901–1906. IEEE (2016)

10. Goncalves, T., Quaresma, P.: A preliminary approach to the multilabel classification problem of Portuguese juridical documents. In: Proceedings of the International Conference on Artificial Intelligence, pp. 435–444, Springer, Heidelberg (2003)

11. Pestian, J., Nasrallah, H., Matykiewicz, P., Bennett, A., Leenaars, A.: Suicide note classification using natural language processing: a content analysis. Biomed. Inform. Insights **2010**(3), 19–28 (2010)

12. Thaseen, S., Kumar, C. A.: An analysis of supervised tree based classifiers for intrusion detection system. In: Proceedings of the International Conference on Pattern Recognition, Informatics and Mobile Engineering, pp. 294–299. IEEE (2013)

13. Wall, D.P., Kosmicki, J., Deluca, T.F., Harstad, E., Fusaro, V.A.: Use of machine learning to shorten observation-based screening and diagnosis of autism. Translational Psychiatry **2**(4), e100 (2012)

14. Zhao, Y., Zhang, Y.: Comparison of decision tree methods for finding active objects. Adv. Space Res. **41**(12), 1955–1959 (2008)

15. Jena, L., Kamila, N.K.: Distributed data mining classification algorithms for prediction of chronic kidney disease. Int. J. Emerg. Res. Manag. Technol. **4**(11), 110–118 (2015)

16. Akben, S.B.: Early stage of chronic kidney disease by using statistical evaluation of the previous measurement results. Biocybern. Biomed. Eng. **36**(4), 626–631 (2016)

17. Basar, M.D., Sari, P., Kilic, N., Akan, A.: Detection of chronic kidney disease by using Adaboost ensemble learning approach. In: Proceedings of the International Conference on Signal Processing and Communication Application, pp. 773–776. IEEE (2016)

18. Chen, Z., Zhang, Z., Zhu, R., Xiang, Y., Harrington, P.B.: Diagnosis of patients with chronic kidney disease by using two fuzzy classifiers. Chemometr. Intell. Lab. Syst. **153**, 140–145 (2016)

19. Quinlan, J.R.: Improved use of continuous attributes in C4. 5. J. Artif. Intell. Res. **4**, 77–90 (1996)

20. Salekin, A., Stankovic, J.: Detection of chronic kidney disease and selecting important predictive attributes. In: Proceedings of the International Conference on Healthcare Informatics, pp. 262–270. IEEE (2016)

21. Soundarapandian, P., Rubini, L.J., Eswaran, P.: UCI Machine Learning Repository. University of California, School of Information and Computer Science, Irvine, CA (2016). (http://archive.ics.uci.edu/ml/datasets/chronic_kidney_disease)

22. Holmes, G., Pfahringer, B., Kirkby, R., Frank, E., Hall, M.: Multiclass alternating decision trees. In: Proceedings of the European conference on machine learning, pp. 161–172). Springer, Heidelberg (2002)

23. Sugumaran, V., Muralidharan, V., Ramachandran, K.I.: Feature selection using decision tree and classification through proximal support vector machine for fault diagnostics of roller bearing. Mech. Syst. Signal Process. **21**(2), 930–942 (2007)

24. Komenda, P., Rigatto, C., Tangri, N.: Screening strategies for unrecognized CKD. Clin. J. Am. Soc. Nephrol. **11**(6), 925–927 (2016)
25. Polat, H., Mehr, H.D., Cetin, A.: Diagnosis of chronic kidney disease based on support vector machine by feature selection methods. J. Med. Syst. **41**(4), 55 (2017)
26. Serpen, A.A.: Diagnosis rule extraction from patient data for chronic kidney disease using machine learning. Int. J. Biomed.Clin. Eng. (IJBCE) **5**(2), 64–72 (2016)
27. Reddy, M., Cho, J.: Detecting chronic kidney disease using machine learning. In: Proceedings of the Qatar Foundation Annual Research Conference, vol. 2016(1), p. ICTSP1534. HBKU Press (2016)
28. Subasi, A., Alickovic, E., Kevric, J.: Diagnosis of chronic kidney disease by using random forest. In: Proceedings of the International Conference on Medical And Biological Engineering in Bosnia and Herzegovina, pp. 589–594. Springer, Heidelberg (2017)
29. Kuncheva, L.I.: How good are fuzzy if-then classifiers? IEEE Trans. Syst. Man Cybern. **30**(4), 501–509 (2000)

Backpropagation and Self-Organizing Map Neural Network Methods for Identifying Types of Eggplant Fruit

Siswanto, Wahyu Pramusinto and Gunawan Pria Utama

Abstract Identification of the image of eggplant fruit is a process aimed at obtaining objects contained in the image of eggplant fruit or dividing the image of eggplant fruit into several regions with each object or region that has similar aribut, such as shape, color, and size. The formulation of the problem in this study is how to get the objects contained in eggplant fruit and how to calculate the accuracy of the results of training on Backpropagation artificial neural networks and Self-Organizing Map (SOM) from an image of eggplants in all directions. The purpose of this study is to compare the accuracy of the process of identifying eggplant fruit types with the process of training Backpropagation and Self-Organizing Map (SOM) artificial neural networks based on size using the MATLAB application. After testing with the backpropagation method one eggplant image class can be classified with one eggplant, so the accuracy produced by the backpropagation network in the testing process is 18.18% and with the Self-Organizing Map (SOM) method, so the accuracy is generated by the network SOM in the testing process is 23.18%.

Keywords Image identification · Artificial neural network · Backpropagation · SOM · Eggplant fruit

Siswanto (✉) · W. Pramusinto · G. P. Utama
Universitas Budi Luhur, Jakarta 12260, Indonesia
e-mail: siswanto@budiluhur.ac.id

W. Pramusinto
e-mail: wahyu.pramusinto@budiluhur.ac.id

G. P. Utama
e-mail: gputama@gmail.com

© Springer Nature Singapore Pte Ltd. 2020
S. Agarwal et al. (eds.), *Machine Intelligence and Signal Processing*,
Advances in Intelligent Systems and Computing 1085,
https://doi.org/10.1007/978-981-15-1366-4_23

1 Introduction

1.1 Background

The progress of the world of information technology that is so fast supported by discovery and innovation has brought many changes in human life. The more things and aspects in life use information technology to run the wheel of activity.

The rapid development of increasingly sophisticated ICT has caused major changes to the habits of its users. These changes are in the use of devices that were initially only used for typing purposes, for programs, electronic transactions, and browsing, so now it is headed to the use that starts to support daily activities, both for work and entertainment. So, with the MATLAB source code, we can identify the type of eggplant fruit with the process of training Backpropagation artificial neural networks and Self-Organizing Map (SOM).

Among the many activities carried out by humans, identification of eggplant fruit is a process aimed at obtaining objects contained in the image of eggplant fruit or dividing the image of eggplant fruit into several regions with each object or area that has similar aribut, such as shape, color and size with the Backpropagation and Self-Organizing Map (SOM) artificial neural network training process.

Identification of the image of eggplant fruit is a process aimed at obtaining objects contained in the image of eggplant fruit or dividing the image of eggplant fruit into several regions with each object or region that has similar aribut, such as shape, color and size.

1.2 The Subject Matter of the Problem

The formulation of the problem in this study is how to get the objects contained in eggplant fruit *and* how to calculate the accuracy of the results of training on Backpropagation artificial neural networks and Self-Organizing Map (SOM) from an image of eggplants in all directions.

1.3 The Purpose of the Research

The purpose of this research was to compare the accuracy of the process of identifying eggplant fruit types with the process of training Backpropagation and Self-Organizing Map (SOM) artificial neural networks based on size using the MATLAB application.

The limitations of the problem in this study are as follows: types of purple eggplant, white eggplant, finger eggplant, green round eggplant, Green Leuncia, and Dutch eggplant, programming used with the MATLAB source code and the process of training artificial neural networks to identify the types of eggplant fruit with the Backpropagation and Self-Organizing Map (SOM) algorithm.

2   Related Works

2.1   Definition of Eggplant

Eggplant is a plant species in the nightshade family Solanaceae. *Solanum melongena* is grown worldwide for its edible fruit [1]. *Eggplants* or aubergines are flowering plants that belong to the nightshade family. Botanically, the eggplant is also considered a berry fruit while culinarily they are considered a vegetable. It originated in India and has been cultivated in both India and China for over 1500 years [2].

Most of us think of eggplant as a large, dark purple, and oblong-shaped vegetable that is always available. It is the perfect vegetable to be used for grilling, roasting, or baking with its slightly bitter taste that turns sweeter as it cooks. Well, it turns that there are a good amount of eggplant varieties other than the basic purple eggplant [3].

There are 13 different types of eggplant fruit, among others [2]:

(1)   African Garden Egg Eggplant
(2)   Bianca Eggplant
(3)   Chinese Round Mauve Eggplant
(4)   Graffiti Eggplant
(5)   Indian Eggplant
(6)   Italian Eggplant
(7)   Japanese White Egg Eggplant
(8)   Little Green Eggplant
(9)   Ping Tung Eggplant
(10) Santana Eggplant/Purple Eggplant
(11) Tango Eggplant/White Eggplant
(12) Thai Eggplant/Green Leuncia
(13) Dutch Eggplant.

2.2   Definition of Backpropagation

Backpropagation is a systematic method for multiplayer neural network training. This method has a strong, objective mathematical basis and this algorithm gets the form of equations and coefficient values in formulas by minimizing the number of squares of error errors through the developed model. This neural network consists

Fig. 1 Backpropagation
neural network architecture

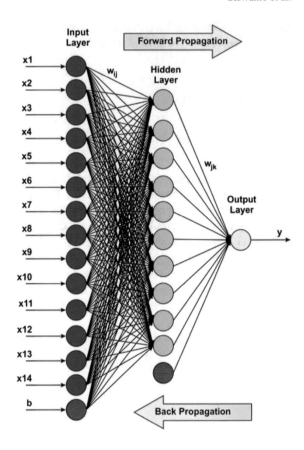

of three layers, namely the input layer, the hidden layer, and the output layer. The
backpropagation neural network architecture is shown in Fig. 1 [4].

Obviously, the error at the first epoch (with randomly selected weights/biases)
will be too large for a good function approximation; therefore, you need to reduce
this error to the acceptable (desired) value called the error limit, which you set at the
beginning of processing. Reducing the network error is done in the backward pass
(also called backpropagation) [5].

The backpropagation neural network algorithm is as follows:

Step 1:   If the condition is not reached, do steps 2–9.
Step 2:   For each training pair, do steps 3–8.

Forward propagation

Step 3:   Each input unit (x_i, $i = 1,\ldots, n$) receives the x_i signal and delivers this signal
             to all the upper layer units (hidden units).

Step 4:   Each unit is hidden (x_i, $i = 1,...,p$), add the weight of the input signal, by Formula (1).

$$z\_in_j = v_{oj} + x\Sigma = ni1_i v_{ij} \tag{1}$$

voj = bias on hidden units apply the activation function to calculate the output signal, $z_j = f(z\_in_j)$ and send this signal to all units in the upper layer (output unit).

Step 5: Each output unit (yea, $k = 1,...,m$) sums the weight of the input signal, by Formula (2).

$$y\_in_k = w_{ok} + z\Sigma = n_{j1j} w_{jk} \tag{2}$$

wok = bias on the output unit k and apply the activation function to calculate the output signal, $y = f(y\_in_k)$.

Backpropagation

Step 6: Each output unit (yea, $k = 1,...,m$) receives a target pattern that is interconnected at the input of the training pattern, calculate the misinformation, by Formula (3).

$$\delta_k = (t_k - y_k) fl (y\_in_k) \tag{3}$$

calculate the weight correction (used to update time later), by Formula (4).

$$\Delta w_{jk} = \alpha \delta_k z_j \tag{4}$$

calculate the correction bias (used to update the wok later) and send kek to the units in the layer below it,

Step 7: Each hidden layer unit ($z_j, j = 1,...,p$) sums up the changes in the input (from the layer units above), by Formula (5).

$$\Delta\_in_j = \Sigma\delta = m_{k1k} w_{jk} \tag{5}$$

multiply by derivating the activation function to calculate the error information, by Formula (6).

$$\delta_j = \delta\_in_j fl(z\_in_j) \tag{6}$$

calculate the weight correction (used to update v_{oj} later),

Step 8: Each output unit (yea, $k = 1...m$) updates the bias and its weight ($j = 0,...,$ p), by Formula (7).

$$w_{jk}(\text{new}) = w_{jk}(\text{old}) + \Delta w_{jk} \tag{7}$$

Each hidden layer unit ($z_j, j = 1,..., p$) updates the bias and its weight ($i = 0,...,$ n), by Formula (8).

$$v_{ij}(\text{new}) = v_{ij}(\text{old}) + \Delta v_{ij} (10) \tag{8}$$

Step 9: Test the stop condition.

2.3 Definition of Self-Organizing Map

Self-Organizing Map is proposed by Kohonen in 1982 which is a non-directed and a two-tier neural network. A brief description of the training process is as Fig. 2 [6].

Assume that the input layer consists of d-dimensional vector data dXR, output layer consists of K neurons, and the number of input neurons is the same with the dimensions of input data. First, to initialize the neural network which mainly determines the learning rate function, the value of initialized connection weight vectors, the size of neighbor domain, and the number of learning T.

At time t, input training sample $X(t)$ to the neural network, and use the competition algorithm about minimum Euclidean distance to select winning neuron c in the neural network, by Formula (9).

$$\|X(t) - w_c\| = \min_{j} \|X(t) - w_j\|, j = 1, 2, \ldots, k \tag{9}$$

Determine the winning neighbor domain $h_{cj}(t)$ and update all of the connection weight vectors of neighbor domain as Formula (10).

Fig. 2 The structure of some neural network

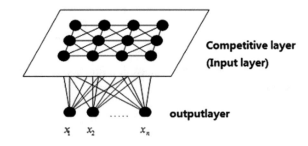

Competitive layer
(Input layer)

outputlayer

x_1 x_2 x_n

$$w_j(n+1) = w_j(n) + a(t)h_{c,j}(t)\big(X - w_j(n)\big) \tag{10}$$

Here, $a(t)$ shows the learning rate at time t, $h_{c,j}(t)$ shows the neighbor domain function of winning neuron c. This process continues until a certain condition or the value of connection weight vectors no longer change.

SOM represents data using nodes as points in the two-dimensional (or three-dimensional) vector space. These SOM nodes have weight vectors which are updated per iteration depending on the input vector from the data set [7, 8].

Generally, the weight vectors are updated as Formula (11).

$$\mathbf{w}_i(t+1) = \mathbf{w}_i(t) + G(t)\alpha_i(t)\|\mathbf{x}(t) - \mathbf{w}_i(t)\| \tag{11}$$

where t represents the iteration number, w_i represents the weight vector of the ith node, $x(t)$ is the input vector chosen randomly from the training set, $\alpha_i(t)$ is the learning rate of the adaptation process, $G(t)$ is a window function which is typically a Gaussian window or a rectangular window, and $\|x(t) - w_i(t)\|$ is the Euclidean distance between $x(t)$ and $w_i(t)$.

3 Research Methodology

The *research* methodology contains the steps used in this study to be well structured. With this, systematic research process can be understood and followed by other parties. Research conducted to design the system is obtained from observing existing data.

The stages *described* in Fig. 4 of the research framework are as follows.

3.1 Identify Problems

The initial stage in this research is to formulate a problem that will be used as the object of research. The formulation of the problem is done by first looking at the actual conditions in the field. After the problem is formulated, the next step is to determine the objectives of the research. The purpose of this study is the target who will want to be realized from the resolution of the problem under study.

3.2 Literature Review

The literature review was conducted to find supporting literature for this research. At this stage, it is explained by studying the applications related to conventional and digital artificial neural networks, theories, for collecting data and tools used by researchers. And, it explained about the method used.

3.3   Data Collection Method

The method used to collect data in this study is two ways, namely by observing the process of artificial neural networks and literature. Observation is used to obtain and collect the data needed. This observation is carried out on conventional neural network processes and artificial neural network processes that have been circulating.

3.3.1   Observation

Observation is done by collecting data and information needed in designing, developing systems, and the process of activities of the artificial neural network processes that are applied.

3.3.2   Study of Literature

While the literature study is used to collect data from previous research, learning from various kinds of literature and documents such as books, journals, and theories that support research, tools that will be used and other supporting data related to artificial neural network processes using the MATLAB application.

3.4   Needs Analysis

The analysis of the needs of the image segmentation process is carried out to determine the user's needs for the image segmentation process developed. This needs to be done so that the process is developed according to user needs. This section also explains who will use this image segmentation process, and what information is used by them.

The process of artificial neural networks can be seen in Fig. 3.

3.5   System Design

At this stage, the author makes the design of the application that will be proposed. The design includes designing the appearance of the user, designing the database for

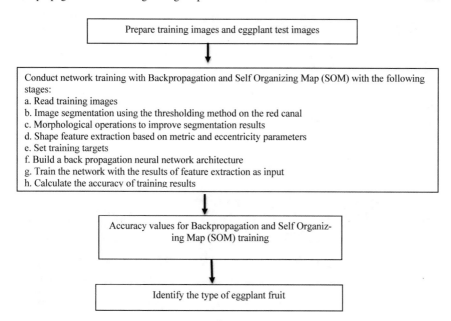

Fig. 3 The process of identifying eggplant fruit types

the application so that the existing file management is more organized, and then the last is designing the coding program of information.

3.6 System Implementation and Testing

Implementation and testing are a process to ensure that the application developed is free from errors, and testing is carried out on the application. At this stage, evaluation of the results of the research conducted will also be carried out. Evaluation includes evaluation of results and benefits of ways by comparing the results obtained with user needs.

3.7 Making Research Reports

The final step of this research is to make a research report. This report contains the things that were done during the research and the results obtained when conducting research.

4 Result and Discussion

4.1 Data Collection

The data needed in this study *are* images of several types of eggplants. And, each type of eggplant has been prepared as many as 100 images can be seen in Figs. 4 and 5.

Fig. 4 Image of a Dutch eggplant

Fig. 5 Image of purple eggplant

4.2   Testing with Backpropragation

In the first trial, MATLAB programming was made to classify the image of eggplant. Images of eggplants are grouped into 11 types. In this example, 33 eggplant images were used which consisted of three images in each type.

The testing steps with the backpropagation method are as follows:

1   Prepare training images and test images. In this example, 33 eggplant images are divided into two parts, namely 22 images for training images and 11 images for test images. The training image consists of 22 images consisting of two images in each class. The test image consists of 11 images consisting of one image in each class.
2   Conduct network training with stages, as follows:
   a.   Read training images.
   b.   Image segmentation using the thresholding method on the red canal.
   c.   Morphological operations to improve segmentation results.
   d.   Shape feature extraction based on metric and eccentricity parameters.
   e.   Setting training targets (type Leunca Green symbolized by number 1, type Leunca Purple with number 2, type of Green Eggplant with number 3, type of White Eggplant with number 4, type of Purple Eggplant with number 5, type of Eggplant Green Netherlands with number 6, the type of Dutch Red Eggplant with number 7, the type of Purple Dutch Eggplant with number 8, the type of Green Round Eggplant with number 9, the type of Purple Round Eggplant with number 10, and the type of Green Index Eggplant with number 11).
   f.   Build a backpropagation neural network architecture.
   g.   Train the network with the results of feature extraction as input.
   h.   Calculate the accuracy of training results.

Process of neural network training can be seen in Fig. 6.

Based on the table, the accuracy produced by the system in the training process is 22/22 * 100% = 100%.

4.3   Testing with SOM

In the second trial, MATLAB programming was made to classify eggplant images. Images of eggplants are grouped into 11 types. In this example, 33 eggplant images were used which consisted of three images in each type.

Testing Steps With the Self-Organizing Map (SOM) method, as follows:

1   Prepare training images and test images. In this example, 33 images of eggplant are divided into two parts, namely 22 images for training images and 11 images for test images. The training image consists of 22 images consisting of two images in each class. The test image consists of 11 images consisting of one image in each class.

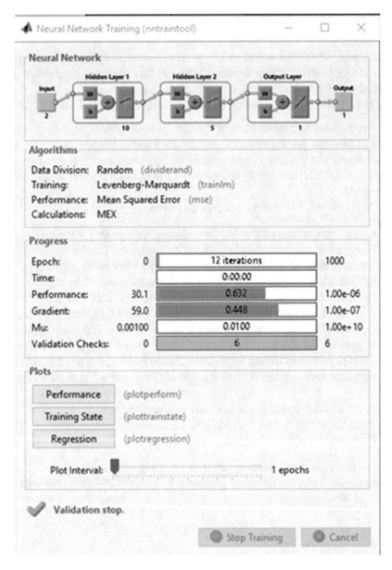

Fig. 6 Neural network training

2 Conduct network training with the following stages:

   a. Read training images.

   b. Image segmentation using the thresholding method on the red canal.

   c. Morphological operations to improve segmentation results.

   d. Shape feature extraction based on metric and eccentricity parameters.

   e. Setting training targets (type Leunca Green symbolized by number 1, type Leunca Purple with number 2, type of Green Eggplant with number 3, type of White Eggplant with number 4, type of Purple Eggplant with number 5, type of

Eggplant Green Netherlands with number 6, the type of Dutch Red Eggplant with number 7, the type of Purple Dutch Eggplant with number 8, the type of Green Round Eggplant with number 9, the type of Purple Round Eggplant with number 10, and the type of Green Index Eggplant with number 11).

   f. Build a Self-Organizing Map (SOM) artificial neural network.

   g. Train the network with the results of feature extraction as input.

   h. Calculate the accuracy of training results.

3 Perform network testing with stages, as follows:

   a. Read test images.

   b. Image segmentation using the thresholding method on the red canal.

   c. Morphological operations to improve segmentation results.

   d. Shape feature extraction based on metric and eccentricity parameters.

   e. Setting training targets (type of Leunca Green symbolized as number 1, type Leunca Purple with number 2, type of Green Eggplant with number 3, type of White Eggplant with number 4, type of Purple Eggplant with number 5, type of Green Eggplant with numbers 6, the type of Dutch Red Eggplant with number 7, the type of Purple Dutch Eggplant with number 8, the type of Green Round Eggplant with number 9, the type of Purple Round Eggplant with number 10, and the type of Green Index Eggplant with number 11).

   f. Call the Self-Organizing Map (SOM) artificial neural network that was created in the training process.

   g. Test the network with the results of feature extraction as input.

   h. Calculate the accuracy of the test results.

Based on the table, the accuracy produced by the system in the training process is $22/22 * 100\% = 100\%$.

In the table below there is one class of eggplant imagery that is wrongly classified so that the accuracy produced by the SOM network in the testing process is 23.18%. Accuracy generated by the system both in the training and testing process shows that the method used can classify the image of eggplant well (Table 23.1).

After testing with the backpropagation method, one eggplant image class can be classified with one eggplant, so the accuracy produced by the backpropagation network in the testing process is 18.18% and with the Self-Organizing Map (SOM) method, so the accuracy is generated by the network SOM in the testing process is 23.18%.

5 Conclusions

The conclusions that can be drawn from this study are:

1 This study uses a sample of 33 eggplant images divided into two parts, namely 22 images for training images and 11 images for test images. The training image consists of 22 images consisting of two images in each class.

2 This research uses test images totaling 11 images consisting of one image in each class.

Table 23.1 Table Testing Procees Results by SOM Network

Test Data

No	Species Class	
	Original	SOM
1	1	1
2	2	1
3	3	5
4	4	1
5	5	5
6	6	1
7	4	1
8	8	6
9	9	6
10	10	3
11	11	1

Information :
1 = Green Leuncia
2 = Leuncia Purple
3 = Green Eggplant
4 = White Eggplant
5 = Purple Eggplant
6 = Green Dutch Eggplant
7 = Red Dutch Eggplant
8 = Purple Dutch Eggplant
9 = Green Round Eggplant
10 = Purple Round Eggplant
11 = Green Index Eggplant

3   After testing with the backpropagation method, one eggplant image class can be classified with one eggplant, so that the accuracy produced by the backpropagation network in the testing process is 18.18% and with the Self-Organizing Map (SOM) method, so that accuracy is SOM network generated in the testing process is 23.18%.

References

1. Wikipedia: Eggplant. https://en.wikipedia.org/wiki/Eggplant, pp. 1–6, 6 July 2019. Last accesed 15 June 2019
2. Popoptiq: 13 different types of eggplant. https://www.popoptiq.com/types-of-eggplant/, pp. 1–6 (2019)
3. Geertsema, C.: The Ultimate Eggplant Varieties Guide. https://www.wideopeneats.com/guide-10-eggplant-varieties/, pp. 1–7, 2 July 2018
4. Mazur, A.: Step By Step Backpropagation Example. https://mattmazur.com/2015/03/17/a-step-by-step-backpropagation-example/, pp. 1–9, 17 March 2015
5. Livshin, I.: Artificial neural networks with java. In: Tools for Building Neural Network Applications, pp. 16–30. Apress, Chicago, IL (2019) ISBN-13 (pbk): 978-1-4842-4420-3 ISBN-13 (electronic): 978-1-4842-4421-0, https://doi.org/10.1007/978-1-4842-4421-0
6. Jianga, X., Liub, K., Yana, J., Chen, W.: Application of improved SOM neural network in anomaly detection. In: 2012 International Conference on Medical Physics and Biomedical Engineering, Physics Procedia , vol. 33, pp. 1093–1099. Elsevier Ltd. (2012).
7. Cordel II, M.O., Azcarraga, A.P.: Fast emulation of self-organizing maps for large datasets. In: 6th International Conference on Ambient Systems, Networks and Technologies (ANT 2015), Procedia Computer Science vol. 52, pp. 381–388 (2015).
8. Chaudhary, V., Bhatia, R.S., Ahlawat, A.K.: A novel Self-Organizing Map (SOM) learning algorithm with nearest and farthest neurons. Alexandria Eng. J. **53**, 827–831 (2014)

Head Pose Prediction While Tracking Lost in a Head-Mounted Display

Himanshu Rohilla and Suneeta Agarwal

Abstract Head-mounted displays are being used nowadays in various industries like defence, training pilots and many more. With these advancements, there is always a need to increase the area of tracking based on cameras or sensors. However, there will still be a limit on the number of cameras on the device for inside-out tracking and number of sensors for outside-in tracking. The proposed deep neural network is a cheaper and easy-to-adopt way for this pose estimation. The major problem of tracking loss occurs in outside-in tracking and is a common issue. In this paper, the proposed approach handles the tracker loss and efficiently predicts the position faster in less than 1 ms. The method exploits the users' behaviour and game environment to follow a particular trace when playing, users can relate it with their gaming patterns. The approach has been tested with NVIDIA's VRWorks SDK for real-time behaviour and it works very well within the timing constraints. Our network achieves an accuracy of nearly 96% with 30% data loss and is better than SHNM and BPNN discussed in [2].

Keywords Virtual reality · Neural network · Machine learning · Head-mounted display

1 Introduction

Virtual reality is an emerging field where advancements keep going on. From wired headsets to wireless headsets, users have seen it all. It catches the interest of a lot of people because of its complete 360° experience that matches what the user is experiencing in-person. Due to this synchronization with the real world, it has applications in a variety of fields like defence, movies, personal training and more.

H. Rohilla · S. Agarwal (✉)
Computer Science & Engineering Department, MNNIT, Allahabad, Prayagraj, India
e-mail: suneeta@mnnit.ac.in

H. Rohilla
e-mail: ruhela.himanshu@gmail.com

© Springer Nature Singapore Pte Ltd. 2020
S. Agarwal et al. (eds.), *Machine Intelligence and Signal Processing*,
Advances in Intelligent Systems and Computing 1085,
https://doi.org/10.1007/978-981-15-1366-4_24

1.1   What is Virtual Reality?

Virtual reality (VR) allows the user to experience simulated real-world experience, which is not similar to what they experience on TVs and hand-held devices. It generally covers three of the five senses that include hearing, touch, and vision using headphones, controllers and display on device, respectively. So, that is why the accuracy of simulation matters a lot for a smoother experience which is what this paper is trying to achieve.

1.2   What is an HMD?

VR headsets are often referred to as head-mounted displays (HMDs) shown in Fig. 1. This particular image is an HMD from Oculus, known as Oculus Rift. Oculus Rift is being used as an HMD device in the whole research process for this paper. Generally, an HMD has two displays, one for each eye giving a 3d experience. In addition to that, for outside-in tracking devices like Oculus, there is a tracker with the HMD that tracks the user's position, while orientation can be measured using gyroscope that is present in Oculus Rift HMD.

There are various VR devices available in the market Oculus Rift, HTC Vive/HTC Vive Pro, Microsoft HoloLens, Sony PlayStation VR, StarVR one, Lenovo Mirage, Varjo and many more. For this research, this paper is targeting wired headset, and Oculus Rift is one of the wired headsets. The VR headsets have a specified field of view (FOV) that determines how much area it can cover at any instant and Rift is

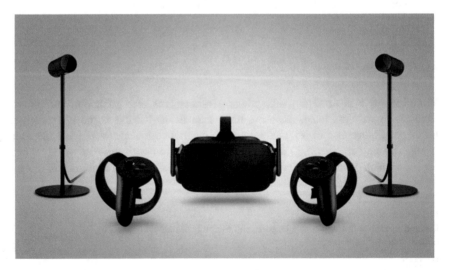

Fig. 1 Oculus Rift with controllers, and sensors

providing approx. 110° of FOV. With this, headsets come with sensors and there are two types of it, outside-in, and inside-out tracking. In this research, the approach is using outside-in tracker HMDs where the trackers are placed outside the HMD, and tracking loss problem originates because of hindrance between sensor and HMD. There should be proper synchronization between head motion in headset and external environment coordinates, otherwise, it results in nausea.

Generally, the sensors and the gyroscope acquire data that consists of six-degrees-of-freedom coordinates (6-DoF). These 6-DoF coordinates combine position and angular motion that is X, Y, and Z coordinates for position, and yaw angle, pitch angle, and roll angle for orientation. Due to various problems like delay, hindrance between the sensor and the device or any other malfunctioning the tracker data is corrupted and resulting in bad accuracy which causes the user to feel nausea because of this data mismatch between the external environment and HMD VR environment. This paper is covering position prediction while tracking is lost in this research and not orientation because the gyroscope is there in the device itself and there can't be any hindrance of this type with gyroscope.

1.3   How Does an HMD Works?

In tethered HMDs, sensors and HMD are connected to the system, Oculus Sensor tracks constellations of IR LEDs to translate user's movements in VR. The player's movements are controlled by the gyroscope in HMD and touch controllers. Now, before the position coordinates actually getting rendered onto the display, the approach logs them to make the dataset and accordingly simulate the tracker loss by modifying the position coordinates in the VRWorks Graphics multiprojection app.

Rest of the paper is organized as follows: Sect. 1 gives a brief introduction about the paper and its components. Sect. 2 provides an overview of the existing techniques for predicting positions and orientations. In Sect. 3, the functionality of the proposed approach and data collection approach is presented. Section 4 presents results and analysis, Sect. 5 concludes the paper, and Sect. 6 provides future work directions.

2   Related Work

Head-mounted displays are for quite some time in the market, they have a good hold and a lot of research is being conducted on how to smoothen the VR experience for the user. For example, manufacturers are shifting from wired headsets to wireless headsets like Oculus Go, things are shifting from outside-in to inside-out tracking and a lot of other stuff to remove the current drawbacks like tracking loss. But, with these, a lot of researchers are working in making the system fault proof. As Tobias, Christopher, and Michael in [1] focused their work in absolute head-to-body orientation by using orientation sensors and not the user's path. They took the assumption that the user will move in the direction in which they are looking or where their head orientation points to, but the suggested approach did not rely on in which direction user is viewing, instead, here the approach used its past behaviour to predict the future, which is a more robust technique and it does not take any assumption. Then Kataria in [2] assembles their setup to track user's position using infrared rays and predicts the missing piece using self-healing neural model (SHNM), but the results are better than them without requiring any extra setup. The model is predicting the whole position that contains x, y, and z coordinate, while they are focusing only on some missing data and not the complete position with all the coordinates in place.

3 Methodology

At the time of conducting the research, various permutation and combinations are tried with the network design, input/output format, and questions like how many consecutive frames can help us understand the user's trajectory to predict user's future positions to be taken? The trained network is used with the application also to see how it behaves in the real-world environment and this trial repeats and repeats until a final approach is achieved. Different sequences and trace length have been tested and also the number of positions to be predicted. After going through an extensive testing phase, the model and the input dataset form are finalized.

3.1 Equipment Used

Various types of equipments are used in the experiment such as Oculus Rift and its components are used as an HMD in the experiment, NVIDIA's TITAN V is used as GPU, CPU is i7-8650U @ 1.90 GHz, and a 1920 × 1080 display used to test the model on the application. Python 3.6 is used to code the network and Anaconda is used as Python distribution. NVIDIA's VRWorks Graphics SDK is used to test the trained model in the real-world scenario using the multiprojection app's Sponza scene. Some tweaks are also made in the app to display some information regarding pose inference and to turn it on/off.

3.2 Data Collection

NVIDIA's VRWorks Graphics SDK's multiprojection app is used to collect data. A total of 50,000 positions are recorded during a continuous app usage for 9–10 min using Oculus Rift, considering an average of 90 frames per second (fps) while recording. These 50,000 positions from frames are saved in a data file, which will later be transformed accordingly to use as input data to train the network. All possible

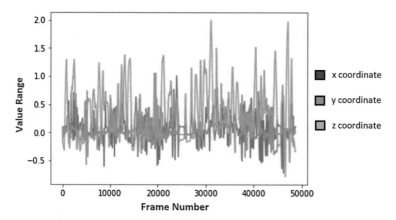

Fig. 2 Graph representing x, y, and z coordinate of collected data

sequences are covered in the scene specified. A position contains x, y, and z coordinates in the space, with (0,0,0) as the centre point of reference in the scene. For training, 80% of the input data is used, and rest 20% data is used for testing out of all acquired data (Fig. 2).

3.3 Head Pose Prediction—The Proposed Approach

Head pose prediction is implemented to remove the glitches at the time of tracking the user's position through sensors. There is a stuttering effect when tracking gets lost, a whole bunch of frames gets lost which results in a very unstable experience and it might result in user to fall because at once the frame suddenly changes and the user might not be expecting it. The whole idea behind the approach is how the user behaves while using the device or while playing any game, for instance, if a person is walking in the game, then there are n number of possibilities in the virtual world to move if only the current frame is considered. However, if one thinks of per game training, then there are repeatable traces in the gameplay (Table 1).

However, predicting the user's next position accurately, just from the current single position is a bit tricky and practically an impossible thing to achieve because there are n number of next positions the user can take, but predicting the whole trace using some x number of previous positions PoseNet can effectively predict the next certain number of positions. Now, either the model can predict just the next position, or can predict some y number of next positions that continues the trace. So, after a lot of experiments and verifying the predicted positions in real time in a VR environment, x is fixed to 10 and y is fixed to 4. Additionally, an experiment with x as 10 input position and y as 1 output position is conducted, but results are better with 10:4 than 10:1 in real time. Remember, that the decision is totally based on the real-time experience and results (Fig. 3).

Table 1 Sample data set containing random positions

X	Y	Z
0.0306912	−0.142518	0.0404296
0.0841651	−0.106357	0.0851185
0.0231342	−0.0195761	−0.172449
0.0524441	−0.00637811	−0.190101
0.101965	0.0312687	−0.265246
−0.0683607	0.0151937	−0.112803
−0.037373	0.0229639	−0.0505158

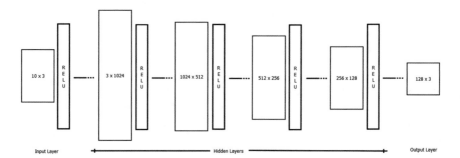

Fig. 3 Illustration of model

This paper proposes a feed forward network that contains 4 hidden layers with $1024, 512, 256,$ and 128 number of neurons in hidden layer 1, 2, 3, and 4, respectively. After each layer, ReLU activation layer is used except the last layer. The input layer consists of 10 input positions with 3 values each i.e. the x, y, and z coordinates of each position. For each predicted position, 10 position values are to be given as input, and further position values will be predicted as output.

3.4 Training

The approach explained above is trained on 80% of available data, with 10 positions as input and 4 expected positions. So, for input 30 values are there and expected value will contain 12 float values representing x, y, and z coordinate of predicted 11th, 12th, 13th, and 14th position. For ease, the coordinate domain set is reduced to $[−1, 1]$ using MinMaxScaler.

For training, TITAN V GPU is used, and at a time batch size of 256 is trained with epoch as 40. While training, permutation is also done on the batch to avoid over-fitting the model with the current dataset. The learning rate is chosen as 0.001, and the cost function is mean squared error (MSE). Adam optimizer is used for the optimization because of batch training, and the aim is minimizing the mean squared error between the actual position and the predicted position using Adam optimizer, TensorFlow is used for the back end.

After training, the learned parameters are saved in a separate data file that will be used in the C++ environment to actually predict positions in a real-time scenario in the VRWorks Graphics SDK's multiprojection app. The approach has reduced the losses to 10^{-5} for mean squared error (MSE), which before training was 0.1735. So, predicted positions match the overall pattern in real-world data for the whole test data.

4 Analysis and Results

When it is about to test how the model works in terms of accuracy and real-world environment, there are a lot of things to prove to make it worthy for use in devices, and for that, extensive tests are conducted on 20% of the total dataset that makes up something approx. 10,000 positions. Using this, it is also seen that how long the future positions can be predicted with accuracy, if suppose the tracking is lost for a considerable amount of time. After training, that model is tested on testing data, and the results are good, predicting the position well with good accuracy. Various permutation and combinations are tested on the input and output data and finalized one that gives the best result in the least time out of them.

Broadly, the tests are categorized into two main subcategories, one that predicts only one position after giving some x number of positions in input, and another one that predicts some y number of positions after providing some x number of positions in the input.

4.1 Single Position Prediction

In this experiment, the approach took a 10:1 ratio for input versus output. The model is trained with 10 consecutive input positions and 1 position that will be the 11th position in that sequence will the output (or label). The network is trained using approx. 40,000 positions after making sequences from them. After training, the model is tested over the test data and loss (MSE) is reduced from 0.25866517 to 6.017256e-05.

Figure 4 represents the predicted versus actual values for the whole test data. In the ideal case, the plotted line should be straight diagonal with slope = 45°. The results can be clearly visualized that the results are approaching towards the ideal scenario. The results shown expects 10 input position as 10 consecutive actual positions from Oculus sensors and then 1 future position will be predicted from the trained model.

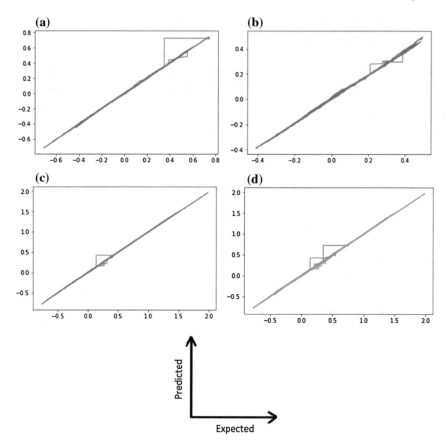

Fig. 4 Graphs representing predicted V/S actual values of **a** x coordinate, **b** y coordinate, **c** z coordinate, and **d** x, y, and z coordinates combined

4.2 Multiple Position Prediction

In this experiment, the approach took a 10:4 ratio for input and output. The model is trained with 10 consecutive input positions and 4 consecutive positions that will be the 11th to 14th position in that sequence will the output (or label). After training, the model is tested the model with testing data and loss (MSE) is reduced from 0.17872562 to 1.6482448e-04.

Figure 5 represents the predicted and actual values of all the test data, where the four graphs are 4 respective predicted positions. All comparisons are made with actual values of position coordinates. Here too, it is expected that previous 10 positions are actual positions from Oculus sensors for the results shown.

PoseNet achieved better accuracy than [2] with both 10 and 28% loss as stated in [2]. Table 2 lists the comparison results well and as it seems, PoseNet achieves better results in 10% and with 25% too.

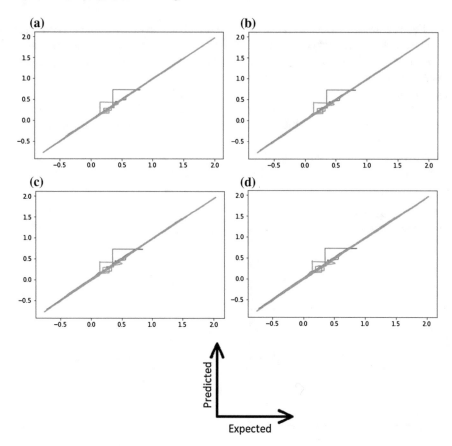

Fig. 5 Graphs representing predicted V/S actual values of **a** 1st position, **b** 2nd position, **c** 3rd position, and **d** 4th position

Table 2 PoseNet comparison with SHNM and BPNN [2]

Model	BPNN		SHNM		PoseNet	
Loss	**10%**	**25%**	**10%**	**25%**	**10%**	**25–28%**
X (%)	91.05	90.20	96.14	95.12	**96.06**	**95.38**
Y (%)	87.70	86.57	92.50	91.34	**95.45**	**95.45**
Z (%)	87.30	86.10	92.80	91.30	**95.66**	**95.48**

Another experiment was conducted that estimates how long the predictions can be done at a time if sensor tracking is lost for a considerable amount of time. In this experiment, a series of positions are predicted such as predict 11th position from previous 10 positions, then 12th using position 2–11 (where 11th is previously predicted position) and so on till 6 positions.

Figure 6 showing the difference between actual and predicted value v/s predicted frame number and as expected as the frames are getting predicted, the error increases. To further add to the results, the network was also tested on the whole test data cumulatively to see how the overall prediction looks like when it comes to global predictions.

As shown in Fig. 7, graph represents the x, y, and z coordinate prediction over the whole test data. There are minor misses in the overlays shown where the predicted is not overlapping with the actual values. The average distance between actual (orange) and predicted graph (blue) is 0.0028644, 0.003031, 0.003811 for Fig. 7a, b, and c calculated over all the 10000 positions in test data. The graphs in Fig. 7 are plotted

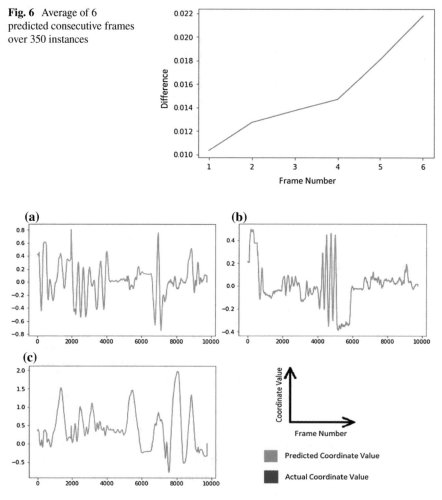

Fig. 6 Average of 6 predicted consecutive frames over 350 instances

Fig. 7 Graphs showing predicted versus actual over test data for **a** X coordinate, **b** Y coordinate, and **c** Z coordinate

Fig. 8 A curve of degree 10 fitting the test data using extrapolation

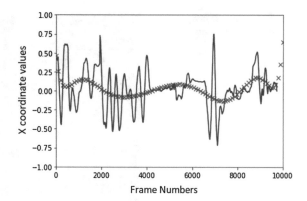

by predicting positions and plotting it against actual data. The approach explained in Sect. 4.2 (multiple position prediction) is used for predictions where 10 positions are taken as input and 4 positions are predicted as output.

Extrapolation is also tried on our data and how its curve will fit the data. So, a curve of degree 10 is plotted along the X coordinate position data and results are shown in Fig. 8. The red curve is the extrapolated curve over the test data, and the blue one represents the x coordinate of data.

So, extrapolation cannot solve the problem as expected because in the gameplay if the character is moving, it is not necessary that he follows a certain path following predefined positions, so extrapolation here fails. However, deep learning handles the job very well as can be seen in Fig. 7.

4.3 Inference

To check the results in real-time app usage, some changes were made in VRWorks SDK multiprojection app, where the model's parameters are loaded and transferred to GPU memory using CUDA APIs, then, CUDA environment is created in the app using CUDA wrappers implemented in C++. After loading, a simulation of tracker loss is created that after 10 frames, 4 predicted frames position will be displayed instead of the tracking positions. The time taken to predict one frame is ~0.2–0.5 ms on TITAN V GPU and that does not affect the performance because considering 90 fps, so 1 frame hardly have 10–12 ms to render, and this prediction is not at all affecting the performance in real time. Some UI tweaks were also made in the app itself to dynamically control the predictions. Timing was a major issue to make it effective in real-life utility and our network and inference handles it well.

In Fig. 9, ENABLE/DISABLE Loss is used to enable/disable predictions for the purpose of simulation for tracking loss. Inc L and Dec L are for increase/decrease, loss of Nth frame means that after N frames, 1 frame (Burst N frame Loss) is getting

Fig. 9 UI changes in Sponza
Scene to toggle features

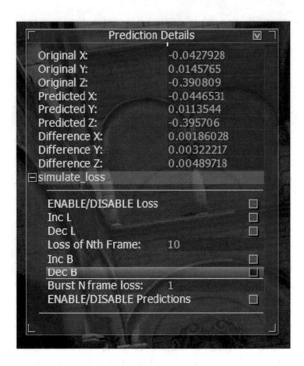

lost. Inc B and Dec B are used to increase or decrease Burst frame loss after loss of Nth frame. All the other labels are self-explanatory.

5 Conclusion

The proposed approach PoseNet (head pose prediction) gives promising results for predicting user's future positions when tracking is lost. The experience while wearing HMD is nausea tic in case tracking gets lost. With this deep learning model, the experience emerges out to be good. The model reduces errors based on game experience with nearly 96% accuracy within 30% data loss. Also, it is important to mention here that there is no need for any external hardware.

6 Future Work

Head-mounted displays come with their own gyroscope these days to track the orientation of the devices using earth's gravity. For now, this paper is covering only predicting the positions and not orientation. So, the next step might be to predict orientations as well to take care of 6-DoF coordinates that is position and orientation. Also, right now the approach is suggesting per game training, but one can also think of building a deeper network with a loss function having more penalties. In this paper, PoseNet is achieving nearly 96% accuracy with per frame position inference in <1 ms.

Secondly, TensorFlow has been used as a back end. However, PoseNet achieved an accuracy of 96% and an inference time of <1 ms, but one can work on other frameworks such as PyTorch to further improve the inference time and results. An important use case for this approach is, while streaming the pose data over the network and some frames get lost due to network congestion.

References

1. Feigl, T., Mutschler, C., Philippsen, M.: Head-to-body-pose classification in no-pose VR tracking systems. In: 2018 IEEE Conference on Virtual Reality and 3D User Interfaces (VR), Reutlingen, pp. 1–2 (2018) https://doi.org/10.1109/vr.2018.8446495
2. Kataria, A., Ghosh, S., Karar, V.: Data prediction of optical head tracking using self healing neural model for head mounted display (2018)

Recommendation to Group of Users Using the Relevance Concept

Vivek Kumar⦿, Saumya Jain⦿ and Vibhor Kant⦿

Abstract Group recommender systems (GRSs) have played an important role in numerous online applications by providing recommendation to the group of users where satisfaction of the entire group is a major concern. In traditional GRSs, the relevance of all the groups and the items is considered equal which does not produce accurate recommendations. In this paper, we propose a formalization of the GRS based on the relevance concept using profile merging scheme where collaborative filtering (CF) is applied on each group profile to generate effective recommendations to the group by considering the ratings of the items, the relevance of the groups and the relevance of the items. Further, our GRS framework provides relevant similarity measures, relevant prediction and recommendation quality measures. The experimental results on the benchmark MovieLens dataset demonstrate the efficacy of our proposed GRS framework.

Keywords Collaborative filtering · Group recommender systems · Quality measures · Relevance

1 Introduction

The online availability of enormous amount of information makes it strenuous for the users to locate the right information which they require. Recommender system (RS) is the technique used to address this information overload problem by analyzing the available information and modelling user preferences [14]. Generally, traditional recommendations are made to a single user but there are innumerable situations where recommendation is required to be given to a group of users such as recommending

V. Kumar (✉) · S. Jain · V. Kant
LNM Institute of Information Technology, Jaipur 302031, India
e-mail: vivekkumar9993@gmail.com

S. Jain
e-mail: saumya.jain507@gmail.com

V. Kant
e-mail: vibhor.kant@gmail.com

© Springer Nature Singapore Pte Ltd. 2020
S. Agarwal et al. (eds.), *Machine Intelligence and Signal Processing*,
Advances in Intelligent Systems and Computing 1085,
https://doi.org/10.1007/978-981-15-1366-4_25

movies to a group of college students or family. Hence, group-based perspective provides a new angle on the traditional recommendation approaches [8]. GRSs have been proposed to provide recommendations to the group of users which must satisfy the whole group [10].

Recommendations to the group of users are generated using mainly two schemes, i.e. by profile merging [3, 4, 10, 11] or by merging recommendations [3, 4, 10, 11]. Among these, the former approach gives better results while considering numerous aspects of the group [10]. GRSs use various filtering algorithms like content-based filtering [2, 6], CF [2, 9], demographic filtering [6], hybrid filtering [7], social filtering [6] and many more for the purpose of recommendations out of which CF is the most prevalent choice [2, 11].

In GRSs, the satisfaction of the entire group is of utmost importance, i.e. each user of the group must be satisfied with the given recommendations. In profile merging-based GRSs, the CF considers all the items and the groups to have equal relevance. This assumption is thought to be debatable as the high-budgeted blockbuster movie would be considered highly relevant in comparison to the little-known movie. Although there might be some groups, who would prefer the little-known movie over the blockbuster movie, most of the groups believe that recommendation error is much more relevant when it is related to a relevant item, i.e. the RSs which make mistakes on the recommendation of relevant items are mistrusted more by the groups in comparison to the mistakes made on the recommendation of non-relevant items.

In this paper, we have incorporated the relevance concept in the GRSs using profile merging scheme. The relevance of the groups and the items assisted in finding the best neighbours of the group which improved the prediction and the recommendation results thus improving the group satisfaction. We have also proposed quality measures for GRSs using the relevance of the groups and the items. While calculating the relevance of the items, we have considered only the ratings given by each group on each item which were further incorporated into traditional similarity and quality measures. While calculating the relevance of groups, we have considered the following two cases:

- Firstly, the groups who have rated numerous items are considered more relevant over the groups who have rated lesser number of items.
- Secondly, the groups who have rated the items in an equitable manner are considered more relevant over the groups who give maximum ratings to most of the items.

The organization of the rest of the paper is as follows: In Sect. 2, background of the RS and some related work is given. In Sect. 3, the proposed approach based on the relevance concept in the GRSs is presented. In Sect. 4, the experimental results are shown. Finally, in Sect. 5, the work is concluded.

2 Literature Survey

The rampant growth of information over the Web demands for an intelligent information system that can provide suggestions about which information is pertinent to users. These systems should include automated Web personalization tools that intelligently transform huge amounts of data and deliver meaningful information to users [1]. Most online shopping sites and other major applications now use RSs which try to assist users in everyday life by reducing the information overload problem, providing personalized recommendations and services to them. RSs generate personalized recommendations to users based on various types of filtering techniques such as content-based [6], collaborative [2] and hybrid [7] filtering. Among these, CF is more prevalent one.

The traditional recommendations are made to a single user but there are numerous situations where recommendation is required to be given to a group of users such as recommending movies to a group of friends or family [3, 10]. GRSs are the ones that generate recommendations for the group of users rather than for individuals about the products that satisfy the whole group. The methods for the recommendation to the group of users are classified into two broad categories [14]:

1. Merging recommendations: In this method, recommendation lists of the users are merged and sorted using some merging algorithm.
2. Profile merging: In this method, a group profile which can be observed as a virtual user in CF is created that represents the entire members of the group and then the individual recommendations can be made to this virtual user.

The merging in GRSs can be visualized at various levels as illustrated in Fig. 1. The merging strategies for the group recommendation on the basis of social choice theory can be classified into four major categories [8, 14]:

1. Consensus-based: fairness strategy and average strategy

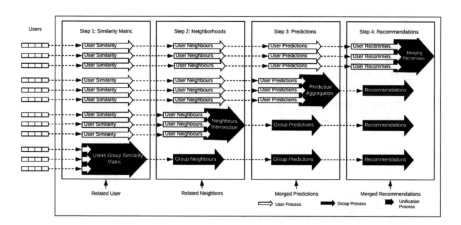

Fig. 1 Different approaches for making recommendations to the group of users [11]

2. Majority-based: plurality voting strategy
3. Borderline: least misery strategy and most pleasure strategy
4. Dictatorship: most respected person.

GRSs have been successfully applied in numerous domains like music (group-fun [13]), movies (MovieLens [11]), news (G.A.I.N [12]) and many more. However, the satisfaction level of the entire group is a major concern. In GRSs, groups and items are considered to have equal relevance while calculating neighbours in CF which does not produce proper recommendations thus affecting the group satisfaction. Therefore, to enhance the satisfaction level of the groups, we have used the relevance concept where the relevance of the groups and the items is calculated using only the ratings given by each group to each item. We have adopted the idea from [5] and the following section shows our proposed method.

3 Proposed GRS Framework

Our proposed GRS framework consists of the following four phases.

3.1 PHASE 1: Formation of Group Profile

Firstly, the groups are formed using the group profile merging scheme. We have considered s users, q groups and t items in our proposed GRS. We have defined the following sets:

$W = \{w \in \text{Natural Numbers} \mid 1 \leq w \leq s\}$, users' set

$L = \{l \in \text{Natural Numbers} \mid 1 \leq l \leq t\}$, items' set

$U = \{u \in \text{Natural Numbers} \mid r_{min} \leq u \leq r_{max}\} \cup \{\bullet\}$, set of possible ratings

Here, r_{min} and r_{max} represents the minimum and the maximum rating that can be given to the item.

$R_w = \{(l, x) \mid l \in L, x \in U\}$, ratings given by user w

$R_g = \{(l, x) \mid l \in L, x \in U\}$, ratings given by group g

$G = \{g \in \text{Natural Numbers} \mid 1 \leq g \leq q\}$, set of groups

$G_{g,l} = \{w \in g \mid r_{w,l} \neq \bullet\}$, where $g \in G$

When the user or the group has not rated the item, $\bullet$ symbol is used.

Let $r_{g,l}$ be the rating of the item l given by the group g:

$$r_{g,l} = \sqrt[|G_{g,l}|]{\Pi_{w \in G_{g,l}} r_{w,l}} \tag{1}$$

Here, $r_{w,l}$ is the rating of the item l given by the user w and $|I|$ represents the cardinality of the set I.

3.2 PHASE 2: Computation of Similar Groups Using the Relevance Concept

After the groups are formed, similarity between the two groups is calculated. We have computed the similarity between the groups by incorporating relevance of the items, relevance of the groups and relevance of the items to the groups which are defined as:

Relevance of the item l

Let $A_l = \{g \in G | r_{g,l} \neq \bullet\}$ be the set of groups who have rated the item l. Then the relevance of the item l (v_l) is defined by the following equation:

$$v_l = \left(\frac{1}{|A_l|} \sum_{g \in A_l} r_{g,l}\right)\left(\frac{|A_l|}{|G|}\right) = \frac{\sum_{g \in A_l} r_{g,l}}{|G|}, r_{g,l} \in [0, 1], v_l \in [0, 1] \quad (2)$$

Relevance of the group g

Let $D_g = \{l \in L | r_{g,l} \in J\}$ be the items' set that have been rated with significant ratings by the group g, and $E_g = \{l \in L | r_{g,l} \in J^c\}$ be the items' set that have been rated with non-significant ratings by the group g. Here, $J = \{r_\alpha, \ldots, r_{max}\}$, $J \subseteq U$ which are considered as significant ratings and $J^c = U - J - \{\bullet\} = \{r_{min}, \ldots, r_\alpha - 1\}$, $J \subseteq U$ which are considered as non-significant ratings.

Then, the relevance of the group g (v_g) is defined as:

$$v_g = \left(1 - \frac{|D_g|}{|D_g| + |E_g|}\right)\left(\frac{|D_g| + |E_g|}{|L|}\right), v_g \in [0, 1] \quad (3)$$

Relevance of the item l for the group g

There are three possible cases in which the relevance of an item l for a group g ($v_{g,l}$) can be defined:

Case (1)

If the group g has rated an item l then we have merged the rating which the group g has given to the item l with the relevance of the group g and the relevance of the item l.

$$v_{g,l} = r_{g,l} v_g v_l \Leftrightarrow r_{g,l} \neq \bullet \wedge v_g \neq \bullet \wedge v_l \neq \bullet, r_{g,l} \in [0, 1], v_{g,l} \in [0, 1] \quad (4)$$

Case (2)

If the group g has not rated an item l, then we have tried to predict the rating by analyzing the ratings which the group g has given over the items similar to the item l.

Let S_l^e be the set of e items that are similar to the item l such that

$$\left|S_l^e\right| = e, \ \forall x \in S_l^e, \ \forall y \notin S_l^e \text{ and } sim(x, l) \geq sim(y, l)$$

Here, similarity between the two items j and m is denoted by $sim(j, m)$.

$$v_{g,l} = v_g v_l \frac{1}{\beta} \sum_{x \in F_g} v_x r_{g,x} sim(x, l) \iff \tag{5}$$

$$r_{g,l} = \bullet \wedge v_g \neq \bullet \wedge v_l \neq \bullet \wedge F_g \neq \phi, \ r_{g,x} \in [0, 1], \ v_{g,l} \in [0, 1],$$

where $\beta = \sum_{x \in F_g} sim(x, l)$ and $F_g = \left\{x \in S_l^e \wedge r_{g,x} \neq \bullet\right\}$.

Here, F_g is the subset of the items of S_l^e that have been rated by the group g.

Case (3)

If the group g has not rated an item l and there are no items similar to the item l which has been rated by the group g, then the prediction is not possible and we have considered that no relevance is assigned to the item l for the group g.

$$v_{g,l} = \bullet \iff r_{g,l} = \bullet \wedge F_g = \phi$$

Similarity measures

We have defined the similarity measures using the various relevance measures defined above. The items on which the relevance of groups j and m can be defined are used to describe the v-similarity measures between the groups j and m, i.e. $B_{j,m} = \{l \in L | v_{j,l} \neq \bullet \wedge v_{m,l} \neq \bullet\}$.

Pearson$_v$ and *Cosine$_v$* v-similarity measures are defined by the following equations:

$$Pearson_v(j, m) = \frac{\sum_{l \in B_{j,m}} \left(v_{j,l} - \bar{v}_j\right)\left(v_{m,l} - \bar{v}_m\right)}{\sqrt{\sum_{l \in B_{j,m}} \left(v_{j,l} - \bar{v}_j\right)^2 \sum_{l \in B_{j,m}} \left(v_{m,l} - \bar{v}_m\right)^2}} \tag{7}$$

where

$$T = \left\{x \in L | v_{e,x} \neq \bullet\right\}, \ \bar{v}_e = \frac{1}{|T|} \sum_{l \in T} v_{e,l}$$

$$Cosine_v(j, m) = \frac{\sum_{l \in B_{j,m}} v_{j,l} v_{m,l}}{\sqrt{\sum_{l \in B_{j,m}} v_{j,l}^2 \sum_{l \in B_{j,m}} v_{j,l}^2}} \tag{3}$$

3.3 PHASE 3: Prediction and Recommendation to Group of Users

Prediction of the rating that would be given by the group g to the item l is done using the ratings of similar groups of the group g on the item l. After the prediction of the ratings is completed, the top N items with the highest predicted values are recommended to the group g. For the prediction and the recommendation process, we have found h-neighbours of the group g using v-similarity measures. If there exists at least h-neighbours, then:

$$K_g \subset G \wedge |K_g| = h \wedge g \notin K_g \forall x \in K_g,$$
$$\forall y \in (G - K_g), \ \text{sim}_v(g, x) \geq \text{sim}_v(g, y)$$

where K_g is the set of h-neighbours of the group g and $\text{sim}_v(j, m)$ is the similarity between the two groups j and m using v-similarity measures.

Then, adjusted weighted aggregation function is used to calculate the predicted rating $p_{g,l}$ for the item l that would be made by the group g.

$$p_{g,l} = \bar{r}_g + \mu_{g,l} \sum_{n \in G_{g,l}^h} \text{sim}_v(g, n)(r_{n,l} - \bar{r}_n) \Leftrightarrow G_{g,l}^h \neq \phi \tag{4}$$

where

$$T = \{x \in L | r_{e,x} \neq \bullet\}, \ \bar{r}_e = \frac{1}{|T|} \sum_{l \in T} r_{e,l}$$

$G_{g,l}^h = \{n \in K_g | \exists r_{n,l} \neq \bullet\}$ is the set of h-neighbours who have rated the item l and $\mu_{g,l} = 1 / \sum_{n \in G_{g,l}^h} \text{sim}_v(g, n) \Leftrightarrow G_{g,l}^h \neq \phi$ is the normalizing factor.

If there is no neighbour of the group g who has rated the item l then $p_{g,l}$ cannot be calculated $i.e. p_{g,l} = \bullet \Leftrightarrow G_{g,l}^h = \phi$.

Then, the top N recommendations for the group g are calculated. Let X_g be the set which contains the items that have not been rated by the group g and for which the ratings that the group g would make can be predicted and let Z_g be the items' set in X_g that have highest predicted rating values. Then,

$$X_g \subset L \wedge \forall l \in X_g, r_{g,l} = \bullet, p_{g,l} \neq \bullet$$
$$Z_g \subseteq X_g \ |Z_g| = N, \ \forall x \in Z_g, \ \forall y \in (X_g - Z_g) : \ p_{g,x} \geq p_{g,y}$$

where we assume that there are at least N recommendations for the items.

3.4 PHASE 4: Proposed Quality Measures for the GRS

After the recommendations are given to the group of users, the quality of the predictions and the recommendations with relevance weighting is calculated using various proposed quality measures:

Prediction quality measures
MAE$_v$
MAE_v is the measure of how efficiently a GRS can make predictions. Let Q be the set of pairs,

$$Q = \left\{ <g, l> | r_{g,l} \neq \bullet \wedge p_{g,l} \neq \bullet \wedge v_l \neq \bullet \wedge v_g \neq \bullet \right\}$$

Then, MAE_v is defined as:

$$MAE_v = \frac{1}{\sum_{(g,l)S \in Q} v_g v_l} \sum_{g \in G} \sum_{l \in L} |r_{g,l} - p_{g,l}| v_g v_l | \langle g, l \rangle \in Q \tag{5}$$

Here, the prediction errors are weighted according to the relevance of the items.

Recommendation quality measures
Recommendation quality measures like *Precision$_v$* and *Recall$_v$* are used to measure the quality of recommendations given by the GRS. An item l is considered recommendable for the group g if $p_{g,l} \geq \alpha$ where α is the threshold constant. Let Z_g be the items' set that are recommended to the group g. Now we will define the following sets:

- True-positives $\left(Y_g\right)$—They are the items' set that are actually recommendable for the group g $\left(r_{g,l} \geq \alpha\right)$ and which are recommended to the group g by the GRS $\left(l \in Z_g\right)$. We denote this by:

$$Y_g = \left\{ l \in Z_g | r_{g,l} \geq \alpha \right\}$$

- False-positives $\left(N_g\right)$—They are the items' set that are not actually recommendable for the group g $\left(r_{g,l} < \alpha\right)$ and which are recommended to the group g by the GRS $\left(l \in Z_g\right)$. We denote this by:

$$N_g = \{ l \in Z_g | r_{g,l} < \alpha \}$$

- False-negatives $\left(O_g\right)$—They are the items' set that are actually recommendable for the group g $\left(r_{g,l} \geq \alpha\right)$ and which are not recommended to the group g by the GRS $\left(l \in Z_g^c\right)$. We denote this by:

$$O_g = \left\{ l \in Z_g^c | r_{g,l} \geq \alpha \right\}$$

Precision$_v$

Pearson$_v$ is defined as the percentage of the results that are relevant. The *Pearson$_v$* for the group g is defined as:

$$k_g = \frac{\sum_{l \in Y_g} v_{g,l}}{\sum_{l \in Y_g} v_{g,l} + \sum_{l \in N_g} v_{g,l}}$$

The *Pearson$_v$* for the whole GRS is defined as:

$$k = \frac{1}{|G|} \sum_{g \in G} k_g \tag{6}$$

Recall$_v$

Recall$_v$ is defined as the percentage of total relevant results that are analyzed correctly by the GRS. The *Recall$_v$* for the group g is defined as:

$$x_g = \frac{\sum_{l \in Y_g} v_{g,l}}{\sum_{l \in Y_g} v_{g,l} + \sum_{l \in O_g} v_{g,l}} \tag{7}$$

The *Recall$_v$* for the whole GRS is defined as:

$$x = \frac{1}{|G|} \sum_{g \in G} x_g \tag{8}$$

4 Experiments and Result Analysis

4.1 Experimental Setup

We have performed several experiments to demonstrate the efficacy of our proposed approach based on the relevance concept in GRSs over the traditional approaches used without the relevancy concept. There is no publically available dataset for GRSs. Therefore, various researchers have used 1 M MovieLens dataset by forming groups of varying sizes in a random manner. We have followed the same setup for our experiments. The detailed description of the dataset is available in [5]. In this way, we got a total of 1022 group profiles. Further, we have applied a threefold cross-validation approach for analyzing its behaviour in a better way. In each fold, we have taken 34% of the items which are selected randomly for the testing data and remaining as the training data. We have performed three experiments to show relative performance of the following approaches:

- To fix the value of h.
- To show performance comparison of traditional similarity measures [2] (*Cosine, Pearson, Correlation*) and proposed similarity measures (*Cosine$_v$, Pearson$_v$*) using proposed prediction quality measures (*MAE$_v$*) and recommendation quality measures (*Pearson$_v$, Recall$_v$*) at fixed value of h and N.
- To show performance comparison of traditional similarity measures (*Cosine, Pearson, Correlation*) and proposed similarity measures (*Cosine$_v$, Pearson$_v$*) using proposed recommendation quality measures (*Pearson$_v$, Recall$_v$*) at fixed value of h and value of N varying from 1 to 20.

4.2 Experiments

Experiment 1

In this experiment, we are observing different values of (*MAE$_v$*) using traditional similarity measures (*Cosine, Pearson, Correlation*) and proposed similarity measures (*Cosine$_v$, Pearson$_v$*) by varying value of h from 1 to 200 to analyze the appropriate value of h.

Analysis of Results

From Figs. 2 and 3, we have observed that the value of (*MAE$_v$*) is getting stable after 160, and therefore, we have taken the value of h to be 160 for our experiments.

Experiment 2

To illustrate the efficacy of our proposed approach, we conducted this experiment to compare the performance of traditional similarity measures (*Cosine, Pearson Correlation*) and proposed similarity measures (*Cosine$_v$, Pearson$_v$*) using proposed prediction quality measures (*MAE$_v$*) and recommendation quality measures (*Pearson$_v$, Recall$_v$*) at $h = 160$ and $N = 3$.

Analysis of Results

Table 1 presents the relative performance comparison of *Cosine, Pearson Correlation* and *Cosine$_v$, Pearson$_v$* using MAE_v, $Precison_v$, $Recall_v$ at $h = 160$ and $N = 3$. The lower values of MAE_v indicate more accurate predictions and it can be clearly observed from the table that MAE_v values have significantly reduced when using proposed similarity measures over traditional similarity measures. The higher values of $Precison_v$ indicate higher number of relevant results and it can be clearly

Fig. 2 MAE_v when using *Cosine* and *Cosine_v* as similarity measure

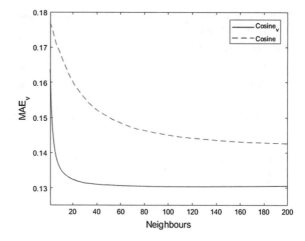

Fig. 3 MAE_v when using *Pearson Correlation* and *Pearson_v* as similarity measure

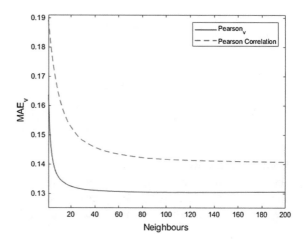

observed from the table that $Precison_v$ values have significantly increased when using proposed similarity measures over traditional similarity measures. The higher values of $Recall_v$ indicate higher number of correctly analyzed relevant results and it can be clearly observed from the table that $Recall_v$ values have significantly increased when using proposed similarity measures over traditional similarity measures.

Experiment 3

To illustrate the efficacy of our proposed approach, we conducted this experiment to compare the performance of traditional similarity measures (*Cosine, Pearson Correlation*) and proposed similarity measures (*Cosine_v, Pearson_v*) using proposed recommendation quality measures (*Precison_v, Recall_v*) at $h = 160$ and value of N varying from 1 to 20.

Table 1 Performance comparison of *Cosine*, *Pearson Correlation* and *Cosine$_v$*, *Pearson$_v$* using *MAE$_v$*, *Precison$_v$* and *Recall$_v$*

Proposed quality measure	*Cosine*	*Cosine$_v$*	*Pearson Correlation*	*Pearson$_v$*
MAE$_v$ ($h = 160$)	0.1431	0.1304	0.1409	0.1304
Precison$_v$ ($h = 160, N = 3$)	0.8461	0.8732	0.8426	0.8738
Recall$_v$ ($h = 160, N = 3$)	0.0286	0.0515	0.0299	0.0513

Fig. 4 *Pearson$_v$* when using *Cosine* and *Cosine$_v$* as similarity measure

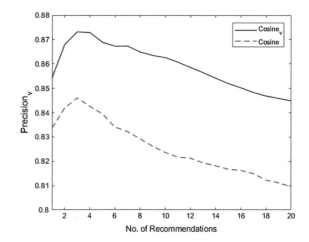

Analysis of Results

Figures 4, 5, 6 and 7 clearly demonstrate the effectiveness of our proposed approach. The *Precison$_v$* and *Recall$_v$* values have improved considerably when using proposed similarity measures over traditional similarity measures for various values of *N*.

5 Conclusions

We have proposed a GRS framework based on the relevance concept. Focusing on the accuracy of GRSs, the major challenge of group satisfaction has been addressed in this framework. To deal with the problem of equal relevance of all the items and the groups, we have generated effective recommendations to the group by considering the ratings given to the items, the relevance of the groups and the relevance of the items. We proposed relevant similarity measures, relevant prediction and recommendation quality measures. Finally, experimental results establish that our proposed similarity and quality measures outperform the traditional measures and produce effective recommendations.

Fig. 5 $Pearson_v$ when using *Pearson Correlation* and *Pearson_v* as similarity measure

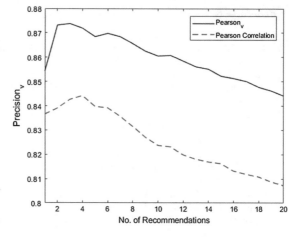

Fig. 6 $Recall_v$ when using *Cosine* and *Cosine_v* as similarity measure

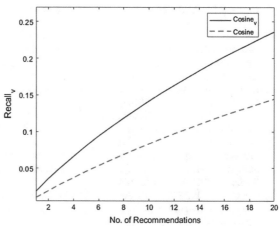

Fig. 7 $Recall_v$ when using *Pearson Correlation* and *Pearson_v* as similarity measure

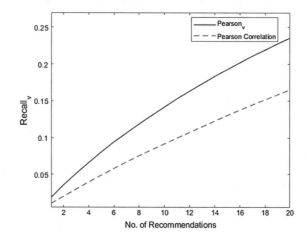

References

1. Adomavicius, G., Kwon, Y.: New recommendation techniques for multicriteria rating systems. IEEE Intell. Syst. **22**(3), 48–55 (2007)
2. Adomavicius, G., Tuzhilin, A.: Toward the next generation of recommender systems: a survey of the state-of-the-art and possible extensions. IEEE Trans. Knowl. Data Eng. **17**(6), 734–749 (2005)
3. Baltrunas, L., Makcinskas, T., Ricci, F.: Group recommendations with rank aggregation and collaborative filtering. In: Proceedings of the Fourth ACM Conference on Recommender Systems (ACM), pp. 119–126 (2010)
4. Berkovsky, S., Freyne, J.: Group-based recipe recommendations: analysis of data aggregation strategies. In: Proceedings of the Fourth ACM Conference on Recommender Systems (ACM), pp. 111–118 (2010)
5. Bobadilla, J., Hernando, A., Ortega, F., Gutiéerrez, A.: Collaborative filtering based on significances. Inf. Sci. **185**(1), 1–17 (2012)
6. Bobadilla, J., Ortega, F., Hernando, A., Gutiéerrez, A.: Recommender systems survey. Knowl. Based Syst. **46**, 109–132 (2013)
7. Burke, R.: Hybrid recommender systems: survey and experiments. User Model. User Adap. Inter. **12**(4), 331–370 (2002)
8. Cantador, I., Castells, P.: Group recommender systems: new perspectives in the social web. In: Recommender systems for the social web, Springer, pp. 139–157 (2012)
9. Herlocker, J.L., Konstan, J.A., Terveen, L.G., Riedl, J.T.: Evaluating collaborative filtering recommender systems. ACM Trans. Inf. Syst. (TOIS) **22**(1), 5–53 (2004)
10. Kompan, M., Bielikova, M.: Group recommendations: survey and perspectives. Comput. Inform. **33**(2), 446–476 (2014)
11. Ortega, F., Hurtado, R., Bobadilla, J., Bojorque, R.: Recommendation to groups of users using the singularities concept. IEEE Access **6**, 39745–39761 (2018)
12. Pizzutilo, S., De Carolis, B., Cozzolongo, G., Ambruoso, F.: Group modeling in a public space: methods, techniques, experiences. In: Proceedings of the 5th WSEAS International Conference on Applied Informatics and Communications, World Scientific and Engineering Academy and Society (WSEAS), pp. 175–180 (2005)
13. Popescu, G., Pu, P.: What's the best music you have? Designing music recommendation for group enjoyment in groupfun. In: CHI'12 Extended Abstracts on Human Factors in Computing Systems, ACM, pp. 1673–1678 (2012)
14. Wang, W., Zhang, G., Lu, J.: Member contribution-based group recommender system. Decis. Support Syst. **87**, 80–93 (2016)

ACA: Attention-Based Context-Aware Answer Selection System

K. Sundarakantham, J. Felicia Lilian, Harinie Rajashree
and S. Mercy Shalinie

Abstract The main goal of question answering system is to develop chatbots capable of answering the questions irrespective of the domain. The system has to provide appropriate answers according to the user queries. The challenge of a question answering system lies in analyzing the question to retrieve the accurate answers from a large amount of data present. The main aim of this paper is to propose a question answering system that analyzes the questions properly and answers according to the type of the question in a precise way. The model uses similarity measures to find the candidate set which are the most relevant answers from the content given. The generated candidate set is then analyzed further using the bidirectional long short-term memory (BiLSTM) to retrieve the appropriate answer. Based on the type of question, we retrieve the exact word/phrase as a response to the user. This is done using an attention mechanism which _nds the exact answer to the query. Results of models without memory and with an attention-based BiLSTM are being compared. Using the attention-based BiLSTM, we found that the retrieved answer had increased accuracy rate by 17% when compared with without-attention models like convolution neural networks (CNN) and recurrent neural networks (RNN).

Keywords Question answer system · Attention mechanism · Context aware · Recurrent neural network · Bidirectional LSTM

1 Introduction

The reading comprehension task understands a given passage by answering the question that is raised. This system is applicable for answering many natural language processing (NLP) applications such as question from medical reports/government reports, questions from product manuals and questions from online articles. A question answering system is confronted with huge data from various sources [1]. These data contain a huge volume of information related to any field ranging from medicine,

K. Sundarakantham (✉) · J. Felicia Lilian · H. Rajashree · S. Mercy Shalinie
Thiagarajar College of Engineering, Madurai, Tamilnadu 625015, India
e-mail: kskcse@tce.edu

© Springer Nature Singapore Pte Ltd. 2020
S. Agarwal et al. (eds.), *Machine Intelligence and Signal Processing*,
Advances in Intelligent Systems and Computing 1085,
https://doi.org/10.1007/978-981-15-1366-4_26

education, law and agriculture. When confronted with such huge data, the working of the system can be enhanced by two phases:

- To analyze the question context properly for matching the phrase from the passage
- Retrieve the candidate set which possibly contains the answer and selects the most appropriate answer.

The necessity to analyze the question rises due to the fact that a properly analyzed question can maximize the opportunities to find a similarity to retrieve the answer. Once the similarity is matched, the deep learning model can be employed to find the most relevant answer from the candidate set. There are various deep learning models used for NLP applications such as RNN, Long Short-Term Memory (LSTM)/Gated Recurrent Unit (GRU), bidirectional LSTMs and Bi-RNNs [2]. The user question may be classified as 'When' implies that the answer should be a date or a time, a question asking 'Where' implies that the answer retrieved has to be a place. For example, consider the below given content, question and answer.

Corpus Lila was fond of cherries. Lila usually bought cherries from the local grocery store. But this time she went to the 'Blossom garden' to pick some cherries for her friends. She was very happy to serve her friends the cherries for dinner at 'Saint Restaurant.'
Question: Where did Lila pick cherries?
Answer: Blossom Garden

From the above given example, the question answering system has to analyze the question to find what is the answer the user is expecting. Only when that step is attended, the next step of arriving at the right answer becomes consequential [3].

In our model, we try to incorporate all the features that are needed in the development of a question answer system for reading comprehension. We use the NLP techniques to analyze the question. We have proposed a novel approach for embedding the text based on the negation words and projected then into the vector space to find the similarity between the words to maintain the consistency among them. We adapt the Word2Vec model to enhance the embedding and use similarity function to generate the proper candidate set. The attention-based BiLSTM attends to generate the most appropriate answer out of the candidate set.

2   Related Work

The rise of deep learning models has paved way for the applications on NLP to emerge with great success by handling huge data [4]. To start with processing, the text we need to focus on its numerical terms. This process of changing the text into their vector form is the foremost step in NLP process, and we call it as word embedding. After vectorization, we input these vectors into the deep learning models and perform the QA task efficiently.

2.1 Word Embedding

Mikolov et al. [5, 6] proposed a neural network-based approach for representing the word as vectors known as Word2Vec model. These representations emerge by considering the semantic and syntactic position of the words. He has developed two architectures, namely skip-gram model (SG) and continuous bag-of-words model (CBOW). They retrieve the local context of the words and retrieve the vector form for each word. Pennington et al. [7] proposed a different approach of representing words known as GloVe, by considering the global bilinear regression model which produces the vector structure along with meaningful substructure. Liu and Zang [8] proposed a transformer recurrent neural network which learns through bidirectional GRU and uses contextual word embeddings. It was specialized for tackling the Chinese machine reading comprehension. Neamah and Saad [9] discussed the various similarity measures and classification techniques employed in a question answering model. Kapashi and Shah [10] proposed a model which uses deep learning networks along with a memory component for QA tasks. It also applies an NLP feature engineering to pass only pruned inputs.

2.2 Question Answer System

The word vectors generated through word embedding are sent into the learning model to understand the question and answer. There are various approaches which have been proposed. Tan and Wei [11] proposed a hierarchical gated recurrent unit to build a context-aware answer selection model. A hierarchical approach was incorporated in order to find matching patterns at both word and sentence levels. The work introduces gates at both the word and sentence levels in order to find the context dependency. Zhang and Ma [12] proposed a coattention-based BiLSTM model for answer selection which captures interactions between the question answers so that it generates different question representations according to the answers. Bahdanu et al. [13] have also proposed an attention network for machine translation to encode the source sentence and used the attention mechanism in the decoder. While Tan et al. [14] and Xiang et al. [15] proposed similar attention-based BiLSTM model to encode and decode. Seo and Kembhavi [16] have proposed a bidirectional attention network along with a hierarchical process to represent context at different levels of granularity.

Based on the survey over various models, we have identified the research gap in dealing with the reading comprehension sentences for question answering system. They are,

- Pre-processing of data and embedding of text has to be enhanced [16]
- Retrieving the needed phrase is based on the factoid type question and not the entire sentence [11]
- Ranking of the responses based on the user's post which has to be relevant with proper candidate set [12]
- Retrieval of data is not consistent as it depends on embedding of text [13].

3 Proposed Work

In this section, we introduce our proposed work 'attention-based BiLSTM model for context-aware answer selection.' The flow of the proposed system is given in Fig. 1.

The input query and the content are processed to obtain their vector form. The processing is performed to handle the negation words and update the input. This is broken down into individual words and they are embedded as vectors. Next, the

Fig. 1 Flow of the proposed work

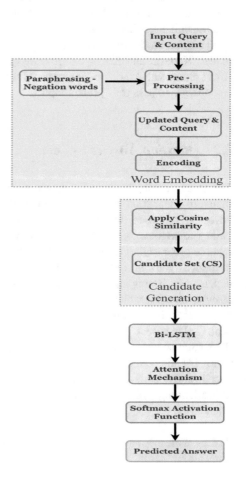

candidate answers are generated by performing the similarity between the question and the content. Then, the BiLSTM model which makes the context aware of the words presents query and answer. Finally, the attention layer decides which words in the context have to be generated. Hence, the major modules are:

1. Word embedding
2. Candidate generation
3. Attention mechanism.

3.1 Word Embedding

The context and the question have to be represented in the form of vectors. The aim of embedding is to make sure semantically similar words which lie in the same space so that semantics of every word is properly captured. In the proposed model, we deploy a model similar to Word2Vec by incorporating the negation word 'Not' [17]. Negation can change the context of a sentence drastically when not considered [18]. For instance,

Sentence 1:   The neighborhood is not friendly.
Sentence 2:   The neighborhood is friendly.

The Sentence 1 conveys that the neighborhood was not very friendly. But, when the negation is not considered, the meaning of the sentence might turn out the other way as given in Sentence 2. To overcome this challenge during the stop word removal pre-processing stage, we have proposed an Algorithm 1 which will handle these negate words and update the corpus to generate word embeddings for each words.

Algorithm 1: Anti-Negation Algorithm

Input: Corpus
Output: Updated Corpus
Parameter: w_i = individual words, n = length of text
begin:
word_tokenize [] = $w_1, w_2, ..., w_n$
for each word w_i in corpus **do**
  if $w_i \neq$ 'not' **then**
  | Update Corpus
  end
  if $w_i ==$ 'not' **then**
    retrieve next word w_{i+1}
    Retrieve Synonym = syn(w_{i+1})
    Generate antonym = ant(w_{i+1})
    Replace 'not' = ant(w_{i+1})
    Remove w_{i+1}
    Update Corpus
  end
end

3.2 Candidate Generation

Once the context and questions word embedding are done, the candidate set of answers is generated based upon the similarity measures. The most similar answers which are considered similar are selected by retrieving the top two or three sentences based on the number of sentences in the passage. In the proposed model, we adapt cosine similarity given in Eq. 1 to generate candidate set. Cosine similarity is a vector-based similarity measure which calculates the cosine of the angle between the two documents, where S_1 and S_2 are the vectors for the generated for the query and each sentence. Based on the similarity score, the top-ranked answers are found to be similar to the context of question are chosen to be as the candidate set.

$$\cos(\theta) = \frac{S_1, S_2}{||S_1|| ||S_2||} \tag{1}$$

3.3 Attention Network

Once the candidates set (CS) is generated, they are encoded using a Word2Vec model based on the context words. The embedded text is sent into the BiLSTM, and the output is represented as $\overline{H}$.

$$\overline{H} = \left[\overrightarrow{h_i} + \overleftarrow{h_i} \right] \tag{2}$$

where $\overline{H}$ is the sum of the forward $\overrightarrow{h_i}$ and backward vectors $\overleftarrow{h_i}$ generated from the BiLSTM layer. The $\overline{H}$ is sent into the attention layer as a vector of length L (sentence length). The output of this attention layer is the weighted sum of $\overline{H}$ vector.

$$T = \tanh\left(\overline{H}\right) \tag{3}$$

$$\beta = \text{softmax}\left(w^L T\right) \tag{4}$$

$$\gamma = \overline{H}\beta^L \tag{5}$$

where $\overline{H} \in R^{WXL}$, W represents the word vector dimension and w^L is the transpose of the trained vectors. The dimensions of β and γ are L and W. The output of the attention layer will be used to retrieve the start and the end of the phrase to generate the answer.

4 Experimental Result

4.1 Dataset and Pre-processing

The dataset that we have considered for our proposed work is the SQUAD dataset.[1] This dataset is developed by Rajpurkar [19] from Stanford University. It consists of questions posted by crowd workers over the Wikipedia article. The answers to the questions are a piece of text or a span of text from the article. It is best suited for reading comprehension [1].

As given in the algorithm 1, the data are pre-processed using the natural language toolkit [20]. The context C is tokenized into words using the word tokenize() function of the toolkit into set of tokens Tokens $= C_1, C_2,…,C_n$. The tokens are scanned to find out the occurrence of 'Not.' When an occurrence is spotted the word next to the 'Not' is checked for the Part of Speech (POS). If it is a verb VB + (VB, VBN, VBG, VBN, VBP, VBZ), then the words are replaced by the antonym in the context using the antonyms() function of nltk. The updated context is embedded into vectors. After figuring out the negate word, the antonym of the word is retrieved and it is being replaced into the content back as shown in Fig. 2.

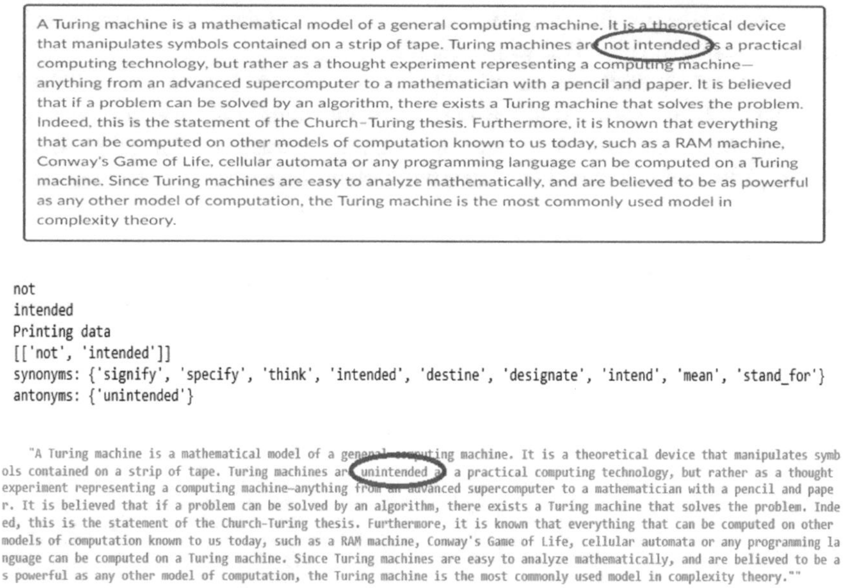

A Turing machine is a mathematical model of a general computing machine. It is a theoretical device that manipulates symbols contained on a strip of tape. Turing machines are not intended as a practical computing technology, but rather as a thought experiment representing a computing machine— anything from an advanced supercomputer to a mathematician with a pencil and paper. It is believed that if a problem can be solved by an algorithm, there exists a Turing machine that solves the problem. Indeed, this is the statement of the Church–Turing thesis. Furthermore, it is known that everything that can be computed on other models of computation known to us today, such as a RAM machine, Conway's Game of Life, cellular automata or any programming language can be computed on a Turing machine. Since Turing machines are easy to analyze mathematically, and are believed to be as powerful as any other model of computation, the Turing machine is the most commonly used model in complexity theory.

```
not
intended
Printing data
[['not', 'intended']]
synonyms: {'signify', 'specify', 'think', 'intended', 'destine', 'designate', 'intend', 'mean', 'stand_for'}
antonyms: {'unintended'}
```

"A Turing machine is a mathematical model of a general computing machine. It is a theoretical device that manipulates symbols contained on a strip of tape. Turing machines are unintended as a practical computing technology, but rather as a thought experiment representing a computing machine–anything from an advanced supercomputer to a mathematician with a pencil and paper. It is believed that if a problem can be solved by an algorithm, there exists a Turing machine that solves the problem. Indeed, this is the statement of the Church-Turing thesis. Furthermore, it is known that everything that can be computed on other models of computation known to us today, such as a RAM machine, Conway's Game of Life, cellular automata or any programming language can be computed on a Turing machine. Since Turing machines are easy to analyze mathematically, and are believed to be as powerful as any other model of computation, the Turing machine is the most commonly used model in complexity theory.""

Fig. 2 Figuring the word 'Not' and replaced the antonym word

---

[1] https://rajpurkar.github.io/SQUAD-explorer/explore/1.1/dev/.

4.2   Proposed Attention-Based BiLSTM Model

The candidate set is generated applying cosine similarity, and the top ranked sentences of 2–3 numbers are retrieved for considering as the closest possible answers. Based on the candidates generated by similarity measures, the BiLSTM model detects the most relevant answer and the attention mechanism works on the specific answer to retrieve the particular phrase out of the answer sentence. The accuracy of the system is measured as the total number of responses that are correctly retrieved for the query to the total number of queries generated. The model is trained for the SQUAD dataset and we used Adam optimizer to optimize the values in the output layer. The accuracy result for our model is compared with the existing models CNN and LSTM and BiLSTM model without attention network. The cross-entropy loss is calculated as shown in Fig. 3, and the words are represented into the vector space through the proposed embedding as shown in Fig. 4.

We resort to Precision and Recall standard metrics in information retrieval and question answering to measure the experimental results. It is based on the confusion matrix as shown in Table 1, and the Precision (P), Recall (R) and Accuracy (Acc) are calculated based on Eqs. 6, 7 and 8. Precision is the measure of accuracy which denotes the relevant answers that have been derived from the possible sentences in the proposed answer context and recall denotes the relevant answers that have been derived out of the possible context area. The retrieval of the exact answer match is performed using the proposed work as shown in Fig. 5.

The proposed attention-based BiLSTM method achieves precision of 92.27% and a recall of 91.03%. We can observe that the attention network influences the answer selection when considering models without attention mechanism. The graph 6 shows the comparison with the existing models based on the Precision, Recall and Accuracy values.

Fig. 3 Calculating the loss over the dataset

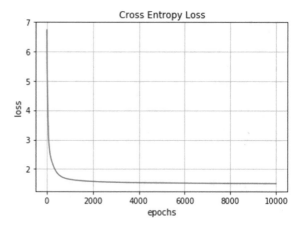

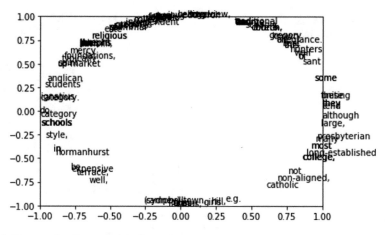

Fig. 4 Representing the words into the vector space

Table 1 Relevant retrieval confusion matrix

	Relevant	Non-relevant
Retrieved	True positive (TP)	False positive (TP)
Not-retrieved	False negative (TP)	True negative (TP)

A Turing machine is a mathematical model of a general computing machine. It is a theoretical device that manipulates symbols contained on a strip of tape. Turing machines are unintended as a practical computing technology, but rather as a thought experiment representing a computing machine—anything from an advanced supercomputer to a mathematician with a pencil and paper. It is believed that if a problem can be solved by an algorithm, there exists a Turing machine that solves the problem. Indeed, this is the statement of the Church–Turing thesis. Furthermore, it is known that everything that can be computed on other models of computation known to us today, such as a RAM machine, Conway's Game of Life, cellular automata or any programming language can be computed on a Turing machine. Since Turing machines are easy to analyze mathematically, and are believed to be as powerful as any other model of computation, the Turing machine is the most commonly used model in complexity theory.

QUESTION: Is turning machine a practical computing technology?

ANSWER: No

QUESTION: What is the term for a mathematical model that theoretically represents a general computing machine?

ANSWER: A Turing machine

Fig. 5 Exact answer match to the question posted

$$\text{Precision } (P) = \frac{TP}{TP + TP} \quad (6)$$

$$\text{Recall } (R) = \frac{TP}{TP + (TP)} \quad (7)$$

$$\text{Accuracy } (\text{Acc}) = \frac{TP + TP}{TP + (TP) + TP + TP} \quad (8)$$

Fig. 6 Comparison with existing models with the proposed model

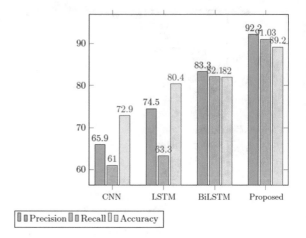

The BiLSTM model outperforms well over the other existing models as shown in Fig. 6. The system also performs well over the different type of factoid questions like when, where, who, what. The accuracy of these types of questions is shown in Fig. 7.

Fig. 7 Accuracy of different type of question

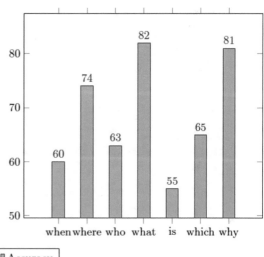

5 Conclusion and Future Work

In our work, we have proposed an attention-based BiLSTM model for question answering in reading comprehension. The attentive mechanism captures the answer precisely, thereby responding to user questions in a specific manner while remaining in the context. The experiment results show that the model achieves an accuracy of 85.2% which is effective but has numerous scopes for improvement. Future work can be carried out by employing gated recurrent units.

References

1. Parikh, S., Sai, A.B., Nema, P., Khapra, M.M.: Eliminet: A model for eliminating options for reading comprehension with multiple choice questions. arXiv preprint arXiv:1904.02651 (2019)
2. Najafabadi, M.M., Villanustre, F., Khoshgoftaar, T.M., Seliya, N., Wald, R., Muharemagic, E.: Deep learning applications and challenges in big data analytics. J. Big Data 2(1), 1 (2015)
3. Ichida, A.Y., Meneguzzi, F., Ruiz, D.D.:. Measuring semantic similarity between sentences using a siamese neural network. In: 2018 International Joint Conference on Neural Networks (IJCNN) IEEE, pp. 1–7, July, 2018
4. Young, T., Hazarika, D., Poria, S., Cambria, E.: Recent trends in deep learning based natural language processing. IEEE Comput. Intell. Mag. 13(3), 55–75 (2018)
5. Mikolov, T., Chen, K., Corrado, G., Dean, J.: Efficient estimation of word representations in vector space. arXiv preprint arXiv:1301.3781 (2013)
6. Almeida, F., Xexo, G.: Word embeddings: A survey. arXiv preprint arXiv:1901.09069 (2019)
7. Pennington, J., Socher, R., Manning, C.: Glove: Global vectors for word representation. In: Proceedings of the 2014 conference on Empirical Methods in Natural Language Processing (EMNLP), pp. 1532–1543 (2014)
8. Liu, S., Zhang, S., Zhanga, X., Wang, H.: R-trans: RNN transformer network for chinese machine reading comprehension. IEEE Access (2019)
9. Neamah, N., Saad, S.: Question answering system supporting vector machine method for hadith domain. J. Theor. Appl. Inf. Technol. 95(7) (2017)
10. Kapashi, D., Shah, P.: Answering reading comprehension using memory networks. Report for Stanford University Course CS224d (2015)
11. Tan, C., Wei, F., Zhou, Q., Yang, N., Du, B., Lv, W., Zhou, M.: Context-aware answer sentence selection with hierarchical gated recurrent neural networks. IEEE/ACM Trans. Audio, Speech, Lang. Process. 26(3), 540–549 (2018)
12. Zhang, L., Ma, L.: Coattention based BiLSTM for answer selection. In: 2017 IEEE International Conference on Information and Automation (ICIA), IEEE, pp. 1005–1011 (2017)
13. Bahdanau, D., Cho, K., Bengio, Y.: Neural machine translation by jointly learning to align and translate. arXiv preprint arXiv:1409.0473 (2014)
14. Tan, M., Dos Santos, C., Xiang, B., Zhou, B.: Improved representation learning for question answer matching. In: Proceedings of the 54th Annual Meeting of the Association for Computational Linguistics (Volume 1: Long Papers), vol. 1, pp. 464–473 (2016)
15. Xiang, Y., Chen, Q., Wang, X., Qin, Y.: Answer selection in community question answering via attentive neural networks. IEEE Signal Process. Lett. 24(4), 505–509 (2017)
16. Seo, M., Kembhavi, A., Farhadi, A., Hajishirzi, H.: Bidirectional attention flow for machine comprehension. arXiv preprint arXiv:1611.01603 (2016)
17. Skovgaard-Olsen, N., Collins, P., Krzyanowska, K., Hahn, U., Klauer, K.C.: Cancellation, negation, and rejection. Cogn. Psychol. 108, 42–71 (2019)

18. Blanco, E., Moldovan, D.: Some issues on detecting negation from text. In: Twenty-Fourth International FLAIRS Conference, Mar, 2011
19. Rajpurkar, P., Zhang, J., Lopyrev, K., Liang, P.: Squad: 100,000+ questions for machine comprehension of text. arXiv preprint arXiv:1606.05250 (2016)
20. Loper, E., Bird, S.: NLTK: The natural language toolkit. arXiv preprint cs/0205028 (2002)

Dense and Partial Correspondence in Non-parametric Scene Parsing

Veronica Naosekpam, Nissi Paul and Alexy Bhowmick

Abstract Scene correspondence, in general, refers to the problem of finding a set of points on one image which corresponds to a set of points in another image. Given the corresponding points on a pair of images, the task of scene correspondence is to align these set of points for categorically labeling the pixels in the image. There are broadly two approaches of finding correspondence between scenes or images: (i) *dense correspondence* and (ii) *partial correspondence*. Dense correspondence is a pixel-by-pixel alignment of the scenes. Partial correspondence is a superpixel, patch, or region-based alignment of the scenes. Partial correspondence has the ability to align a large, cohesive group of pixels simultaneously. This paper provides a compact review of the state-of-the-art methods in the past few years for scene correspondence in the non-parametric scene parsing.

Keywords Scene correspondence · Scene parsing · Dense correspondence · Partial correspondence · Superpixels

1 Introduction

Scene correspondence refers to ascertaining which parts of one image correspond to which parts of another image. It is one of the modules of the non-parametric scene parsing pipeline and is a prerequisite for scene parsing. Scene correspondence involves matching features of the images and to find similarity. The importance of scene correspondence in scene parsing is highlighted by an increasing number of applications developed for inferring knowledge from imagery such as self-driving vehicles, human–computer interaction, and virtual reality.

Non-parametric scene parsing is a data-driven approach for scene parsing. This approach incurs almost no training at all. Instead, it involves retrieval of a small set of similar images and transfer of labels from the retrieved images to the query

V. Naosekpam (✉) · N. Paul · A. Bhowmick
Department of Computer Science & Engineering, School of Technology, Assam Don Bosco University, Guwahati, Assam, India
e-mail: venaosekpam11@gmail.com

© Springer Nature Singapore Pte Ltd. 2020   339
S. Agarwal et al. (eds.), *Machine Intelligence and Signal Processing*,
Advances in Intelligent Systems and Computing 1085,
https://doi.org/10.1007/978-981-15-1366-4_27

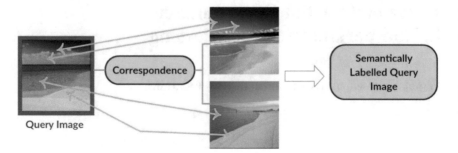

Fig. 1 Scene correspondence followed by scene labeling in non-parametric scene parsing

image. In the data-driven approach, the dataset size can vary at any point of time, and therefore, it is scalable to a large number of categories and works on "Open Universe" datasets. There are three major steps involved in nonparametric scene parsing: (i) Scene Retrieval, (ii) Scene Correspondence, and (iii) Markov Random Field (MRF) inference.

A typical correspondence step in the non-parametric scene parsing pipeline is shown in Fig. 1. The query image is the image to be labeled. The two images are the images retrieved in the scene retrieval step whose semantic labels are to be transferred to the query image via a correspondence algorithm. The output is a coarsely semantically labeled query image.

This paper focusses on the scene correspondence module of the non-parametric scene parsing pipeline. There are two approaches for scene correspondence: Dense correspondence (holistic approach) [1] and Partial correspondence (superpixels or patch-/region-based approach) [2]. The rest of the paper is organized as follows: Sect. 2 starts with the compact overview of related works so far done in the field of dense and partial scene correspondence in the non-parametric scene parsing pipeline. Section 3 discusses the evaluation of the state-of-the-art approaches of scene parsing using dense and partial correspondence on benchmark datasets. An overview of Markov Random Field (MRF) is discussed in Sect. 4. Finally, a conclusion of the research is drawn in Sect. 5.

2 Related Works

2.1 Dense Correspondence

Many solutions have been offered for establishing dense correspondences such as by changing the design approach and considering the nature of the scene. The dense descriptors are used for dense image correspondence and label transfer. In this section, we discuss the different approaches for establishing non-parametric dense correspondence.

SIFT Flow: Liu et al. [1] introduced the concept of dense scene alignment using SIFT flow [3], which is analogous to the optical flow algorithm. The SIFT flow algorithm consists of matching densely sampled SIFT features between the two images, while preserving spatial discontinuities. The aim of SIFT flow dense correspondence [3] is to find correspondence between a pair of images present in a three-dimensional scene that shares some scene characteristics. In SIFT flow algorithm, it is assumed that image matching is performed without much scale change. Let $p = (x, y)$ be the grid coordinate of images, $w(p) = (u(p), v(p))$ the flow vector at a position p, and s_1 and s_2 be two SIFT images that are to be matched. $s_i(p)$ denotes a SIFT descriptor at position p of SIFT image s_1. ε contains all spatial neighborhoods. The energy function for SIFT flow is defined as—

$$E(w) = \sum_p \min\left(\|s_1(p) - s_2(p)\|_1, t\right) + \sum_p \eta(|u(p)| + |v(p)|)$$

$$+ \sum_{p,q \in \varepsilon} \min(\alpha|w(p) - w(q)|, d) \tag{1}$$

The first term (data term) constrains the SIFT descriptors to be matched along with the flow vector $w(p)$, the second term (small displacement) constrains the flow vectors to be as small as possible, and the third term (smoothness) constrains the flow vectors of adjacent pixels to be similar. It has been found that the ordinary SIFT flow implementation [3] took more than two hours to process image pair of size 256×256. To overcome the performance drawback, coarse-to-fine SIFT flow was introduced which consists of estimating the flow at the coarse level and propagate and gradually refine the flow. It took 31 seconds per image execution which is a very speeded up improvement.

Scale-space SIFT Flow [4]: It explores the large object scale differences automatically. This method has obtained a significant improvement while dealing with large-scale differences images using "scale-space" theory [5]. When matching a pair of images, the second image is set at its own scale. In the first image, a scale factor is assigned for every position which is denoted by $\sigma(p) \in R$. The new SIFT flow field of position p of first image of image as is $s_1(p, \sigma(p))$ where $\sigma(p)$ is the scale space. The smoothness term of Eq. 1 is changed to $\sum_{p,q \in \varepsilon} \min(\beta|\sigma(p) - \sigma(q)|, \tau)$ called scale smoothness term where β and τ are used for adjusting the weight of scale smoothness term and truncating scale smoothness term.

Dense, Scale-less Descriptors [6]: The Scale-Less SIFT (SLS) [6] is a descriptor used for conversion of a subspace to a point representation. A subspace is computed from SIFT extracted at multiple scales at a single image location which can be represented as a high-dimensional point. SLS has the ability to capture multi-scale information, and it can be extracted anywhere in the image to represent multi-scale image appearances.

Dense Segmentation-aware Descriptors: Segmentation-aware descriptors [7] are invariance to occlusions and background changes. Basically, the aim is to reduce or eliminate the effects of background changes during features extraction obtained by combining mid-level segmentation module obtained by using three techniques (spectral clustering, generalized boundaries, and structured edge forests) with SIFT and scale-invariant descriptor (SID).

Approximate Nearest Neighbor (ANN) Bilateral Matching: In image-to-image approximate nearest neighbor bilateral matching scheme [8], the query image is matched to the retrieved images by establishing dense correspondence at pixel level. There are four main steps involved: inner-propagation, random search, intra-propagation, and mismatch rejection which are shown in Fig. 2. Prior to these steps, for two images, say A and B, feature vectors f_a and f_b are initialized randomly with uniformed sampled offset vectors from the valid offset vector space. Taking two pixels (a, b) of image A, and (c, d) of image B, the main steps are:

- Inner-propagation: If a was matched to c and b was matched to d, b will check whether c is a better match than d, if yes, b updates its match to c.
- Random search: If a was matched to c, it will search around c to find a better match by testing a sequence of randomly selected position.
- Intra-propagation: If a was matched to c and b was matched to a, when the matching distance between b and a is smaller than that of a and c, the matching correspondence of a will be updated to b.
- Mismatch rejection: If a was matched to c and c was matched to b, when the shift between a and b is bigger than a predefined small range, the correspondence a and c will be rejected.

The algorithm converges when the average matching distance of matched pairs keeps unchanged. The matching distance for correspondence computed as:

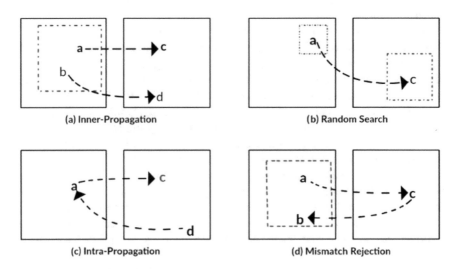

(a) Inner-Propagation   (b) Random Search

(c) Intra-Propagation   (d) Mismatch Rejection

Fig. 2 ANN bilateral matching

$$D(a, c) = \begin{cases} \|D_a - D_c\|^2, \text{if } C_c \in R = \Phi \\ +\infty, \text{else} \end{cases} \tag{2}$$

where D_a^i and D_c^j are the feature descriptors of a and c. C_c is the category label of c. R is the image level prior about the categories contained in A. This approach outperformed the dense correspondence approach by Liu et al. [1] in terms of global accuracy and category-wise accuracy.

Adaptive Window approach: Collage parsing [9] is an approach that extracts mid-level content-adaptive windows from the retrieved images and the query image using objectness algorithm by Alexe [10] to extract the content-adaptive window. The label transfer is performed by matching the query image's content-adaptive windows with retrieval set images' content-adaptive windows. For describing the content-adaptive window, the features used are the Histogram of Oriented Gradient (HOG). For matching for correspondence, a pair of windows with the same HOG feature dimensions is compared. This approach encourages the matching to come from spatially similar regions in the respective images. During one run of the algorithm, scene correspondence is performed twice, firstly with the coarse retrieval set and secondly with the semantically labeled retrieval set.

2.2 Partial Correspondence

While establishing partial correspondence at the superpixel level, the first step involved is feature extraction of the segmented regions or superpixels which is followed by superpixel matching. Some of the superpixel features used are shape, location, texture, appearance, color, SIFT descriptors, Histogram of Oriented Gradient (HoG) features, etc. [1, 9, 11–13].

In this section, we discuss the different approaches of partial correspondence in the nonparametric scene parsing pipeline.

Superpixels-based Correspondence: In Superparsing [2], bottom-up segmentation is used to obtain the superpixels. The superpixel features are assumed to be independent of each other and by using the Naïve Bayes theorem, the likelihood ratio for class c and test superpixel s_i is given as:

$$L(s_i, c) = \frac{P(s_i|c)}{P(s_i|\bar{c})} = \Pi_k \frac{P(f_i^k|c)}{P(f_i^k|\bar{c})} \tag{3}$$

where $\bar{c}$ is the set of all classes excluding c, and f_i^k is the feature vector of the kth type for s_i. The labeling of the test image is obtained by assigning to each superpixel the class that maximizes Eq. 3. Though this system outperforms the approach of Liu et al. [1] in terms of accuracy, this approach somewhat deviates from the typical pipeline of a nonparametric scene parsing which relies heavily on the retrieval set obtained. Instead the higher accuracy obtained in this approach is mainly because the likelihood ratio calculated is dependent on the whole annotated training set. Nevertheless, many more works have been carried by taking this approach as a baseline.

Correspondence using adaptive neighborhood set [12] extends Superparsing [2] where per-descriptor weights for each superpixels' segment are learnt in order to minimize classification error. Rare classes which are not present in the retrieval set while performing global context similar image retrieval are further added back to the retrieval set using an indexing scheme which is based on semantic context descriptor of a superpixel. Leave-one-out classification on each image in the training set is done to index the rare classes, and index each superpixel whose class occurs below a threshold of r times in the global retrieval set. Using this approach, the labeling accuracy improved.

Scene parsing using regions and per-exemplar detector [14] is done using super-parsing as a baseline algorithm. First, the region-based parsing is performed using likelihood ratio of superparsing [2]. The score of the region-based data term E_R for each pixel p and class c is—

$$E_R(p, c) = L(s_p, c) \tag{4}$$

where s_p is the region containing p. It is followed by the per-exemplar detector by using the framework of [15]. For each labeled per object instance, a per-exemplar detector is trained.

Other Superpixels-based Partial Correspondence Approaches: Rare classes are added back to the retrieval set for a more balanced superpixels classification in Yang et al. [16]. In order to refine the superpixel matching, the global semantic information and the local semantic context information are set in a feedback manner. The retrieval set's label distribution could be noisy due to the presence of irrelevant images. The distribution of the retrieval is partitioned into head and tail parts based on the 80%–20% Pareto rule, and define the classes in the tail as "rare" L_r while the other classes in the head as "common" L_c. The label subset of retrieved superpixels can be partitioned into $L' = (L'_r \cap L'_c)$ and populate the superpixels in the classes of L'_r with exemplars. It has the potential to improve the appearance based image retrieval along with semantic scene description. Together with the local appearance descriptors, contextually consistent classification results have been obtained with increased in per-class performance.

Partial scene correspondence can be performed by combining various classifiers (ensemble approach) as in [17]. In this approach, likelihood scores of various classi-fiers are combined to label each superpixel of the query image with a balanced score. The weights of combining the scores are learnt through likelihood normalization. Significant improvement in the coverage of foreground classes in a given scene is obtained with this approach. Another highlight of this approach is that it skips the scene retrieval step and directly works on the entire annotated training set, and hence, the label costs are learned from the global contextual correlation of labels. A boosted decision tree (BDT) [18] model is used to obtain the label likelihood. This approach encapsulates multiple classifiers to create an ensemble. The final cost of assigning the label c to superpixel s_i of the query image can be represented as combination of the likelihood scores of all classifiers.

Scene correspondence via adaptive relevance set [11] calculates the appearance likelihood of the query image as a whole as—

$$P(A|L) \approx \prod_{i=1}^{S} P(a_i|l_i) \tag{5}$$

where A is the appearance vector and L is the likelihood of the labels. The individual label likelihood $P(a_i|l_i)$ for a superpixel is obtained using a k-NN method. The highlight of this approach is that the individual superpixels' features are assigned weights and are adjusted to get the best match for each individual superpixel of the query image. Most of the algorithms developed for partial scene correspondence uses a fixed value of k nearest neighbors for assigning label to the query superpixels. However, the k-value can be varied [13] and can be adaptively set as the minimum number of nearest neighbor which is required to get the best category prediction of the query superpixel. The nearest neighbors of a sample is specific to the sample.

Stable scene layout of road scenes is parsed [19] by adding spatial prior in terms of histogram distribution of each object prior to the non-parametric scene parsing pipeline. Spatial prior histogram $hist_c(p)$ for class $c \in C$ at pixel p is estimated from the training set T as the ratio of the number of pixels at location p in the training set whose semantic labels are c, and the number of pixels at p in the training set. The spatial prior term, which indicates the probability that object category c appears at p, is given by—

$$E_{sp}(y_p) = -\log hist_c(p) \tag{6}$$

The labeling of the query scene is performed using the approach of [2]. Label transfer via efficient filtering [20] is an approach that performs parsing by first sampling of superpixels and transferring labels via efficient filtering using a Gaussian kernel $k(x_i, x'_j)$ that encodes how similar two superpixels are in terms of their feature vector x_i and x'_j. The Gaussian filtering operation consists of three step: *Splatting* which is the mapping of the training data to the permutohedral lattice and then computing the values at the vertices of the lattice; *Blurring* which locally approximates the Gaussian filter and updates the values at the vertices of the lattice; *Slicing* which is the mapping of the query superpixels to the lattice and computing their values as a linear combination of values of a few vertices.

Patch-Based Partial Correspondence: Patch-based is another approach of partial correspondence based on the aggregation of pixels in the form of a grid. Super-PatchMatch [21] is based on the PatchMatch algorithm [22] developed for matching of irregular structures between images and transferring the colors of the matched superpatches. Superpatch refers to a patch formed by aggregation of superpixels that lie below a given threshold radius R. It has three steps: *Initialization* step consists of associating randomly a patch of the image X with a patch of the image Y, which leads to an initial Approximate Nearest Neighbor (ANN) field. The following two steps are performed iteratively to improve the correspondences. *Propagation* assumes that if a patch in X corresponds to a patch in Y, then the adjacent patches in X should also match the adjacent patches in Y. *Random Search* consists of sampling around the current approximate nearest neighbor (ANN) to escape from possible local minima. The distance measurement between the superpixels is computed using Euclidean distance.

Region-Based Partial Correspondence: In this partial scene correspondence approach, a graphical model [23] is constructed in which training image regions (V^t) and test image regions (V^q) are nodes of a graph. An undirected graph $G = (V, E)$ is constructed where nodes are shown by $V = \{V^t \cup V^q\}$ and the three types of edges are represented by $E = \{E^t, E^q, E^{tq}\}$. By using the relationships between nodes in the defined graphical model, a quadratic convex cost function is defined over regions likelihood which is further optimized. Finally, for achieving the semantic label of each region in a query image, maximum a posterior probability (MAP) estimation is used. This approach does not use pairwise patch matching (Markov Random Field) for label smoothing during the label aggregation stage and it does not limit to the general datasets used for evaluating non-parametric scene parsing.

3   State-of-the-Art in Dense and Partial Correspondence

As discussed, dense correspondence is a pixel-by-pixel alignment of the query image with the retrieval set using dense local features, such as Scale Invariant Feature Transform (SIFT). The per-pixel local information is transferred from an image in the retrieval set to the query image. Dense correspondence estimation has mostly been motivated by two specific problems: stereo vision and optical flow. Parsing a scene based on dense correspondence is a traditional approach in the non-parametric scene parsing.

The steps in dense correspondence include:

- Extraction of features from each pixel of the query image and the images belonging to the retrieval set.
- Matching of the query image and the annotated retrieval set using the features extracted using a similarity function or a classifier.
- Finding out the nearest neighbor annotated pixels for label transfer.

Table 1 gives the summary of dense correspondence approaches reported using various datasets resulting in evaluation score, i.e., per-pixel accuracy, found in the complete non-parametric scene parsing pipeline. α and β are the parameters used for inferencing in MRF using graph cuts. λ represents the weight of the pairwise energy. k is the number of nearest neighbor images during retrieval. The higher accuracy is obtained in Tung et al. [9] because of use of content-adaptive windows which select only relevant objects in a scene.

Table 1 A brief summary of representative works on dense correspondence in non-parametric scene parsing

Dense Correspondence in Scene Parsing

Reference	Datasets	MRF Parameters	Accuracy (%)
Liu et al. [1]	LabelMe Outdoor SUN Dataset	$\alpha = 0.1, \beta = 60,$ $\lambda = 0.7$	74.57 64.45
Zhang et al. [8]	Google Street View Polo	N/A	73.1 82.5
Tung et al. [9]	SIFT Flow	$k = 400, \alpha = 0.1,$ $\gamma = 0.38, \lambda = 0.01$	79.8

It has been found that dense correspondence is complex and computationally expensive in terms of parameters like efficiency, memory consumption rate, time of execution, etc. as every pixel has to be taken into consideration and evaluation has to be done at pixel level. Because of these bottlenecks, most of the recent researches in the scene correspondence have been diverted toward the superpixel level matching.

On the other hand, partial correspondence is the alignment of the query image to the images in the retrieval set by considering a group of pixels at a time. Partial correspondence is further classified into (i) superpixels correspondence, (ii) patch-based correspondence, and (iii) region-based correspondence. Among the three, the most popular approach is the superpixels-based correspondence. Superpixel refers to a cluster of homogeneous pixels. The ability of the partial correspondence is to label cohesive groups of pixels together without loss of geometrical information of the image, and hence, the number of elements to process is significantly reduced. The steps for partial correspondence include:

- Oversegmentation of the query image and the retrieval set into superpixels/patch/regions.
- Extraction of features of the annotated training and the query superpixels.
- Superpixels or patch or region-based matching using the features extracted for calculation of the likelihood of each query segment to belong a particular class label of the retrieved set segments.

Table 2 gives the summary of partial correspondence approaches reported using various datasets resulting in evaluation score, i.e., per-pixel accuracy, found in the complete non-parametric scene parsing pipeline. α, β, λ, and k represents the same parameter as in Table 1. v refers to the number of top ranked similar images and r denotes the nearest neighbor points in the rare segments index. There is a significant increase in accuracy of 81.7% on SIFT flow data in Yang et al. [16] mainly because their approach expands the contents of the rare classes in the retrieval set by using exemplar on the rare class labels. A very high accuracy is achieved in Shuai et al. [19] because of the use of road scenes datasets where the number of class category is relatively less compared to other datasets.

It has been seen that label transfer via partial correspondence allows for more variation between the layout of the query image and the retrieved images in the retrieval

Table 2 A brief summary of representative works on partial correspondence in non-parametric scene parsing

Partial correspondence in scene parsing

Reference	Datasets	MRF parameters	Accuracy (%)
Eigen et al. [12]	Stanford	$v = 200, k = 10,$	75.3
	SIFT Flow	$r = 200$	76.8
Tighe et al. [14]	SIFT Flow	$\lambda = 16$	78.6
	LM + SUN		61.4
Singh et al. [11]	SIFT Flow	$k = 9, \lambda = 0.4$	77.2
	SUN09		49.5
Nguyen et al. [13]	SIFT Flow	$\lambda = 16, k = 1000$	78.9
	SUN		82.7
George et al. [17]	SIFT Flow		79.8
	LM-SUN		60.6
Yang et al. [16]	SIFT Flow	$\alpha = 0.7, \lambda = 0.6$	81.7
	LM-SUN	$\alpha = 0.9, \lambda = 0.6$	61.2
Di et al. [19]	ROAD1	$\alpha = 0.3, \lambda = 6$	94.7
	ROAD2		98.4
	ROAD3		98.1
Najafi et al. [20]	SIFT Flow		83.1
	LM-SUN		69.3
Talebi et al. [27]	LM-SUN	$\lambda = 1$	82.4
Razzaghi et al. [23]	SIFT Flow	$k = 15$	75.84
	LM-SUN	$k = 5$	44.05

set. Moreover, it gives better spatial support for combining features that belong to the same object, thus leading to high computational efficiency. The only drawback of partial correspondence is it cannot efficiently utilize the shape information.

4 MRF Inference

Both the dense and partial correspondence matching results are smoothened and final labeling is obtained using Markov Random Field (MRF) [24, 25] energy function. The initial labeling of the pixels at the scene correspondence module may be noisy and they need to be refined further using global context information. For example, a region marked as "car" cannot be surrounded by "water". The global semantic labeling is performed by minimizing the pairwise MRF energy function where the associated energy is factorized as a sum of potential functions defined on cliques of order strictly less than three. A pairwise energy function consists of a graph with G with a set $(\theta(.))_{i \in V}$ of unary potentials defined on single variables and a set $(\theta(.))_{i,j \in \mathcal{E}}$ of pairwise potentials defined on a pair of variables. The general MRF energy has the following form:

$$E_x = \sum_{i \in V} \theta_i(x_i) + \sum_{i,j \in \varepsilon} \theta_{ij}(x_{ij})$$ (7)

The first term in Eq. 7 is called the Data Term and the second term is the Smoothing Term. Semantic labeling is obtained by performing MAP inference on the energy function E using alpha/beta swap algorithm [26].

5 Discussions and Conclusion

In this paper, the different state-of-the-art approaches available in the scene correspondence module of the non-parametric scene parsing pipeline and their performances are discussed. It is found that the dense correspondence is less efficient and computationally expensive. Since partial correspondence has the ability to label cohesive groups of pixels simultaneously, it gives better spatial support for aggregating features that could belong to the same object. Hence, the partial correspondence is a preferred approach for most of the state-of-the-art researches and it is also highlighted in this paper. Scene correspondence by detecting salient regions of an image along with superpixels segmentation for matching of similar regions for label transfer will be a good research direction.

References

1. Liu, C., Yuen, J., Torralba, A.: Nonparametric scene parsing: label transfer via dense scene alignment. Institute of Electrical and Electronics Engineers (2009)
2. Tighe, J., Lazebnik, S.: Superparsing: scalable nonparametric image parsing with superpixels. In: European Conference on Computer Vision, pp. 352–365. Springer, Berlin, Heidelberg, 5 Sep 2010
3. Liu, C., Yuen, J., Torralba, A.: Sift flow: dense correspondence across scenes and its applications. IEEE Trans. Pattern Anal. Mach. Intell. 33(5), 978–994 (2011)
4. Qiu, W., Wang, X., Bai, X., Yuille, A., Tu, Z.: Scale-space sift flow. In: IEEE Winter Conference on Applications of Computer Vision, pp. 1112–1119. IEEE, 24 Mar 2014
5. Witkin, A.P.: Scale-space filtering. In: Readings in Computer Vision, pp. 329–332. Morgan Kaufmann 1 Jan 1987
6. Hassner, T., Mayzels, V., Zelnik-Manor, L.: Dense, scale-less descriptors. In: Dense Image Correspondences for Computer Vision 2016, pp. 51–70. Springer, Cham (2016)
7. Trulls, E., Kokkinos, I., Sanfeliu, A., Moreno-Noguer, F.: Dense segmentation-aware descriptors. In: Proceedings of the IEEE Conference on Computer Vision and Pattern Recognition 2013, pp. 2890–2897 (2013)
8. Zhang, H., Fang, T., Chen, X., Zhao, Q., Quan, L.: Partial similarity based nonparametric scene parsing in certain environment. In: CVPR 2011, pp. 2241–2248. IEEE 20 Jun 2011
9. Tung, F., Little, J.J.: Collageparsing: nonparametric scene parsing by adaptive overlapping windows. In: European Conference on Computer Vision, pp. 511–525. Springer, Cham, 6 Sep 2014
10. Alexe, B., Deselaers, T., Ferrari, V.: Measuring the objectness of image windows. IEEE Trans. Pattern Anal. Mach. Intell. 34(11), 2189–2202 (2012)

11. Singh, G., Kosecka, J.: Nonparametric scene parsing with adaptive feature relevance and semantic context. In: Proceedings of the IEEE Conference on Computer Vision and Pattern Recognition, pp. 3151–3157 (2013)
12. Eigen, D., Fergus, R.: Nonparametric image parsing using adaptive neighbor sets. In: IEEE Conference on Computer Vision and Pattern Recognition, pp. 2799–2806. IEEE, 16 Jun 2012
13. Nguyen, T.V., Lu, C., Sepulveda, J., Yan, S.: Adaptive nonparametric image parsing. IEEE Trans. Circuits Syst. Video Technol. **25**(10), 1565–1575 (2015)
14. Tighe, J., Lazebnik, S.: Finding things: image parsing with regions and per-exemplar detectors. In: Proceedings of the IEEE Conference on Computer Vision and Pattern Recognition, pp. 3001–3008 (2013)
15. Rubinstein, M., Liu, C., Freeman, W.T.: Joint inference in weakly-annotated image datasets via dense correspondence. Int. J. Comput. Vision **119**(1), 23–45 (2016)
16. Yang, J., Price, B., Cohen, S., Yang, M.H.: Context driven scene parsing with attention to rare classes. In: Proceedings of the IEEE Conference on Computer Vision and Pattern Recognition 2014, pp. 3294–3301 (2014)
17. George, M.: Image parsing with a wide range of classes and scene-level context. In: Proceedings of the IEEE Conference on Computer Vision and Pattern Recognition 2015, pp. 3622–3630, 2015
18. Collins, M., Schapire, R.E., Singer, Y.: Logistic regression, AdaBoost and Bregman distances. Mach. Learn. **48**(1–3), 253–285 (2002)
19. Di, S., Zhang, H., Mei, X., Prokhorov, D., Ling, H.: Spatial prior for nonparametric road scene parsing. In: 2015 IEEE 18th International Conference on Intelligent Transportation Systems, pp. 1209–1214. IEEE, 2015 Sep 15
20. Najafi, M., Taghavi Namin, S., Salzmann, M., Petersson, L.: Sample and filter: nonparametric scene parsing via efficient filtering. In: Proceedings of the IEEE Conference on Computer Vision and Pattern Recognition, pp. 607–615 (2016)
21. Giraud, R., Ta, V.T., Bugeau, A., Coupé, P., Papadakis, N.: SuperPatchMatch: an algorithm for robust correspondences using superpixel patches. IEEE Trans. Image Process. **26**(8), 4068–4078 (2017)
22. Barnes, C., Shechtman, E., Finkelstein, A., Goldman, D.B.: PatchMatch: A randomized correspondence algorithm for structural image editing. In: ACM Transactions on Graphics (ToG), vol. 28, Issue No. 3, p. 24. ACM, 27 July 2009
23. Razzaghi, P., Samavi, S.: A new fast approach to nonparametric scene parsing. Pattern Recogn. Lett. **1**(42), 56–64 (2014)
24. Boykov, Y., Kolmogorov, V.: An experimental comparison of min-cut/max-flow algorithms for energy minimization in vision. IEEE Trans. Pattern Anal. Mach. Intell. **1**(9), 1124–1137 (2004)
25. Kolmogorov, V., Zabih, R.: What energy functions can be minimizedvia graph cuts? IEEE Trans. Pattern Anal. Mach. Intell. **1**(2), 147–59 (2004)
26. Boykov, Y., Veksler, O., Zabih, R.: Fast approximate energy minimization via graph cuts. In: Proceedings of the Seventh IEEE International Conference on Computer Vision 1999, vol. 1, pp. 377–384. IEEE (1999)
27. Talebi, M., Vafaei, A., Monadjemi. S.A.: Nonparametric scene parsing in the images of buildings. Comput. Electr. Eng. **70**, 777–788 (2018)

Audio Surveillance System

Kedar Swami, Bhardwaz Bhuma, Semanto Mondal and L. Anjaneyulu

Abstract Surveillance systems are omnipresent in our day-to-day life as these systems ensure our safety, most of the surveillance systems are based on video sensory modality and machine intelligence. Video surveillance systems in real time are very expensive because they require very high processing. Most of the times, these lack reliability and robustness in real-time systems. To confront this issue, we have come up with real-time audio surveillance system. Where the audio signals are picked up using highly sensitive microphones and then the signal is amplified to a great extent. The picked-up signals contain a lot of other noise apart from the desired information. Hence, the procured audio signals are denoised with the help of signal processing technique called spectral subtraction. After denoising, there are chances of the mixing of audio signals so we are applying the speech separation techniques to get separate speech signals from the multi-speaker audio signal, which will help us to get clear insights of what a monitored person is speaking about.

Keywords Audio surveillance · Denoising · Deep neural network speech separation

K. Swami (✉) · B. Bhuma · S. Mondal · L. Anjaneyulu
National Institute of Technology, Warangal, India
e-mail: skedar@student.nitw.ac.in

B. Bhuma
e-mail: bbhuma@student.nitw.ac.in

S. Mondal
e-mail: msemanto@student.nitw.ac.in

L. Anjaneyulu
e-mail: anjan@nitw.ac.in

© Springer Nature Singapore Pte Ltd. 2020
S. Agarwal et al. (eds.), *Machine Intelligence and Signal Processing*,
Advances in Intelligent Systems and Computing 1085,
https://doi.org/10.1007/978-981-15-1366-4_28

351

1   Introduction

Perception of safety is a state of the mind concept yet the magnitude of demands and the rising costs of surveillance systems against the complexities of threats that the people face today has demanded new improved capabilities and innovative approaches. In modern-day society, surveillance is one of the most primary things among the masses as the rate of criminal activities by individuals and menace by terrorist groups has been intensified in recent times. Surveillance systems in many environments are widespread and common. These systems have played a major role in vouching for security at casinos, banks, correctional institutions, and airports. The surveillance systems are mainly based on video signal processing but these systems are very costly and require a lot of computational power.

Presently, video surveillance is widely used. Still, there are several difficulties that must be administrated in video surveillance like the field of view of a sensor network for capturing the entire region as well as the fact that several scenes may look normal even though an atypical situation is in progress [1]. The acoustic modality can be very much helpful in detecting the audio signals from all the directions as the microphones are omnidirectional. The advantages of acoustic sensors over visual sensors are: (a) lower computational requirements during information processing and (b) illumination conditions of the surrounding where the system is employed do not have an immediate effect on audio signals [2].

The paper is divided into the following sections. Section 2 represents audio surveillance system block diagram, audio amplifier circuit, denoising algorithm, and multi-speaker speech separation using deep learning network. Sections 3, 4, and 5 include implementation, results, and conclusion, respectively.

2   Audio Surveillance System

This section introduces the real-time audio surveillance system which first collects the audio signals from surroundings and then amplifies the signal to a great extent so that very weak signals can also be recognized. Post-amplification is done with the help of the denoising algorithm to eliminate and then the speech separation is done to get the insights of individual being monitored. The block diagram of the overall system is shown in Fig. 1.

2.1   Audio Signal Capturing and Amplification

This section gives details of the amplifier design with the audio capturing. For collecting audio from surrounding, very sensitive omnidirectional microphones are used.

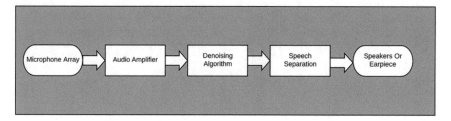

Fig. 1 Block diagram for audio surveillance system

The signals collected by microphones are then fed to the amplifier circuit. The amplifier circuit is used to increase the signal strength of weak signals. The audio amplifier is designed with LM386 IC. LM386 is a low-power audio amplifier. It consumes very less power (mostly batteries of 9v). The IC package is 8-pin mini-DIP [3]. The stereo amplifier is designed with the overall gain of 20 dB. The amplifier design is shown in Fig. 2.

2.2 Denoising

The performance of the digital processors which are used for the applications such as speech recognition, compression, and authentication is degraded by the acoustically added background noise [4, 5]. In this section, an exclusive noise suppression algorithm is discussed which is used for lessening spectral effects of noise superimposed to the speech acoustically. Spectral subtraction [6] is a few of the effective methods used for efficiently examining the speech signal. This also offers computationally efficient and processor independent approaches. The assumption in the method is that the noise is considered remains stationary. The stationary noise is subtracted from the speech signal by subtracting the bias calculated for spectral noise using non-speech activity. Residual noise remained after the spectral subtraction is attenuated using secondary procedures.

Considering the place of usage of our system, we can say that the background noise environment remains locally stationary which implies the spectral magnitude of noise prior to speech and during speech activity, remains the same. If the environment changes to a new stationary state, then we will have enough time to evaluate noise spectral magnitude in the new background before speech activity commences.

Firstly, the speech is low pass filtered with the filter of the bandwidth of 5 kHz and then it is analyzed by windowing data from half-overlapped input data buffers. The magnitude spectra of both the noise spectra during non-speech activity and the windowed data are calculated. The spectra of the noise signal are subtracted from the signal. After subtraction, if the resulting amplitudes are negative, then these coefficients are zeroed out. The remaining residual noise is attenuated using the secondary noise separation methods. A time waveform of the output denoised signal

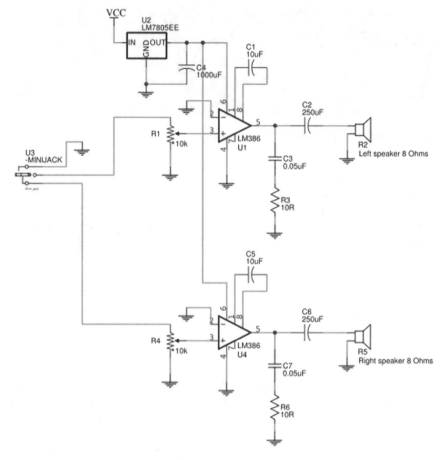

Fig. 2 Circuit diagram of audio amplifier

is obtained after the above procedures. The denoising block diagram is shown in Fig. 3.

Assume that the signal $y(k)$ is the sum of $n(k)$ and $m(k)$, windowed noise signal and the windowed speech signal, respectively. Then

$$y(k) = m(k) + n(k) \tag{1}$$

After converting the signals to the frequency domain, the equation becomes

Fig. 3 Block diagram of denoising algorithm used for removing the background noise

$$Y(e^{jw}) = M(e^{jw}) + N(e^{jw}) \tag{2}$$

where

$$y(k) \leftrightarrow Y(e^{jw}) \tag{3}$$

$$y(k) = \frac{1}{2\pi} \int_{-\pi}^{\pi} Y(e^{jw}) e^{jwk} dw \tag{4}$$

To represent the spectral subtraction, the magnitude $|N(e^{jw})|$ of $N(e^{jw})$ is replaced by its average value taken during non-speech activity so that the spectra of $N(e^{jw})$ can be readily measured. The phase $\alpha N(e^{jw})$ of $N(e^{jw})$ is replaced by the phase $\alpha_y(e^{jw})$ of $Y(e^{jw})$. The substitutions did above result in the spectral subtraction estimator $M(e^{jw})$:

$$M(e^{jw}) = [|Y(e^{jw})| - \mu(e^{jw})] e^{j\alpha_y^{jw}} \tag{5}$$

$$\mu(e^{jw}) = E\{|N(e^{jw})|\} \tag{6}$$

The spectral error can be given as follows:

$$\delta(e^{jw}) = \hat{M}(e^{jw}) - M(e^{jw}) = N(e^{jw}) \mu(e^{jw})(e^{j\alpha_y}) \tag{7}$$

Some simple modifications like magnitude averaging and half-wave rectification are applied for reducing auditory effects of the spectral error.

Local averaging of spectral magnitudes of non-speech activity is used to reduce the error, since the spectral error is differencing between the noise spectrum N and its mean μ. Replacing $|Y(e^{jw})|$ with $\overline{|Y(e^{jw})|}$

$$\overline{|Y(e^{jw})|} = \frac{1}{P} \sum_{i=0}^{M-1} |Y_i(e^{jw})| \tag{8}$$

$|Y(e^{jw})| = $ ith time windowed transform of $y(k)$ gives

$$M_A(e^{jw}) = \left[\overline{|Y(e^{jw})|} - \mu(e^{jw})\right] e^{j\alpha_y(e^{jw})} \tag{9}$$

The reason behind averaging is that the spectral error becomes approximately

$$\delta(e^{jw}) = M_A(e^{jw}) - M(e^{jw}) \approx \overline{|N|} - \mu \tag{10}$$

where

$$\overline{N(e^{jw})} = \frac{1}{P} \sum_{i=0}^{P-1} N_i(e^{jw}) \tag{11}$$

Thus, the sample mean of $N(e^{jw})$ will converge to $\mu(e^{jw})$ when the longer average is taken into consideration.

For each frequency w where the noisy signal spectrum magnitude $Y(e^{jw})$ is less than the average noise spectrum magnitude $\mu(e^{jw})$, the output is set to zero, and this is similar to half-wave rectification. Thus, the effect of half-wave rectification is to bias the magnitude spectrum by the noise bias determined at that frequency at each frequency w.

After subtracting the noise bias and half-wave rectification, we get the spectrum which will have very good SNR compared with input signal so inverse fast Fourier transform (IFTT) is taken and the signal is converted to the time domain.

2.3 Multi-Speaker Speech Separation

The output signal of the denoising script will be free from noise but the audio signal may have many speakers speaking at the same time. We are often confronted with the problem of selectively attending the object whose features are intermixed with one another in the incoming sensory signal. In audio, there is a corresponding problem known as auditory scene analysis [7–9], which seeks to identify the components of audio signals corresponding to individual sound sources in a mixture signal.

The problem of acoustic source separation is addressed with deep learning framework called "deep clustering." Instead of estimating the signals with masking functions, a trained deep neural network model from Hershey et al. [10] is used. The network proposed in the experiments has four layers of bidirectional long short-term memory (BLSTM), followed by one feedforward layer. The input to the network is encoding of the audio signal generated by STFT. Respective BLSTM layer has 600 hidden cells and feedforward layer corresponds with embedding dimension, stochastic gradient descent with momentum 0.9 and a fixed learning rate 10–5 was used for training. In each updating step, Gaussian noise with zero mean and 0.6 variance was added to the weight. The network was trained on the dataset of three males and three females which produces spectrogram embeddings that are discriminative for partition labels given in training data. The audio duration per person was 30 min; the training set was generated by mixing the two audios with Python script. The given deep learning framework performs outstandingly well in speech separation of two speakers at a time.

During testing, the speech separation was accomplished by constructing the time domain signals depending on particular time–frequency masks per speaker. The time–frequency masks for all source speakers were obtained by clustering the row embedding vectors W, where W was outputted from the introduced model for each segment similar to the training set. The number of clusters shows the number of speakers in the signal. The trained model works perfectly for the speech separation of three-speaker mixtures also simply by changing the clustering step to three from two clusters.

3 Implementation

The microphones are hanged out at a place where one needs to surveil. The designed audio amplifier amplifies the perceived signal and simultaneously, the signal is sent to the Raspberry Pi model. The denoising script is then executed for denoising the signal and after that speech, separation will be completed. The proposed system is shown in Fig. 4.

Fig. 4 Audio surveillance system implementation

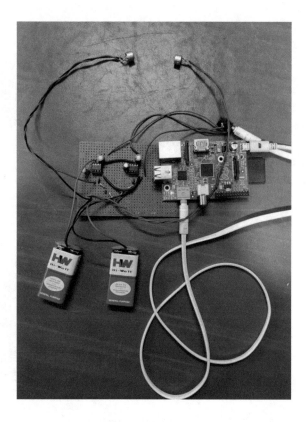

4   Results

The proposed system works very efficiently in real time. The spectrum of signal with and without noise is shown in Fig. 5 and the time domain signal representation of an original and denoised signal is shown in Fig. 6.

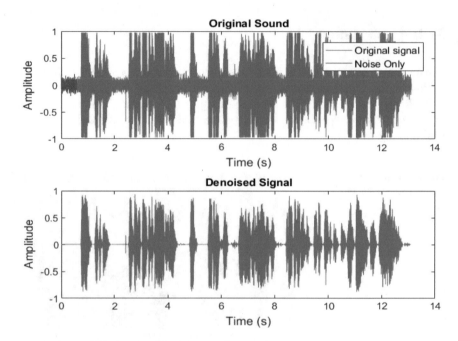

Fig. 5   Time domain signal representation of original and denoised signal

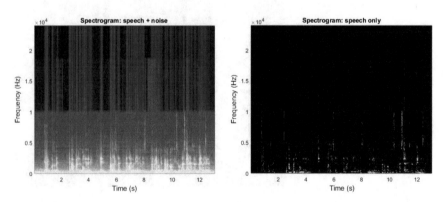

Fig. 6   Spectrum of original signal amplified by amplifier block and the spectrogram of signal after applying proposed denoising algorithm

Fig. 7 Time domain representation of original signal amplified by amplifier block and the spectrogram of signal after applying proposed denoising algorithm

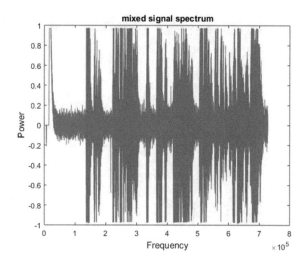

The multi-speaker speech separation also works perfectly with two speakers at a time audio signal irrespective of the gender of the speaker. The spectrograms in Figs. 7 and 8 show the speech separation performed by our proposed system.

The results of the system cannot be very reliable if the background environment changes frequently and we are working on eliminating the effects of the randomly changing background.

5   Conclusion

The paper aims at an old, but an interesting problem which has a lot of practical applications now. The problem of denoising is very complex as noise is highly random but our system works for stable background conditions if random background variations are there and the performance of the system is not up to the expectations. The speech separation works perfectly for two speakers. This research work aims at improving the surveillance systems and the addition of audio surveillance system will improve the overall efficiency of surveillance systems.

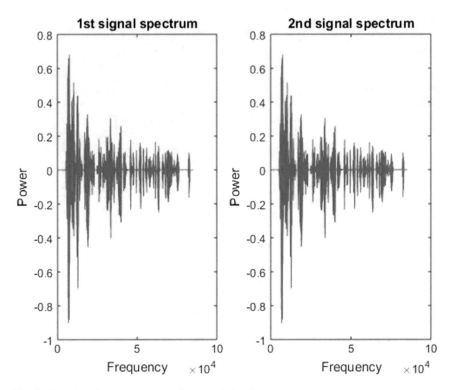

Fig. 8 Time domain representation of separated signals

References

1. Ntalampiras, S.: Audio surveillance. WIT Trans. State Art Sci. Eng. **54**. ISSN 1755-8336
2. Crocco, M., Cristani, M., Trucco, A., Murino, V.: Audio surveillance: a systematic review. ACM Comput. Surv. (CSUR) (2016)
3. Vimal Raj, V., Aravindan, S., Jayaprakash, Agnishwar, Harith, M., Manivannan, R., Bhuvaneshwaran, G.: Comparative study of audio amplifiers. IOSR-JECE **12**(1), 32–36 (2017). ver. I
4. Sambur, M.R., Jayant, N.S.: LPC analysis/synthesis from speech inputs containing quantizing noise or additive white noise. IEEE Trans Acoust. Speech Signal Proc. (ASSP) **24**, 488–494 (1976)
5. Boll, S.: Suppression of acoustic noise in speech using spectral subtraction. IEEE Trans. Acoust. Speech Signal Process. **27**(2), 113–120 (1979)
6. Bregman, A.S.: Auditory Scene Analysis: The Perceptual Organization of Sound. MIT press, Cambridge (1994)
7. Darwin, C.J., Carlyon, R.P.: Auditory grouping. In: Moore, B. (ed.) Hearing. Elsevier (1995)
8. Foote, J.T.: An overview of audio information retrieval. ACM-Springer Multimedia Syst. **7**(1), 2–11 (1999)
9. Gold, B.: Robust speech processing. M.I.T. Lincoln Lab. Tech. Note 1976-6, DDC AD-A012 P99/0. 27 Jan 1976
10. Hershey, J.R., Chen, Z., Le Roux, J., Watanabe, S.: Deep clustering: discriminative embeddings for segmentation and separation. CoRR (2015)

MOPSA: Multiple Output Prediction for Scalability and Accuracy

B. K. Dhanalakshmi, K. C. Srikantaiah and K. R. Venugopal

Abstract Cloud computing is entailed for elasticity, scalability, and accuracy for the multi-sharing systems; predicting resource for the multi-sharing system in an existing system is not accurate which leads to under or over-provisioning of resources and increases higher cost for on-demand resources. In this paper, we develop a prediction model called Multiple Output Prediction for Scalability and Accuracy (MOPSA). The key characteristic of this model is to improve accuracy of prediction of resources with multiple outputs in multi-sharing system by using gradient descent method. Experimental results exhibit that the proposed model improves the prediction accuracy of resources and provides scalability in multi-sharing system for minimizing cost by reducing number of on-demand resources.

Keywords Cloud computing · Gradient descent · Multi-sharing · Prediction

1 Introduction

In today's world, we are surrounded by large amounts of data and resources in cloud environment and processing those amount of data is very difficult. The resources in cloud are shared among many users which is said to be multi-sharing system. The allocation of resources dynamically to multi-users in multi-sharing system is indispensable in terms of elasticity, scalability, and accuracy. The transactions and activities of multi-users are stored in log files; extracting and analyzing these log files are tedious [1]; and it can be simplified by using preprocessing techniques. The information about each user like resource usage, time logs, past information, etc,

B. K. Dhanalakshmi (✉) · K. C. Srikantaiah
S J B Institute of Technology, Visvesvaraya Technological University, Bangalore 560064, Karnataka, India
e-mail: malathi.dl@gmail.com

K. C. Srikantaiah
e-mail: srikantaiahkc@gmail.com

K. R. Venugopal
Bangalore University, Bangalore 560001, Karnataka, India

© Springer Nature Singapore Pte Ltd. 2020
S. Agarwal et al. (eds.), *Machine Intelligence and Signal Processing*,
Advances in Intelligent Systems and Computing 1085,
https://doi.org/10.1007/978-981-15-1366-4_29

361

is extracted from log files by using preprocessing techniques such as data cleaning and filtration, data cube construction, on-line analytical processing (OLAP). The preprocessing is used for converting unstructured log files to structured log files.

The structured log files are to use to compute the work-flow and workload. Work-flow is an activity of each user, workload is defined as resources required to execute the activity of an user, and it is formed by using work-flows of corresponding users. The work-flows are grouped based on the types of workloads like static workload, periodic workload, once-in-a-lifetime workload, unpredictable workloads, and continuously changing workloads [2]. Prediction of resources like CPU, memory, bandwidth, and disk accurately for workloads is a meticulous job. In an existing model, the prediction models predict one set of value as single output in multi-sharing system based on reservation plans, where reservation plans are on basis of daily, weekly, monthly, half-yearly, and yearly. Single output like CPU, memory, and bandwidth are one set of predicted output which is not dependable on number of jobs. Suppose, if allocation of resources are not utilized completely or over utilized which leads to over provisioning and under provisioning respectively and on-demand resources uncertainty was not known in prior and it incurs more cost. So, to overcome this problem prediction model Multiple Output Prediction for Scalability and Accuracy (MOPSA) is proposed.

MOPSA is designed by using a gradient descent method [3] for predicting resources accurately for the multi-sharing system in different reservation plans. The multiple outputs are predicted based on number of jobs for multi-users in multi-sharing system and one of the outputs is selected based on the previous usage of resource history on same resources type by using this information administrator reserves the resources to respective users. The goal of the model is to achieve higher accuracy of prediction resources that in turn minimizes the cost by reducing on-demand resources and avoids over-provisioning or under-provisioning of resources in multi-sharing system.

The rest of the paper is organised as follows: Sect. 2 discusses the related works and covers necessary background by defining the concept of multi-resource sharing. Section 3 defines the problem definition. Section 4 describes the system architecture and calculates the mathematical model. Section 5 represents the system model and the experimental results. Section 6 concludes the paper.

2 Related Work

Da silva et al. [4] proposed a prediction model for resource utilization by using regression and queuing network-based approaches based on workload arrangement to predict the resources like CPU, memory, disk, and network usage. Pietri et al. [5] designed a periodical work-flow for scientific and business analysis field by applying integer programming model and precedence tree-based heuristic (PTH) to determine the sufficient amount of resources for many periodical work-flow operations and minimization of cost and portray the different applications of log file analysis. Pham et al. [6] discussed some fundamental techniques of log file analysis and three case studies to illustrate a different application of analyzing log files and the log files are most important to calculate the resource utilization.

Zheng et al. [7] explored the provisioning strategies to gratify forthcoming resource demands in cloud hosting platforms to develop a prediction model using neural network and linear regression methods. Comer et al. [8] proposed a model by using hierarchical temporal memory (HTM) to execute the work-flow tasks by detecting anomalies to satisfy upcoming resource demands in cloud hosting platforms to develop a prediction model on real-time devices by collecting continuously monitoring the consumed resources. Vora et al. [9] proposed a method using hidden Markov modeling (HMM) to predict variations of workload patterns efficiently for provisioning computing resources in the cloud and cloud manager analyses the capabilities and predicting workload on the virtual machines (VMs).

3 Problem Definition

Consider a set U of n users, where $U = \{$ A, B, C, ..., Z and the set of R resources where $R = \{$ CPU, Memory, Bandwidth, Disk, IO Devices} and set of P Reservation Plans where $P = \{$ Week, Month, Year, Half-Yearly, Yearly} and resources usage of users are extracted from log files D.

Our objective is to design a prediction model called (MOPSA) to minimize the cost by reducing number on-demand resources and to avoid over-provisioning or under-provisioning of resources.

4 Proposed Model

Consider 'n' number of users in multi-sharing system and the resources are shared among users and during utilizing the shared resources, their activities are stored in the unstructured data. To convert unstructured log files into structured log files, the data mining preprocessing techniques are applied. By processing these structured log files (D), the transaction and activity of each user are extracted. An activity is drawn as work-flow (Wf) which represents the resource usage of a user, and these work-flows are grouped into each type of workloads (Wd) based on the resource types r_i. The users exceed their resource usage from reserved resources; then, such user is said to be on-demand user and the actual resources usage $R_{actual,}$ and reserved resources are also extracted from log files. The number of resource usage is computed by actual and reserved resources by using following formula: $R_{usage} = R_{actual} - R_{reserved}$.

where R_{actual} is actual resources utilized by each user and $R_{reserved}$ is resources reserved for each user.

Table 1 On-demand request

User resources	Reserved on-demand	Request
A	3	1
B	2	0
C	1	1
D	2	1
E	3	0
F	1	1

If the R_{usage} value is positive, then user has requested for on-demand resources and assigns value 1 (yes) else assign 0 (not requested for on-demand resources) as shown in Table 1.

The number of on-demand user is obtained by adding the on-demand request column and gives the total number of on-demand users.

The on-demanded users are classified into class 1 and class 2 as based on the amount of resource usage; if on-demand users have utilized all the resources in R, then we classify into class 2, otherwise classified as class 1. Once it is classified as class 2, the prediction is done only once because all on-demand users have used same type of resources, and for class 1, we need to apply prediction for every type of resources. This process is done to reduce time and on-demand users are considered for prediction of future resource usage.

Next, predict the resource usage (R_{Predi}) of ith resources in $R_{(ri)}$ by actual resources (R_{actual}) using gradient descent method which is an optimization algorithm to find the optimal values of intercept a and slope line b which reduces the prediction error and defined as:

$$a = \left(\sum\right)\left(\sum x^2\right) - \left(\sum x \sum xy / n \left(\sum x^2\right) - \sum x\right)^2 \tag{1}$$

$$b = n\left(\sum xy\right) - \left(\left(\sum x\right)\left(\sum y\right)/n\left(\sum x^2\right) - \sum x\right)^2 \tag{2}$$

where x-axis is users and y-axis is actual usage of resources and n is number of users.

$$R_{\text{predi}} = a + b\, R_{\text{actuali}} \tag{3}$$

where $1 < i < n\, r_i$ and R_{Pred} is predicted resource for the future. The dotted line gives the resource usage (R_{Pred}) of given resources(R_{actual}) and bold line gives the actual resource usage from obtained data actual resources (R_{actual}). The variation between R_{actual} and R_{Pred} (drawn by the dotted lines) is the prediction error (E) as shown in Fig. 1.

Now, to find optimal a, b that minimizes the error between actual and predicted values of resource usage by using sum of squared error (SSE) metric as shown in Eq. (4):

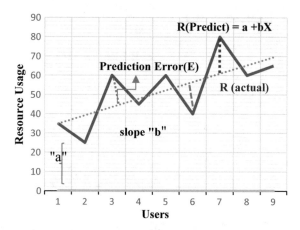

Fig. 1 Comparison of actual and predicted resources

$$SSE = 1/2 \, \text{Sum}\left(R_{\text{actuali}} - R_{\text{predi}}\right)2 \tag{4}$$

is defined as in Eq. (5) and (6):

$$\partial SSE/\partial a = -\left(R_{\text{actuali}} - R_{\text{predi}}\right) \tag{5}$$

$$\partial SSE/\partial b = -R_{\text{actuali}} - R_{\text{predi}})r_i \tag{6}$$

$\partial SSE/\partial a$ and $\partial SSE/\partial b$ are the gradients, and they show the direction of the movement of a, b with respect to SSE as shown in Eq. (5).

Here,

$$SSE = 1/2 \, \text{Sum}\left(R_{\text{actual}} - R_{\text{pred}}\right)2 \tag{7}$$

Generation of multi-outputs using gradient descent as defined:

$$\text{Hypothèses} = h_{(\Theta)}(x) = \Theta^t x = \Theta_0 x_0 + \Theta_1 x_1 + \Theta_2 x_2, \ldots, \Theta_n x_n$$

where X_0 is resource-type CPU, X_1 is resource-type bandwidth, and so on; nr_i is number of resource types in R, and m is the number of output options.

$$J(\Theta) = \frac{1}{2m} \sum_{i=1}^{m} h_{\Theta} x^{(i)} - y^{(i)})^2 \tag{8}$$

$$\Theta_0 = \Theta_0 - \alpha \frac{1}{m} \sum_{i=1}^{m} h_{\Theta} x^{(i)} - y^{(i)} x_0^{(i)} \tag{9}$$

Gradient descent

$$\Theta_j := \Theta_j - \alpha \frac{\partial}{\partial \Theta_j} J(\Theta) \tag{10}$$

When $(n = 1)$,

$$\Theta_0 = \Theta_0 - \alpha \frac{1}{m} \sum_{i=1}^{m} (h_\Theta x^{(i)} - y^{(i)}) \tag{11}$$

$$\Theta_1 = \Theta_{01} - \alpha \frac{1}{m} \sum_{i=1}^{m} (h_\Theta x^{(i)} - y^{(i)}) \tag{12}$$

This process repeats till we obtain prediction single valve. When $(n \geq 1)$,

$$\Theta_1 = \Theta_1 - \alpha \frac{1}{m} \sum_{i=1}^{m} h_\Theta x^{(i)} - y^{(i)} x_j^{(i)} \tag{13}$$

$$\Theta_1 = \Theta_1 - \alpha \frac{1}{m} \sum_{i=1}^{m} h_\Theta x^{(i)} - y^{(i)} x_1^{(i)} \tag{14}$$

$$\Theta_0 = \Theta_0 - \alpha \frac{1}{m} \sum_{i=1}^{m} h_\Theta x^{(i)} - y^{(i)} x - 0^{(i)} \tag{15}$$

$$\Theta_1 = \Theta_1 - \alpha \frac{1}{m} \sum_{i=1}^{m} h_\Theta x^{(i)} - y^{(i)} x - 1^{(i)} \tag{16}$$

The prediction of multiple outputs is obtained by using gradient descent method by equations from (8) to (16); the entire model is tuned to scale up and/or scale down the resources which provides scalability to multi-sharing system.

MOPSA: Multiple Output Prediction for Scalability and Accuracy.

Algorithm 1 MOPSA the proposed algorithm

Purpose: To predict resources for on-demand users
Input: Log files
Output: Multiple output predictions
Step 1: Obtain structured log file D from the given log file using data mining preprocessing techniques.
Step 2: Compute work-flows and workloads from D
Step 3: For each user $ui \in U$ extract $R_{actuali}$ and $R_{reservedi}$
Step 4: Compute $R_{usagei} = R_{actuali} - R_{reservedi}$
if R_{usagei} is positive, then on-demand request [i] $= 1$ else on-demand request [i] $= 0$

Step 5: For each user $ui \in U$ for which on-demand request [i] = 1, compute the number of resource utilization (nr_i); if $(nr_i =$ number of resource types in R), i.e., user u_i utilized all resource types in R, then classify user u_i as class 2 else class 1
Step 6: Compute the values a and b using Eqs. (1) and (2)
Step 7: Calculate the R_{predi} using Eq. (3)
Step 8: Calculate the gradients by using SSE from Eq. (4)
Step 9: Repeat the step 6 and 8 till the error reduced significantly
Step 10: Produced m number of multiple output options for multi-sharing system for multi-users using Eqs. (8, 16).

5 Experimental Setup

We carried out the experiment for predicting accurate resources for multi-sharing system by dividing into two phases: In the first phase, log files are extracted using preprocessing techniques. Log files contain all information about each user. Information is selected carefully by preprocessing the log files. The three parameters to be extracted by preprocessing log files are namely (i) cost of each transaction, (ii) CPU utilization per transaction, and (iii) time taken to execute each transaction. In the second phase, the software application Comindware Tracker is used to generate work-flows and convert it into a workload. The system is configured as follows: CPU (32-bit) processor intel $i7$ CPU (2.9 GHz) with 32 GB RAM and 2 TB hard disk. The IDE of choice is Net beans 8.1 editor, and it is coded in Java language, MOPSA is implemented by using a cloud simulator tool of version 3.03, and the size of the workload is **1 GB** of data.

5.1 Evaluation Criteria

We measure the accuracy of the MOPSA models generated from different learning metrics based on root mean square error (RMSE) and sum of squared error (SSE) metrics. The root mean square error (RMSE) metric is defined as [10], and SSE is defined in Eq. (2):

$$\text{RMSE} = \sqrt{\sum_n^{i=1} \left| R_{\text{actual}} \wedge R_{\text{predi}} / nr_i \right|} \tag{17}$$

where R_{actual} is the actual output, R_{predi} is the predicted output and nr_i, the number of resource types in R. A lesser RMSE value proves a more effective prediction model.

Table 2 Comparison between metrics for SOP and MOP accuracy

Metrics	Single output prediction (%)	Multi-output prediction (%)
RMSE	2.1	1.1
SSE	2.1	1.1

Table 3 Comparison between existing model and proposed model

Model	Cost	Accuracy	On-demand resources
RPMRS	Low	Not accurate	Yes
MOPSA	Very low	Accurate	No

5.2   Results and Discussion

Effect of Accuracy: The prediction accuracy of resource requirement in single output prediction (SOP) using RPMRS model and in multiple output prediction (MOP) using MOPSA model is shown as shown in Table 2. From the table, prediction error is less in SSE is because of using gradient descent method.

Effect of Cost: The existing model and proposed model are compared with three parameters: cost, on-demand request, and accuracy. In the RPMRS model, the cost is low, but it is not accurate due to single output prediction, so it leads to on-demand resources. In the MOPSA model, the cost is very low compared to RPMRS model due to no on-demand resources because it predicts multiple outputs as shown in Table 3.

6   Conclusion

Predicting accurate resource utilization for future usage in multi-sharing system is tedious. It overcomes the downsides of single output prediction with one set of value which leads to over-provisioning and under-provisioning of resources and a greater number of on-demand resources which incurs more cost. In this paper, an accurate prediction MOPSA model is proposed for multi-sharing system based on reservation plans for multi-user. So, MOPSA is designed by using a gradient descent method for predicting resources accurately for the multi-sharing system. The multiple outputs are predicted based on number of jobs for multi-users in multi-sharing system, and one of the outputs is selected based on previous usage of log files on same resources type. Hence, it is concluded that it minimizes the cost by reducing number of on-demanded resources and achieves higher accuracy of prediction resources and avoids over-provisioning or under-provisioning of resources in multi-sharing system. Predictions can be still more accurate by using random forest method for multi-sharing system.

References

1. Valdman, J.: Log _le analysis, Technical Report, DCSE/TR-2001-04, pp. 51 (2001)
2. Rodriguez, M.A., Kotagiri, R., Buyya, R.: Detecting performance anomalies in scientific workflows using hierarchical temporal memory. Futur. Gener. Comput. Syst. **88**, 624635 (2018)
3. Dhanalakshmi, B.K., Srikantaiah, K.C., Venugopal, K.R.: Reservation policy for multi sharing resources in heterogeneous cloud user. In: Computing and Net Work Sustainability- IRSCNS, Springer International Conference, Goa (2018)
4. da Silva, R.F., Juve, G., Rynge, M., Deelman, E., Livny.: Online task resource consumption prediction for scientificWorkows. Parallel Process. Lett. **25**, 1541003 (2015)
5. Pietri, I., Juve, G., Deelman, E., Sakellariou, R.: A performance model to estimate execution time of scientific workflows on the cloud
6. Pham, T.P., Durillo, J.J., Fahringer, T.: Predicting workflow task execution time in the cloud using a two-stage machine learning approach. IEEE Trans. Cloud Comput. **7161**, 113 (2017)
7. Zheng, Z., Wu, X., Zhang, Y., Member, S., Lyu, M.R., Wang, J.: QoS ranking prediction for cloud services. IEEE Trans. Parallel Distrib. Syst. (2012)
8. Comer, D.K., Clark, C.R., Canelas, D.A.: Writing to learn and learning to write across the disciplines: Peer-to-peer writing in introductory-level MOOCs. Int. Rev Res. Open Distance Learn. **15**, 2682 (2014)
9. Vora, M.N.: Predicting utilization of server resources from log data. Int. J. On Comput. Theory Eng. **6**, 118123 (2014)
10. Islam, S., Keung, J., Lee, K., Liu, A.: Empirical prediction models for adaptive resource provisioning in the cloud. Futur. Gener. Comput. Syst. **28**, 155–162 (2012)

Impact of Cluster Sampling on the Classification of Landsat 8 Remote Sensing Imagery

Vikash Kumar Mishra and Triloki Pant

Abstract Remote sensing images are rich enough in terms of both qualitative and quantitative information. Extraction of different land features on earth surface from imagery is highly required. Since satellite imagery includes very large data sets, it is quite difficult to deal with such large data sets, specially extracting the land features. Classification techniques help researchers and analyst to extract such features available on earth surface which further depends on the sampling methods applied for testing purpose. In this study, the focus is on sampling methods, how the samples should be collected so that the classification of satellite imagery can be done effectively and efficiently. Out of many sampling schemes stratified random sampling and cluster sampling Congalton RG, Green K (Assessing the accuracy of remotely sensed data: principles and practices. CRC Press, 2008 [1]), Hashemian MS, Abkar AA, Fatemi SB (Int. Congr. Photogramm. Remote. Sens. 2004 [2]) and Vieira CAO, Santos NT (Analysis of sampling methods and its influence on image classification process of remotely sensed images through a qualitative approach, pp. 6773–6780, 2009 [3]) are used to classify Landsat 8 imagery, and it is found that cluster sampling outperforms with an accuracy up to 95.8%.

Keywords Remote sensing · Supervised classification · Unsupervised classification · Confusion matrix and sampling

1 Introduction

An image is worth more than billion of words but satellite imagery is worth more than million acres of land which consist of different earth surface features. An image is the collection of pixels, i.e., picture elements and the process of sorting pixels into a finite number of individual classes, or categories of data, based on their spectral response is image classification. The purpose of image classification is to replace the visual analysis of the image data with quantitative techniques for automizing the

V. K. Mishra (✉) · T. Pant
Indian Institute of Information Technology, Allahabad, India
e-mail: rsi2018505@iiita.ac.in

© Springer Nature Singapore Pte Ltd. 2020
S. Agarwal et al. (eds.), *Machine Intelligence and Signal Processing*,
Advances in Intelligent Systems and Computing 1085,
https://doi.org/10.1007/978-981-15-1366-4_30

feature identification in an image. A satellite image is actually composed of multiple bands of electromagnetic spectrum; all the bands together form an image but that only gives spectral information. So converting this spectral information into some land use land cover information in classes is the process of classifying an image.

Information classes are categorical classes, such that pixel represents a water area, urban area, forest area, or some other area. So based on the spectral response of the pixel, it can be inferred what the pixel is going to represent. The process of classification involves two stages classification and labeling. Classification is the process to collect similar pixels representing same land cover on the basis of similarity or dissimilarity measures, whereas labeling is the process to assign some name to these collections of similar pixels.

The classification system must possess the following criteria to classify an image accurately and efficiently in appropriate classes [1–3]:

a. There should be minimum 85% accuracy obtained to interpret the land use land cover categories using remote sensing imagery. The accuracy of classification or categorization should be approximately equal for all the classes; it should not be the case that one class has 85% accuracy and other is having 50% [4]. Big differences are not acceptable.
b. Repetitive results should be obtainable from one interpreter to another and from one time of sensing to another. The classification system should be applicable over extensive areas. The classification system should be suitable for use with remote sensor data obtained at different times of the year.
c. Effective use of subcategories that can be obtained from ground surveys or from the use of larger scale or enhanced remote sensor data should be possible, aggregation of categories must be possible, comparison with future land use data should be possible, and multiple uses of land should be recognized when possible.

To classify the land covered area into categorical classes based on land use, we need to follow the following steps:

a. Choose a study region and define the purpose of classification, why is it needed and how it will be used.
b. Decide if it is really required and check for its feasibility.
c. Design and develop a classification method or choose the existing one.
d. Find out remote sensing imagery corresponding to your study region and preprocess the imagery for classification.
e. Collect ancillary data, for example, samples for supervised classification, and classify the imagery.

The United States Geological Survey (USGS) proposed nine major categories of earth surface feature which can be further subcategories at lower levels; Table 1 shows USGS Earth Surface categories up to level-II [4].

Table 1 Land use and land cover classes of remote data by USGS

Level-I	Level-II
1. Urban or Built-up Land	(1.1) Residential. (1.2) Commercial and Services. (1.3) Industrial. (1.4) Transportation, Communications, and Utilities. (1.5) Industrial and Commercial Complexes. (1.6) Mixed Urban or Built-up Land. (1.7) other Urban or Built-up Land
2. Agricultural Land	(2.1) Cropland and Pasture. (2.2) Orchards, Groves, Vineyards, Nurseries, and Ornamental Horticultural Areas. (2.3) Confined Feeding Operations. (2.4) other Agricultural Land
3. Rangeland	(3.1) Herbaceous Rangeland. (3.2) Shrub and Brush Rangeland. (3.3) Mixed Rangeland
4. Forest Land	(4.1) Deciduous Forest Land. (4.2) Evergreen Forest Land. (4.3) Mixed Forest Land
5. Water	(5.1) Streams and Canals. (5.2) Lakes. (5.3) Reservoirs. (5.4) Bays and Estuaries
6. Wetland	(6.1) Forested Wetland. (6.2) Non-forested Wetland
7. Barren Land	(7.1) Dry Salt Flats. (7.2) Beaches. (7.3) Sandy Areas other than Beaches. (7.4) Bare Exposed Rock. (7.5) Strip Mines, Quarries and Gravel Pits. (7.6) Transitional Areas. (7.7) Mixed Barren Land
8. Tundra	(8.1) Shrub and Brush Tundra. (8.2) Herbaceous Tundra. (8.3) Bare Ground Tundra. (8.4) Wet Tundra. (8.5) Mixed Tundra
9. Perennial Snow or Ice	(9.1) Perennial Snowfields. (9.2) Glaciers

1.1 Classification Methods

For the classification of remote sensing images, generally three approaches are used: supervised classification, unsupervised classification, and hybrid approach of classification [5–7].

1.1.1 Supervised Classification

This approach of classification is implemented in three steps, viz. training, classification, and testing.

Training: On the basis of a known area also called as "training area" or "sample area" different "numerical interpreter key" is computed. Each numerical interpreted key describes spectral attribute of each earth surface feature.

Classification: Each pixel from the data set is compared to numerical interpreter key categorically and labeled with the name of category it looks most like.

Testing: Test samples from the classified image are selected and matched with the pixel values of original image to verify what number of pixels are properly classified and what number of pixels are misclassified; on the basis of these numbers, confusion matrix is formed. This accuracy matrix is used to find the accuracy of classification method.

1.1.2 Unsupervised Classification

This is a pixel-based automated classification approach. This approach completes in two stages classification and testing.

Classification: User specifies the number of classes and spectral groupings that are formed based on available data set. Clustering algorithms are used to determine the natural, statistical grouping of the data. The pixels are grouped together into classes based on their spectral similarity.

1.1.3 Hybrid Approach of Classification

This is the combined approach of supervised and unsupervised classification. The intension of using hybrid approach is to increase the efficiency and accuracy of classification methods.

2 Sampling

Sampling is the process of selecting some data points from whole set in such a way that on the basis of selected data points the whole set could be predicted. In remote sensing classification, some data from each existing classes is needed so that spectral information could be converted to LULC information. Such data is referred as samples or training data. On the basis of these samples, spectral signatures of the earth features are computed which help to classify the imagery.

2.1 Sampling Rule

Selection of sample is very important and is the critical part of classification. Inaccurate sampling may lead to misclassification of remote sensing imagery. One should take care of the following facts to avoid inaccurate sampling [1–3]:

a. The reference data and satellite data should belong to same time period (Temporal resolution should be same).
b. We should have enough number of samples (η) and it could be decided through Eq. 1:

$$\eta = z \times \frac{pq}{e^2} \qquad (1)$$

$z = 2 \times$(standard deviation), $p =$ standard accuracy, i.e., between 85 and 95%, $q = 1-p$, $e =$ elloyable error $\approx 100\%\text{-}z$

c. If the area is large (more than million acres) or land use category is more than 12, then the number of samples should be increased to 75–100.
d. More samples can be taken in more important categories and less in less important categories.
e. The minimum number of pixel samples by class can be obtained by Eq. 2:

$$\text{Samples} \equiv 30 \times N \times C \qquad (2)$$

N is the number of discriminate variables, and C is the number of classes.

2.2 Sampling Methods

Samples from a remote sensing imagery can be selected through any of the following methods [1–3].

2.2.1 Simple Random Sample (RS)

Simple random sample method has no restrictions on the randomization. This means that all sample points are selected with equal probability and independently from each other. Figure 1 shows the complete study area as a whole having five classes represented by dotted lines; the disadvantage of this sampling is that some of the region may be ignored due to the randomness, right corner is ignored, and it is a valid separate class.

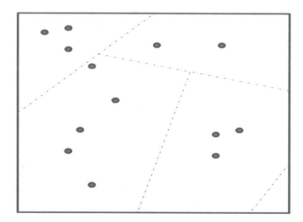

Fig. 1 Samples chosen randomly as simple random sampling

Fig. 2 Samples chosen randomly by dividing the region in stratum

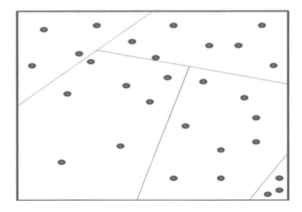

2.2.2 Stratified Random Sampling (SRS)

In this method, the complete study area is divided into small parts; each part is called as strata and then samples are randomly chosen from each stratum. The advantage of this method is that the study area has predefined classes, and when samples are chosen, no class is ignored whether small or big. Figure 2 shows five classes, and samples are chosen from each class even lower right corner which is a small class also sampled properly.

2.2.3 Systematic Sampling (SS)

This method divides the complete area into square grids, and samples are taken as one sample per square grid. Samples are collected at some equal interval over space (Fig. 3).

Fig. 3 Samples chosen systematically

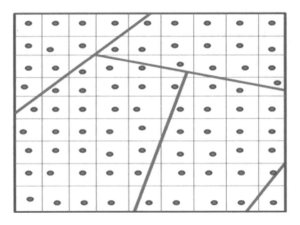

Fig. 4 Samples chosen as
cluster

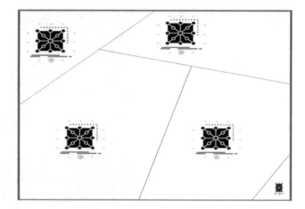

Fig. 4 Samples chosen as cluster

2.2.4 Cluster Sampling (CS)

It is a method of sampling in which one pixel is chosen and nearby pixels is also counted to group them into a single sample. Figure 4 shows the cluster sampling, where one pixel in center is chosen and neighboring pixels are included to form one cluster sample.

3 Study Area and Data Set

The study area covers the district of Allahabad in the state of Uttar Pradesh, India. The Triveni Sangam or Sangam is the place where two rivers Ganges and Yamuna used to meet, this region majorly focused to carry out the study. Landsat 8 imagery from the United States Geological Survey (USGS) is downloaded free from USGS Web site to carry out the study [8].

Landsat is an earth observation series of satellites by United States of America, launched and managed by USGS. Landsat 8 have 11 different bands out of which initial nine bands are known as operational land imagery (OLI) and band 10 and band 11 are known as Thermal Infrared Sensors (TIRS) [11]. So for the observation of different earth surface feature only OLI bands are used. Mishra and Pant [9, 10] found that band 7 is most suitable for the identification of water region from multiple sources. So band 7 is used for the study of sampling and classification of the earth features near Sangam region (Fig. 5).

4 Proposed Method

The proposed method is explained through a flow chart in Fig. 6.

Fig. 5 Band 7 of Landsat 8
from 2018

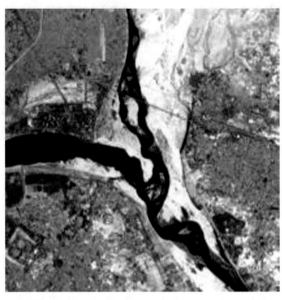

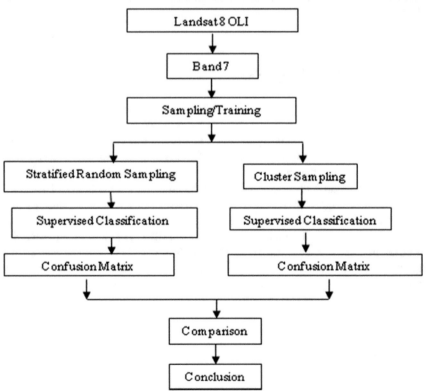

Fig. 6 Proposed Method

5 Result

From the ground survey, it is observed that the study area is covered by five major earth features, viz. water, sand, land, vegetation, and urban. The samples are chosen on the basis of area covered by the feature; more the coverage area more the number of samples, e.g., the water region in study area is more so more number of water samples are chosen, and the urban area is comparatively lesser so less samples are taken. Out of available sampling schemes stratified random sampling and cluster sampling are considered for the selection of samples.

In both the schemes, samples in each class are constant, and on the basis of these samples, supervised classification is performed. The overall accuracy is measured on the basis of confusion matrix. The number of samples is 75, 65, 60, 60, and 50 in water, sand, land, vegetation, and urban classes (Fig. 7).

Similarly, data sets from 2013, 2014, 2015, 2016, and 2017 are used for training the samples using stratified random sampling (SRS) and cluster sampling CS. Then the study area is classified into five classes, viz. water (red), sand (green), land (blue), vegetation (yellow), and urban (cyan). On the basis of trained samples, the accuracy assessment of the classified images is performed. In Table 2, the overall accuracy of the two sampling schemes over the different data sets from different years is tabulated (Fig. 8).

(a)   **(b)**

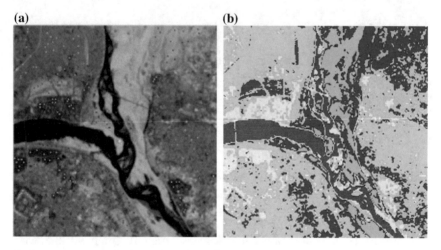

Fig. 7 **a** Training samples of 2018 data set using stratified random sampling and **b** Classified imagery due to stratified random sampling

Table 2 Overall accuracy of classification obtained in two different sampling approaches for six different data of different years

Year	Overall accuracy (%)	
	SRS	CS
2013	81.29	85.16
2014	86.77	90.96
2015	86.12	86.77
2016	88.32	90.47
2017	90.00	95.80
2018	83.87	91.94

(a)   **(b)**

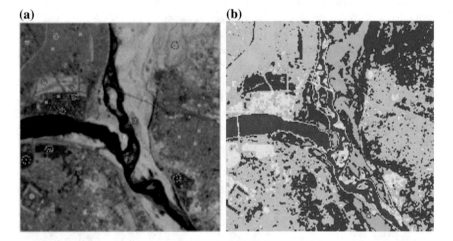

Fig. 8 **a** Training samples of 2018 data set using cluster sampling and **b** Classified imagery due to cluster sampling

6   Conclusion

In the present study, a supervised classification has been applied over six different data sets from 2013 to 2018. Two sampling approaches are used stratified random sampling and cluster sampling. The study region is having five earth surface features; hence, it is divided into five stratum and samples are chosen from each stratum using SRS and CS. The overall accuracy for SRS varies from 81.29% to 90.00%, while for CS overall accuracy varies from 85.16 to 95.80%. It can be concluded that CS gives better results for the classification of Landsat 8 imagery. Random sampling is not used because all five feature categories are important; none can be skipped or ignored and the number of samples in each stratum is fixed in all the six data sets. The surface features in study area are frequently varying so systematic sampling is not preferred. These two sampling approaches can be used for more possibilities in future.

References

1. Congalton, R.G., Green, K: Assessing the Accuracy of Remotely Sensed Data: Principles and Practices. CRC Press (2008)
2. Hashemian, M.S., Abkar, A.A., Fatemi, S.B.: Study of sampling methods for accuracy assessment of classified remotely sensed data. Int. Congr. Photogramm. Remote. Sens. (2004)
3. Vieira, C.A.O., Santos, N.T.: Analysis of sampling methods and its influence on image classification process of remotely sensed images through a qualitative approach Leonardo Campos de Assis Rômulo Parma Gonçalves 2. Anais XIV Simpósio Brasileiro de Sensoriamento Remoto, Natal, Brasil, 25-30 April 2009, INPE, pp. 6773-6780
4. Anderson, J.R.: A Land Use and Land Cover Classification System for Use with Remote Sensor Data, vol. 964. US Government Printing Office (1976)
5. Lillesand, T., Kiefer, R.W., Chipman, J.: Remote Sensing and Image Interpretation. Wiley (2015)
6. Mather, P.M., Koch, M.: Computer Processing of Remotely-Sensed Images: An Introduction. Wiley (2011)
7. Lu, Dengsheng, Weng, Qihao: A survey of image classification methods and techniques for improving classification performance. Int. J. Remote Sens. **28**(5), 823–870 (2007)
8. https://libra.developmentseed.org/
9. Mishra, V.K., Pant, T.: Application of classification techniques for identification of water region in multiple sources using landsat-8 OLI imagery. In: 2019 URSI Asia-Pacific Radio Science Conference (AP-RASC). IEEE (2019)
10. Mishra, V.K., Pant, T.: Monitoring the change in water class of two rivers in Sangam Region, Prayagraj, India using Landsat-8 OLI Imagery. In: IEEE Geoscience and Remote Sensing Symposium, 28 July to 02 August 2019, Yokohama, Japan
11. Barsi, J., et al.: The spectral response of the Landsat-8 operational land imager. Remote. Sens. 6(10), 10232–10251 (2014)

Deep Neural Networks for Out-of-sample Classification of Nonlinear Manifolds

Tissa P. Jose and Praveen Sankaran

Abstract In information processing domains, high dimensional data poses several challenges in terms of storage, visualization, and retrieval. There are several instances where, even though the data points are of high dimension, the data actually resides on a lower dimensional space. Dimensionality reduction methods attempt to find meaningful representation of data which is present in the lower dimension leading to better visualization, removal of noisy features and redundant information. Traditional linear dimensionality reduction techniques are incapable of dealing with nonlinear datasets. Nonlinear dimensionality reduction methods like Isomap work well for such datasets, but suffer from the issue of out-of-sample extension. In this paper, a solution for out-of-sample problem of Isomap method and extended Isomap method is put forward by employing neural networks. Out-of-sample extensions of Isomap and extended Isomap using deep neural network (DNN) are proposed. The proposed method is tested using AT&T face database, Yale face database, and MNIST handwritten digit database. The proposed technique is compared with the existing out-of-sample extension method using general regression neural network (GRNN).

Keywords Extended Isomap · Nonlinear dimensionality reduction · Deep neural networks

1 Introduction

Researches and innovations in every field depend on collection, storage, analysis, and visualization of data. For many computer vision and pattern recognition applications, the data available will be high dimensional. Dimensionality is defined as

T. P. Jose (✉) · P. Sankaran
National Institute of Technology Calicut, Kozhikode, India
e-mail: tissapjose@gmail.com

P. Sankaran
e-mail: psankaran@nitc.ac.in

© Springer Nature Singapore Pte Ltd. 2020
S. Agarwal et al. (eds.), *Machine Intelligence and Signal Processing*,
Advances in Intelligent Systems and Computing 1085,
https://doi.org/10.1007/978-981-15-1366-4_31

the number of features or coordinates used to specify any data point in a space or in a dataset. As dimension increases, (i) computational complexity increases, (ii) there is a loss of meaningful distinctness between similar and dissimilar objects, and (iii) visualization of data becomes a hard task. Reducing the dimensionality of data, without much loss of relevant information, seems to be a viable solution for this problem. Dimensionality reduction is the process of transforming data from higher dimension to a lower dimension in which the data has a meaningful representation [1]. The intrinsic dimensionality of data is the minimum number of dimensions or attributes needed to correctly represent the observed properties of the data.

Linear discriminant analysis (LDA), principal component analysis (PCA) [2], and multidimensional scaling (MDS) [1] are the most important linear dimensionality reduction techniques. If the data is nonlinear in nature, linear methods are incapable of finding the underlying structure of data. Tenenbaum et al. [3] developed a nonlinear dimensionality reduction method called Isomap. The idea behind Isomap is to transform data points from high-dimensional input space to low-dimensional space representation of a nonlinear manifold. The main difference between MDS and Isomap is in the computation of pairwise distances between data points in the input space. In MDS, the pairwise distances are calculated in the input Euclidean space, whereas in Isomap, it is in the geodesic space of the nonlinear manifold. The other important nonlinear dimensionality reduction methods in the literature include Laplacian Eigenmap, Locally Linear Embedding(LLE) [4], etc.

The issue with Isomap algorithm and other nonlinear dimensionality reduction methods is that these methods cannot embed an out-of-sample data point [5]. Bengio et al. [5] developed a kernel solution for this problem. In that method, the algorithms are considered to be learning eigenfunctions of a kernel. The drawback is that the final map should retain all data points and the corresponding low dimensional embedding, which in turn is the problem to be addressed. Locality Preserving Projections (LPP) were developed by He et al. [6] but are limited to LLE and Laplacian criteria. Another idea was proposed by Zhang et al. [7] in which Gaussian RBF is used to model the forward and inverse maps of nonlinear dimensionality reduction methods. In their method, there is no need to store all original samples, only the cluster centers need to be stored which is not a compact enough expression yet.

Although Tenenbaum et al. [3] discussed the applicability of neural networks to model Isomap embedding, there are not many methods based on it. Gong et al. [8] proposed a multilayer perceptron model where the input is the high dimensional data and the output of neural network gives the low dimensional embedding. But in their approach, a meaningful quantitative measure to evaluate the quality or accuracy of the learned low dimensional representation, in terms of classification, has not been proposed. Out-of-sample extension using general regression neural network (GRNN) was used in [9]. It is the most widely used out-of-sample method, as the number of variable parameters is minimal.

All the abovementioned works focus on getting an accurate low dimensional representation of data, but do not concentrate on the classifiability of the data after dimensionality reduction. It has been well proved that dimensionality reduction, in most of the cases, leads to better classification of data [1]. Based on this fact, many subspace approaches like eigenfaces and fisherfaces [10] were used for data classification tasks like face recognition.

Isomap algorithm is a good choice for this among nonlinear methods, in reconstruction point of view, but is not very appropriate for classification tasks. Yang [11] proposed extended Isomap for classifications tasks such as face recognition. In this method, after the computation of geodesic distances, LDA is used to find the low dimensional representation of data. A similar approach was adopted by Wu et al. [12] in face recognition based on relevant component analysis (RCA).

Recently, many techniques for obtaining dimensionality reduced data which is discriminative in nature have been developed. A large number of such methods use autoencoders, since it is already proved that autoencoders can be used for dimensionality reduction [13]. We had, in fact, applied it on the same classification problem setup by incorporating the idea of discriminative autoencoders. But it resulted in very poor accuracy values, based on which we came to the conclusion that a method based on autoencoders is not suitable for this particular multi-class discriminant problem.

2 Proposed Approach

The issue with most of the nonlinear dimensionality reduction techniques is in finding the embedding of new test samples. This leads to the need to find a method or a system which can mimic the performance of this nonlinear dimensionality reduction technique. A good choice of implementation would be using neural networks. The multilayer perceptron model proposed by Gong et al. [8] does not cater to the needs of data classification task. The model is not based on class separability. We propose a method for dimensionality reduction of out-of-sample data using DNN which ensures class separability. In this method, the low dimensional data is further used for classification of original high dimensional data. The performance of the proposed system is measured based on two performance evaluation criteria–classification accuracy and fitting error method.

2.1 Methodology

Manifold learning finds a low dimensional representation $\{y_i\}$ for $\{x_i\}$ by any of the approaches like Isomap, LLE, etc. In the proposed method, first $x_i \rightarrow y_i$ representations are found out using Isomap algorithm or extended Isomap, and these (x_i, y_i) pairs are used to train a deep neural network. x_i is given as input and y_i is given as desired output (target) to train the deep neural network to obtain optimized network weights.

A neural network (for example, MLP or DNN) trained using high dimensional input and its corresponding low dimensional embedding can be used for out-of-sample extension of testing samples. The high generalization ability of neural networks is the main reason for choosing them for out-of-sample solution. In this work, DNN with different number of hidden layers is analyzed.

A general DNN model is as follows. Consider a DNN with n_l layers, with input layer being the first layer and output layer being the n_lth layer. Let W be the set of weights, and b be the vector of bias values. Let f be the activation function (e.g., sigmoid function). z_i^l is the input to ith activation node in layer l, and a_i^l is its output. $W_{ij}^{(l)}$ is the weight from node i in layer l to node j in layer $(l+1)$. $b_i^{(l)}$ is the bias to ith node in layer l. Let $\mathbf{x} = \begin{bmatrix} x_1 & x_2 & \ldots & x_D \end{bmatrix}^T$ be the input and $\mathbf{y} = \begin{bmatrix} y_1 & y_2 & \ldots & y_d \end{bmatrix}^T$ be its corresponding output.

For the first layer (input layer),

$$a_i^1 = x_i \tag{1}$$

For all hidden layers,

$$z_j^l = \sum_k W_{kj}^{(l)} . a_k^{(l-1)} \tag{2}$$

$$a_i^l = f(z_i^l) \tag{3}$$

For the output layer,

$$y_i = a_i^{n_l} \tag{4}$$

The proposed approach uses a DNN with number of hidden nodes chosen moderately small to ensure generalization. The activation function used is sigmoid function. Training algorithm used is scaled conjugate gradient. The desired output of the system is the accurate Isomap embedding or extended Isomap embedding of the high dimensional input. To evaluate the performance of the system, we need to know what is the desired output, which is unknown for the test points. Therefore, there is no straightforward method to evaluate the performance of this network. Two different approaches are used to assess the performance. First evaluation criterion is classification accuracy. In this way, class information is incorporated into the system. The second evaluation criterion is based on the fitting error, which is described in detail in Sect. 3.2.

2.2 Out-of-sample Extension of Extended Isomap

In extended Isomap algorithm, geodesic distances of test points to all other points have to be calculated, which is computationally expensive and time-consuming. Similar to Isomap, out-of-sample extension is an issue in extended Isomap method also. It is desirable to have a system which will mimic the performance of extended Isomap. A DNN is trained to learn extended Isomap technique, similar to DNN modeling of Isomap. The proposed method is compared to out-of-sample method using GRNN [9].

2.3 Selection of DNN Hidden Layer Structure

Assigning small number of hidden units might result in high training errors and may lead to underfitting, whereas high number of hidden units will give low training errors but will result in overfitting. Many rules or general procedures have been suggested by the scientific community for finding the best neural network structure. One such rule insists on the size of the hidden layer to be between number of input units and number of output units. Another rule states that number of hidden neurons should not be more than twice of input dimension [14]. Practically, the number of hidden nodes depends not only on the input and output layer size, but also on the number of training points, on the complexity of the function to be learned by the network, on the activation functions used, and on the training algorithm.

3 Experimental Analysis and Results

Two sets of experiments are done as part of this paper. First experiment is the out-of-sample extension of Isomap method. As a subpart, a comparison of DNN modeling of Isomap and DNN modeling of PCA is also undertaken. Second experiment deals with the proposed out-of-sample extension of extended Isomap algorithm. The performance of the proposed method is done on the basis of two evaluation approaches, namely classification accuracy and fitting error approach. The proposed out-of-sample extension method is also compared with existing out-of-sample method.

3.1 Experiment 1—DNN Modeling of Isomap

This experiment is done to prove that DNN is a good candidate for out-of-sample extension problems. For experimental analysis, Swiss roll dataset with 10,000 points is taken. The data is divided randomly into training (60%), validation (20%), and testing (20%) sets. The number of output classes is chosen as ten. Two sub-experiments are done. In the first part, the number of hidden neurons in each layer and the number of hidden layers are varied and each network is trained. For training, three-dimensional Swiss roll dataset is used as the input and its corresponding Isomap embedding (two-dimensional) is taken as the desired output. Training continues until error gradient goes below 10^{-6}. The whole experiment is repeated for 100 iterations to generalize the results and the average accuracy values are noted.

During testing phase, each test data point is passed through the network, and an approximate Isomap embedding of test point is obtained at the output layer of the network. The test data is classified to that class whose centroid (in low dimensional space) is closest to the obtained output. Original Isomap training accuracy is 96.04%. The subpart of the first experiment is a comparison of DNN modeling of PCA and

Table 1 Performance of DNN modeling of Isomap and DNN modeling of PCA (for Swiss roll dataset)

Hidden layers	DNN modeling PCA		DNN modeling Isomap	
	Training accuracy (%)	Testing accuracy (%)	Training accuracy (%)	Testing accuracy (%)
[10, 5]	42.96	42.75	94.25	93.93
[10, 5, 5]	44.00	43.58	91.66	91.65
[10, 5, 5, 5]	44.55	44	94.21	93.94
[10, 10, 10, 10, 10]	43.35	41.90	91.00	90.80
[5, 5]	44.10	42.89	91.50	89.29

DNN modeling of Isomap, with nonlinear data as input. Network design is fixed and only the training data is changed. The experimental results for DNN modeling PCA are given in Table 1. From Table 1, it can be observed that DNN modeling of Isomap gives better classification accuracy than DNN modeling of PCA. This is due to the highly nonlinear nature of the dataset used.

3.2 Experiment 2—DNN Modeling of Extended Isomap

The dimensionality of the reduced space for extended Isomap and LDA is determined by the number of classes (i.e., $C - 1$), where C is the number of classes in the extended Isomap method. For pure Isomap, the most optimal reduced dimension is found out empirically.

Face recognition–AT & T database The AT&T face database (formerly known as ORL database) consists of 10 images each of 40 subjects [15]. The images are resized to 23×28 pixels. For evaluating the classification performance, leave-one-out strategy is adopted The training images and the test image are projected to low dimension using projection matrix. Class of the test data is found using nearest neighbor classifier. This procedure is repeated by taking each of the images as test image. For AT&T database, number of classes is 40. The reduced dimension for extended Isomap and LDA is 39. The reduced dimension for pure Isomap is chosen to be 45 (the reduced dimension giving the best performance is taken as the most optimum one). Training accuracy of LDA obtained is 100%, and testing accuracy of LDA is 92.5%. The optimal value of neighborhood parameters ϵ and k is chosen such that the classification accuracy is the highest. Tables 2 and 3 show the variation in classification accuracy for different ϵ and k values, respectively. From Table 2, it can be seen that the optimal ϵ value is 1.88. The optimal k value is chosen to be 85, based on Table 3.

During testing, extended Isomap (at optimal neighborhood values) outperforms LDA and Isomap [11]. For finding the testing accuracy of extended Isomap, geodesic distance vector of the testing points needs to be found out. All training data needs

Table 2 Classification accuracy of extended ϵ-Isomap for different ϵ values (for AT&T database)

Epsilon (ϵ)	Extended Isomap	
	Training accuracy (%)	Testing accuracy (%)
1	100	89.00
1.10	100	87.00
1.30	100	93.25
1.50	100	96.75
1.70	100	98.25
1.75	100	98.00
1.80	100	98.00
1.85	100	98.25
1.87	100	98.50
1.88	100	98.75
1.89	100	98.50
1.90	100	98.25
1.95	100	98.50
2	100	98.50
5	100	98.50
10	100	98.50
20	100	98.50

to be available during testing, which is a disadvantage. This indicates the need for out-of-sample extension method for extended Isomap.

DNN is used to mimic extended Isomap. The network is trained using input face images (dimension $23 \times 28 = 644$) and their respective extended Isomap projections (in reduced dimension 39). The number of hidden layers is varied starting with one. A number of hidden neurons are selected to be between input dimension and output dimension [14]. The number of hidden layers is fixed and the number of hidden neurons is increased. From Table 4, it can be seen that while increasing hidden neurons from 100 to 400, the performance is slightly deteriorating. Two hidden layers are used which give considerably better performances compared to single hidden layer. In two hidden layer structure, [400 400] gives the best performance. Further increase in number of hidden neurons ([500 500]) leads to high training accuracy but poorer test accuracy. This may be due to overfitting. Three hidden layer networks give better performance than two hidden layer networks. But, further increase in number of hidden layers leads to overfitting. This is why [100 100 100 100] has high training accuracy, but low testing accuracy.

The proposed out-of-sample extension using DNN is compared with the existing method using GRNN [9]. In GRNN, the only variable parameter is spread factor σ, which is usually found using hold out method. From Table 5, it can be seen that the variation in σ affects the classification accuracy. σ values equal to or above 1 give

Table 3 Classification accuracy of extended k-Isomap for different k values (for AT&T database)

k	Extended Isomap	
	Training accuracy (%)	Testing accuracy (%)
5	100	90.50
10	100	91.00
15	100	86.25
20	100	90.00
25	100	88.50
30	100	88.75
35	100	88.50
40	100	90.25
45	100	90.75
50	100	90.50
55	100	88.50
60	100	91.00
65	100	88.25
70	100	92.00
75	100	90.25
80	100	92.50
85	100	92.75
90	100	92.50
95	100	90.00
100	100	90.50

very low accuracy values, whereas for σ values below 0.2, the accuracy is 97.5%. This might be due to the fact that a training data point that is very similar to the test data might be well within the Gaussian curve centered at the testing sample, even if the spread is very less. This sometimes can lead to misclassification in the case of boundary points. It is therefore preferred to keep the spread constant neither too small nor too large. In the case of AT&T database, the average distance between two images belonging to the same class is 2. Ideally, spread constant should be in that range. The comparison between out-of-sample extension using DNN and GRNN is given in Table 6.

Another performance evaluation of out-of-sample extension using DNN is done, similar to that given in Gong et al.'s paper [8]. The entire dataset is taken and its extended Isomap is done resulting in low dimensional representation of full data. Dataset (A) is split into training (A1) and testing (A2) and extended Isomap embedding of training data is found out. The difference between embeddings of training samples in A and A1 is measured. The average difference for each iteration is calculated. A1 is used to train the DNN and embedding of A2 is obtained during testing. The difference between embeddings of testing samples in A and A1 is measured. The

Table 4 Classification performance of DNN modeling of extended Isomap for different hidden layer structures (for AT&T database)

Hidden layer	Training accuracy (%)	Testing accuracy (%)
[100]	91.86	74.75
[200]	92.54	74.75
[300]	88.71	61.25
[400]	85.47	55.50
[100, 100]	95.47	81.75
[200, 200]	97.52	92.25
[200, 100]	96.31	87.50
[300, 300]	97.78	91.50
[350, 350]	97.83	92.00
[400, 400]	97.95	94.00
[500, 500]	97.85	89.75
[100, 100, 100]	95.78	88.75
[200, 200, 200]	98.44	94.25
[300, 300, 300]	98.77	96.00
[400, 400, 400]	98.76	95.25
[100, 100, 100, 100]	95.39	85.50

Table 5 GRNN for out-of-sample extension (for different values of spread factor σ)

Spread factor (σ)	Training accuracy (%)	Testing accuracy (%)
5	2.51	1
2	2.54	1
1	2.69	1
0.8	2.9	1.25
0.5	24.86	4
0.2	100	97.25
0.1	100	97.5
0.05	100	97.5

average difference for each iteration is calculated. This is termed as experimental error. The experimental error is compared with the error in training samples for each iteration as shown in Fig. 1. It is clear that experimental error is comparable with the error associated with training samples which implies the network is working well in the aspect of dimensionality reduction.

Face recognition - Yale database Yale database consists of face data of 15 people, 11 images per person (Available at http://vision.ucsd.edu/content/yale-face-database). The images are resized to 32×32 pixels for computational convenience and all pixels are normalized to lie between 0 and 1. The number of classes is 15. Reduced

Table 6 Comparison of out-of-sample extension using DNN and GRNN in AT&T database

Method	Testing accuracy (%)
Extended Isomap (using geodesic feature vectors of test points)	98.75
DNN (Hidden layer [300 300 300])	96.00
GRNN ($\sigma = 0.1$)	97.50

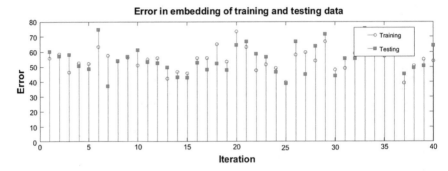

Fig. 1 Comparison of experimental error (testing data) and error in training sample embedding for 40 iterations

dimension of extended Isomap is 14. The classification performance is found out using leave-one-out strategy (explained in Sect. 3.2). The most optimum ϵ value is found out by trial and error analysis given in Table 7, and the chosen value is 15. The proposed DNN is used to mimic the performance of extended Isomap.

Different two, three, four, and five layered networks are tried out considering the aspects of underfitting and overfitting. Best performance is given by the hidden structure [75 75 75], as shown in Table 8. This accuracy level (66.67%) is in comparison with the best performance of existing GRNN method with $\sigma = 1$ (given in Table 10). Considering the fact that the average distance between data points belonging to same class is found to be 8.8 and maximum distance is 11.6, the spread constant value (σ) should be in that range. But at those σ values, the existing method fails as shown in Table 9. This makes the proposed method a better alternative.

Handwritten digit recognition—MNIST database For further analysis, the proposed method is applied for handwritten digit recognition using MNIST database [16]. MNIST database consists of handwritten digits divided into a training set of 60,000 examples and a testing set of 10,000 examples. For computational easiness, a subset of MNIST database with 1500 training samples and 250 testing samples was taken. The experiment is repeated for five iterations by taking different non-overlapping training and testing sets and the average results are calculated. Number of classes is ten and the reduced dimension is nine. Table 11 shows the variation in performance with different ϵ value (neighborhood parameter). The most optimum

Table 7 Classification accuracies of extended ϵ- Isomap for different ϵ values for Yale database

Epsilon	Training accuracy (%)	Testing accuracy (%)
2	99.07	6.67
4	97.60	25.45
6	100	46.06
8	100	41.21
10	100	58.79
12	100	67.27
14	100	78.18
15	100	80.00
16	100	78.18
17	100	78.79
18	100	78.79
19	100	78.79

Table 8 Classification performance of DNN modeling of extended Isomap for different hidden layer structures (for Yale database)

Hidden layer	Training accuracy (%)	Testing accuracy(%)
[50 50]	84.14	55.15
[75 75]	86.14	56.36
[100 100]	85.58	46.67
[50 50 50]	84.13	55.15
[75 75 75]	87.45	66.67
[90 90 90]	88.05	62.42
[100 100 100]	88.43	56.97
[300 300 300]	82.62	43.03
[500 100 50]	88.37	61.82
[75 75 75 75]	86.13	60.61
[100 100 100 100]	88.41	63.64
[200 200 100 100]	88.69	61.21
[100 100 100 100 100]	86.59	56.97

value for $\epsilon = 15$ is chosen for further experiments done on the dataset. Different hidden layer DNN structures are used and their performances in terms of classification accuracy are given in Table 12. The maximum optimum performance is shown by two DNN with hidden layer structures [400 400 400 400] and [600 400 200 100], respectively. The existing GRNN method is performed on the same dataset and the spread factor σ is varied and the most optimum value is $\sigma = 0.5$. The average distance between datasetdata points belonging to same class is 9.18 and maximum distance is 12.61. Ideally, σ should be in this range, but since the performance in these cases

Table 9 GRNN for out-of-sample extension (for different values of spread factor σ) - Yale database

Spread factor (σ)	Training accuracy (%)	Testing accuracy (%)
10	8	0.6
5	28.90	7.27
2	100	61.82
1.2	100	65.45
1	100	66.06
0.7	100	65.45
0.5	100	65.45
0.1	100	64.85
0.01	100	64.85

Table 10 Comparison of out-of-sample extension using DNN and GRNN in Yale database

Method	Testing accuracy (%)
Extended Isomap (using geodesic feature vectors of test points)	80.00
DNN (Hidden layer [75 75 75])	66.67
GRNN ($\sigma = 1$)	66.07

is very poor, σ is chosen to be very small which makes the Gaussian curves centered at training (known) points (in GRNN output equation) to be very steep. This is a major drawback in the existing method. In MNIST dataset, the proposed method gives better classification accuracy compared to that of existing method, as shown in Tables 13 and 14.

Reason for choosing GRNN method for comparison Most of the existing works which deal with out of sample extension of nonlinear dimensionality reduction techniques focus on the embedding part. (Examples: [5], [8]). The drawbacks of these existing methods are: Bengio's paper [5] is based on Nystrom approximation which does not work for classification tasks. It has been proved to give very low classification accuracy levels. That is why it has not been used in the comparison studies. Gong's paper [8]) also deals with mimicking Isomap using neural networks. It fails to give a meaningful quantitative measure of its performance. Also, it is not based on classification point of view. In classification point of view, Nystrom approach does not work. Nystrom approach fails in out-of-sample extension of data for classification. As a part of the experiment, the proposed method was already compared with existing Nystrom method also. The accuracies obtained were well below 30%. That is why only the comparison with GRNN method has been given in this paper.

Table 11 Classification accuracies of extended ϵ- Isomap for different ϵ values for MNIST database

Epsilon	Training accuracy (%)	Testing accuracy (%)
2	100	11.11
3	100	19.11
4	100	12.89
5	100	37.78
6	100	46.22
7	100	31.56
8	100	15.11
9	100	28.89
10	100	29.33
11	100	30.22
12	100	61.33
13	100	80.89
14	100	89.33
15	100	91.56
16	100	90.67
17	100	90.67
18	100	90.67
19	100	90.67
20	100	90.67

Table 12 Classification performance of DNN modeling of extended Isomap for different hidden layer structures (for MNIST database)

Hidden layer	Training accuracy (%)	Testing accuracy (%)
[300 300 300]	96.81	83.11
[400 400 400]	96.89	85.78
[500 500 500]	96.67	84.44
[600 400 20]	97.63	86.22
[700 700 700]	96.52	82.67
[300 300 300 300]	96.96	86.22
[400 400 400 400]	97.11	88.00
[600 400 200 100]	97.04	88.00
[700 400 200 50]	97.26	83.56
[500 500 500 500]	95.48	82.67
[400 400 400 400 400]	96.81	84.00
[500 500 500 500 500]	96.52	85.33

Table 13 GRNN for out-of-sample extension (for different values of spread factor σ) - MNIST database

Spread factor (σ)	Training accuracy (%)	Testing accuracy (%)
10	11.11	11.11
5	43.48	31.56
1	100	83.56
0.8	100	83.56
0.5	100	84
0.1	100	84
0.01	100	84

Table 14 Comparison of out-of-sample extension using DNN and GRNN in MNIST database

Method	Testing accuracy (%)
Extended Isomap (using geodesic feature vectors of test points)	91.56
DNN (Hidden layer [600 400 200 100])	88.00
GRNN ($\sigma = 0.5$)	84.00

4 Conclusion

Out-of-sample extension of Isomap and extended Isomap using DNN are proposed based on classification accuracy. The essence of extended Isomap is incorporated well in DNN, but with less computational complexity compared to extended Isomap, since there is no geodesic computation in DNN approach during testing.

The basic premise and focus of this work are how extended Isomap can be used for a multi-class discriminant embedding and classification problem. So, the results would only be maximum as good as the accuracies obtained in extended Isomap (given in Yang's paper [11]). Comparing current state-of-the-art methods with this particular case would be meaningless, as in this method, we are checking particularly on how out-of-sample problem of extended Isomap could be dealt with. The importance here is that we have put a framework for handling test data in a general nonlinear manifold learning algorithm for a multi-class classification problem.

The proposed method of out-of sample extension of extended Isomap is used for the classification of different high dimensional data. The datasets used are AT&T face database, Yale face database, MNIST handwritten digit dataset. In the out-of-sample extension scenario, for MNIST database, the proposed DNN method gives better classification accuracy. In Yale and AT&T databases, both DNN and GRNN approaches result in comparable performances in terms of accuracy. DNN approach has certain advantages over GRNN method. In GRNN, all training points and their low dimensional embeddings need to be available during testing time, whereas in DNN, once the network is trained, there is no need to store the training data. This is a great advantage as lesser storage space is required. DNN offers more generalization. From the experimental results in Table 4, it can be seen that the classification accuracies are almost stable over the range of different network structures which have been designed, whereas in GRNN method, the performance of the system is highly dependent on spread factor σ value. In DNN approach, a critical value like σ need not be taken care of. It can be concluded that DNN is a more generalized and robust method compared to GRNN for out-of-sample extension of extended Isomap.

References

1. Wang, J.: Geometric Structure of High-Dimensional Data and Dimensionality Deduction. Springer (2011)
2. Martínez, A.M., Kak, A.C.: Pca versus lda. IEEE Trans. Pattern Anal. Mach. Intell. **23**(2), 228–233 (2001)
3. Tenenbaum, J.B., De Silva, V., Langford, J.C.: A global geometric framework for nonlinear dimensionality reduction. Science **290**(5500), 2319–2323 (2000)
4. Roweis, S.T., Saul, L.K.: Nonlinear dimensionality reduction by locally linear embedding. Science **290**(5500), 2323–2326 (2000)
5. Bengio, Y., Paiement, J.-F., Vincent, P., Delalleau, O.,Roux, N.L., Ouimet, M.: Out-of-sample extensions for lle, isomap, mds, eigenmaps, and spectral clustering. Adv. Neural Inf. Process. Syst., 177–184 (2004)
6. He, X., Yan, S., Hu, Y., Niyogi, P., jiang Zhang, H.: Face recognition using laplacianfaces. IEEE Trans. Pattern Anal. Mach. Intell. **27**, 328–340 (2005)
7. Zhang, J., Li, S.Z., Wang, J.: Nearest manifold approach for face recognition. In: Proceedings Sixth IEEE International Conference on Automatic Face and Gesture Recognition, 2004, pp. 223–228. IEEE (2004)
8. Gong, H., Pan, C., Yang, Q., Lu, H., Ma, S.: Neural network modeling of spectral embedding. BMVC 227–236 (2006)
9. Balasubramanian, V.N., Ye, J., Panchanathan, S.: Biased manifold embedding: a framework for person-independent head pose estimation. In: IEEE Conference on Computer Vision and Pattern Recognition, 2007. CVPR'07. pp. 1–7. IEEE (2007)
10. Belhumeur, P.N., Hespanha, J.P., Kriegman, D.J.: Eigenfaces versus fisherfaces: recognition using class specific linear projection. IEEE Trans. Pattern Anal. Mach. Intell. **19**(7), 711–720 (1997)
11. Yang, M.-H.: Extended isomap for classification. In:Proceedings 16th International Conference on Pattern Recognition, 2002, vol. 3, pp. 615–618. IEEE (2002)
12. Wu, Y., Chan, K.L., Wang, L.: Face recognition based on discriminative manifold learning. In: Proceedings of the 17th International Conference on Pattern Recognition, 2004, vol. 4, pp. 171–174. ICPR 2004. IEEE (2004)
13. Wang, Y., Yao, H., Zhao, S.: Auto-encoder based dimensionality reduction. Neurocomputing **184**, 232–242 (2016)
14. Xu, S., Chen, L.: A novel approach for determining the optimal number of hidden layer neurons for fnn's and its application in data mining. In: International Conference on Information Technology and Applications: iCITA, pp. 683–686 (2008)
15. Samaria, F.S., Harter, A.C.: Parameterisation of a stochastic model for human face identification. In: Proceedings of the Second IEEE Workshop on Applications of Computer Vision, 1994, pp. 138–142. IEEE (1994)
16. LeCun, Y., Bottou, L., Bengio, Y., Haffner, P.: Gradient-based learning applied to document recognition. Proc. IEEE **86**(11), 2278–2324 (1998)

FPGA Implementation of LDPC Decoder

Shruti and Barathram Ramkumar

Abstract Low-density parity-check (LDPC) codes are the most powerful error-correcting codes used in the transmission of signals and also provide a near-optimal performance that can approach to capacity for a lot of different channels. LDPC codes are preferred nowadays as it is bandwidth-efficient and provide less Bit Error Rate (BER). In this paper, LDPC decoder using log-belief propagation algorithm which reduces the complexity is implemented with Xilinx System Generator (XSG) blocks combined with Simulink blocks and finally dumped on Virtex-5 FPGA board for hardware, and device utilization summary was also found.

Keywords Low-density parity check (LDPC) · Log domain algorithm · FPGA · Tanner graph

1 Introduction

In wireless communication, the reception of the data at the destination depends on the channel noise. The external noise that gets added in the transmitted signal creates interference to the signal by adding errors while transmitting the signal by wireless medium. In digital transmission of signal, the nature of channel is the main cause of distorting the signal. Shanon published his paper, "A Mathematical Theory of Communication" around 60 years back in which he showed that efficient communication is possible when the data rate is smaller than capacity of the channel. He also showed that coding is the only way to transmit the data in a efficient and reliable way. Channel coding is a process of transforming a signal by encoding in such a way that it has less error when the signal gets distorted [1]. Coded modulation is a process by which coding techniques and modulation techniques combine together to

Shruti · B. Ramkumar (✉)
School of Electrical Science, IIT Bhubaneswar, Bhubaneswar, India
e-mail: barathram@iitbbs.ac.in

Shruti
e-mail: sh20@iitbbs.ac.in

give better performance which is a bandwidth-efficient scheme than uncoded modulation. There is trade-off in channel coding between error performance, bandwidth and power transmitted. The process of adding the extra bits to the actual message bits for error detection and correction of the signal is called as error correction codes (ECC). There is no need of retransmission of messages as it causes delay and adds cost to the system and also waste the bandwidth. There are various applications of error-correcting codes like it is used in wireless and mobile communication, satellite communication, military communication, deep space communication and for data storage. Low-density parity-check (LDPC) codes are the most powerful error-correcting codes used in the transmission of signals and also provide a near-optimal performance that can approach to capacity for a lot of different channels. LDPC codes are preferred nowadays as it is bandwidth-efficient and provide less bit error rate (BER). LDPC codes are used in Wi-Fi, Wi-Max, Ethernet, 4G, Digital Video Broadcasting (DVB), etc. This paper proposes FPGA implementation of LDPC decoder for hardware.

The rest of this paper is organized as follows. In Sect. 2, we present the types, representation and parameters to design LDPC codes. In Sect. 3, we discussed the log-domain belief propagation decoding algorithm. In Sect. 4, we presented FPGA implementation of LDPC decoder. This paper ends with a conclusion in Sect. 5 and references.

2 Low-density Parity-Check Codes

Low-density parity-check codes were first developed by Robert Gallager in his Ph.D. thesis at MIT in 1962 [2]. But at that time, these codes were ignored for more than 35 years because they were not useful as it was too complex to decode such codes using the existing technologies of that time. The hardware systems which were used that time were not compatible with the encoding and decoding of LDPC codes. Afterward, these codes were again recaptured by Mackay and Neal in 1996 when the capacity of computers was made vast along with the development of powerful algorithm like belief propagation (BP) algorithm.

LDPC codes are called as a linear block codes and are introduced as error correction codes that allow reliable transmission on noisy channels and approaches Shanon capacity bound. This section provides the basic concept of LDPC codes, their classification, representations, parameters and structure of LDPC codes.

LDPC codes are defined as (n, w_c, w_r) where, 'n' is the size of the codeword, 'w_c' is the number of 1's in each column of the parity-check matrix, 'w_r' is the number of 1's in each row of parity-check matrix.

2.1 Types of LDPC Codes

There are two types of LDPC codes:

1. Regular Codes: In Regular LDPC codes, there are equal number of 1's in each column and row of the parity-check matrix. The regularity of LDPC codes is defined if w_c and w_r are smaller than the codelength 'n' and the numbers of rows in LDPC matrix respectively.
2. Irregular Codes: Irregular codes are those codes which have unequal number of 1's in each column and in each row of the LDPC matrix. The degree of variable nodes and check nodes are allocated based on the degree distribution polynomials.

2.2 Representation of LDPC Codes

There are two ways of representing LDPC codes:

1. Matrix representation: In matrix representation, there is sparse parity-check matrix 'H' which has very less numbers of 1's as compared to the numbers of 0's. It is called low-density LDPC matrix if it satisfies two conditions:

(a) $w_c \ll n$
(b) $w_r \ll m$

To meet these conditions, LDPC matrix should be large.

2. Graphical representation: LDPC codes can be represented by tanner graph introduced by Tanner in 1981 [3] and also known as bipartite tanner graph. This graph is used in decoding algorithm.

$$H = \begin{bmatrix} 1 & 1 & 0 & 0 & 0 & 1 & 0 & 1 & 0 & 1 \\ 0 & 1 & 1 & 0 & 0 & 1 & 0 & 0 & 1 & 0 \\ 0 & 0 & 1 & 1 & 0 & 0 & 1 & 0 & 0 & 1 \\ 0 & 0 & 0 & 1 & 1 & 1 & 0 & 0 & 1 & 0 \\ 1 & 0 & 0 & 0 & 1 & 0 & 1 & 0 & 1 & 0 \end{bmatrix}$$

The tanner graph for the above parity-check matrix is shown in Fig. 1. Tanner graph consists of two sets of nodes:

(a) Variable or bit nodes which comprise of 'n' nodes for the codeword bits.
(b) Check nodes which comprise of 'k' nodes for the parity-check equations.

The lines joining variable nodes and check nodes are called as edge. An edge is connected to a particular variable and check node if the bit is '1' in the LDPC matrix. The number of 1's in the matrix is the total number of edges in tanner graph.

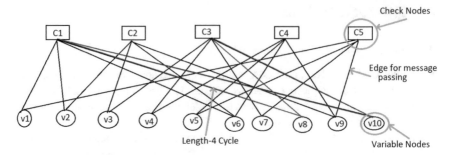

Fig. 1 Tanner graph corresponding to H matrix

The variable nodes refer to the number of columns and check node refers to the number of rows. Two sets of nodes are connected to each other by undirected edges so it is called undirected bipartite graph. The number of edges connected to variable nodes or check nodes is called as variable node degree and check node degree, respectively.

2.3   Parameters for LDPC Code Design

There are various parameters to design LDPC code which affects the performance and implementation of LDPC encoder and decoder.

1. Code size: Code size refers to the number of bits in the messages. Long codes perform better than shorter ones. But the implementation of very long codes requires more resources like memory, processing nodes, power, etc.
2. Code rate: Code rate is the ratio of bits containing information to the total length of codeword. If the code rate is high, there will be high throughput but the performance of decoder will be low. If the code rate is low, there will be low throughput but the performance of decoder will be good.
3. Code structure: Code structure tells how the connection between variable node and check node is made. Code structure defines the connection between the rows and columns in the matrix and in tanner graph. This connection defines the hardware complexity of LDPC encoder and decoder.
4. Number of iterations: In LDPC decoding algorithm, the messages are decoded by repeatedly passing the messages between the nodes. If the number of iterations is more, then there is chance of getting more accurate results and thus decoder performance will increase. But this causes more delay and power consumption of decoder.

3 LDPC Decoding

Gallager has proposed decoding algorithm which is optimal and later on many modifications were done to the algorithms independently for different types of applications. First, design parity-check matrix H which has 'v' variable nodes and 'c' check node. The entries in H matrix are '1' if a particular variable node is connected to a particular check node. These decoding algorithms can be divided into two types:

1. Hard decision decoding: It is the decision made by the decoder based on the received information whether it is 0 or 1. Bit flipping algorithm comes under this category.
2. Soft decision decoding: If the decoder can discriminate between the values that lie between 0 and 1, then it is called as a soft decision decoder. Probability domain, log domain and simplified log domain (min-sum) come under this category.

Log domain is preferred because in probability domain SPA multiplications are associated which could become numerically unstable. So by using log domain, complex multiplications are converted to additions which are less costly to implement. Now consider some log-likelihood ratios (LLR):

$$L(c_i) = \log \frac{P_r(c_i = 0|y_i)}{P_r(c_i = 1|y_i)} \tag{1}$$

where $P_r(c_i = 0|y_i)$ is the probability that the sent codeword is 0 when the received codeword is given.

$$L(v_{ji}) = \log \frac{v_{ji}(0)}{v_{ji}(1)} \tag{2}$$

where $L(v_{ji})$ is the LLR of message going from check node to variable node.

$$L(c_{ij}) = \log \frac{c_{ij}(0)}{c_{ij}(1)} \tag{3}$$

where $L(c_{ji})$ is the LLR of message going from variable node to check node.

$$L(R_i) = \log \frac{R_i(0)}{R_i(1)} \tag{4}$$

where $L(R_i)$ is used for making the hard decision. Now we consider two channels:

$$L(c_{ij}) = L(c_i) = \begin{cases} +\infty, & \text{when } y_i = 0 \\ -\infty, & \text{when } y_i = 1 \quad (BSC) \\ 0, & \text{when } y_i = \epsilon \end{cases} \tag{5}$$

$$L(c_{ij}) = L(c_i) = 2y_i/\sigma^2 \quad (BI - AWGN) \tag{6}$$

where y_i is the received codeword. Now from [3] we can write:

$$1 - v_{ji}(1) = \prod_{i' \in V_j} (1 - 2c_{i'j}(1))$$

Use the rule—$\tanh \left[\dfrac{1}{2} \log(p_0/p_1) \right] = p_0 - p_1 = 1 - 2p_1$

The above equation can be written as:

$$\tanh \left(\frac{1}{2} L(v_{ji}) \right) = \prod_{i' \in V_j \backslash i} \left(\frac{1}{2} L(c_{i'j}) \right) \tag{7}$$

Now we can split $L(c_{i'j})$ into its sign and its absolute value.

$$L(c_{ij}) = \alpha_{ij} \beta_{ij} \tag{8}$$

$$\alpha_{ij} = \text{sign}[L(c_{ij})] \tag{9}$$

$$\beta_{ij} = |L(c_{ij})| \tag{10}$$

So the above equation can be rewritten as:

$$\tanh \left(\frac{1}{2} L(c_{ji}) \right) = \prod_{i' \in V_j \backslash i} \alpha_{i'j} . \prod_{i' \in V_j \backslash i} \tanh \left(\frac{1}{2} \beta_{i;j} \right) \tag{11}$$

we can write,

$$L(v_{ji}) = \prod_{i'} \alpha_{i'j} . 2\tanh^{-1} \left(\prod_{i'} \tanh \left(\frac{1}{2} \beta_{i'j} \right) \right)$$

$$= \prod_{i'} \alpha_{i'j} . 2\tanh^{-1} \log^{-1} \log \left(\prod_{i'} \tanh \left(\frac{1}{2} \beta_{i'j} \right) \right)$$

$$= \prod_{i'} \alpha_{i'j} . 2\tanh^{-1} \log^{-1} \sum_{i'} \log \left(\tanh \left(\frac{1}{2} \beta_{i'j} \right) \right)$$

$$L(v_{ji}) = \prod_{i' \in V_j \backslash i} \alpha_{i'j} \phi \left(\sum_{i' \in V_j \backslash i} \phi(\beta_{i'j}) \right) \tag{12}$$

where, $\phi(x) = -\log[\tanh(x/2)] = \log \left(\dfrac{e^x + 1}{e^x - 1} \right)$

We have considered $\phi^{-1}(x) = \phi(x)$ when $x > 0$.

From [3] we can write:

$$L(c_{ij}) = L(c_i) + \sum_{j' \epsilon C_i \backslash j} L(v_{j'i}) \tag{13}$$

Similarly, Eq. (4) can be rewritten as:

$$L(R_i) = L(c_i) + \sum_{j' \epsilon C_i} L(v_{j'i}) \tag{14}$$

4 FPGA Implementation of LDPC Decoder

Here the implementation of LDPC decoder is discussed using Xilinx System Generator (XSG) along with Simulink which is a MATLAB tool. Using XSG blocks combined with Simulink blocks, LDPC system can be easily ported on FPGA board.

The decoding algorithm which we have implemented in this paper is belief propagation (BP) algorithm with log-likelihood Ratio (LLR) which minimizes the complexity. Figure 2 shows the tanner graph of (10, 5) code which is used for the FPGA implementation. The XSG implementation of message calculation block from check node to variable node is given in Fig. 3.

Figure 4 shows the variable node calculation blocks where variable nodes calculate the messages which are sent by the connected check nodes and update its current values and then sent it to the connected check nodes. Figure 5 shows the overall implementation of LDPC decoder in XSG. The first block is the variable node blocks that will calculate the messages which are sent to the connected check nodes. The second and fourth blocks are the memory blocks which are used to store the messages which are transmitted from variable node to check node and then again check node to variable node. The third block is the check node blocks that will calculate

Fig. 2 Tanner graph of (10, 5) code

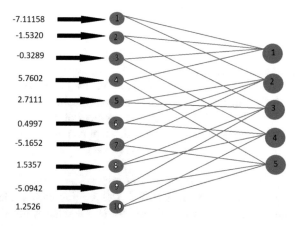

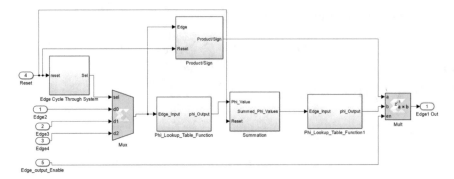

Fig. 3 Check node calculation

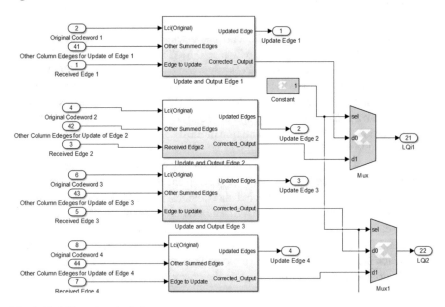

Fig. 4 Variable node calculation

the messages which are transmitted to the connected variable nodes. Figure 6 shows the implementation of full LDPC decoder in XSG. The threshold blocks are used to hard clipped the values which are received after completing the all iterations. Control/Timing block is used to enable and disable some blocks and when to reset values to zero. It contains the maximum number of iterations to be performed. After dumping the design on Virtex-5 board, we obtain the hardware co-simulation block which is shown in Fig. 7. Virtex5xc5vlx110t-1ff1136 board and point-to-point Ethernet are used.

Figure 8 shows the picture of Virtex-5 board which was used to implement LDPC decoder, and it includes USB-JTAG programming cable. The design is synthesized and implemented using Xilinx ISE 10.2 and the synthesizer used in Xilinx Synthesis

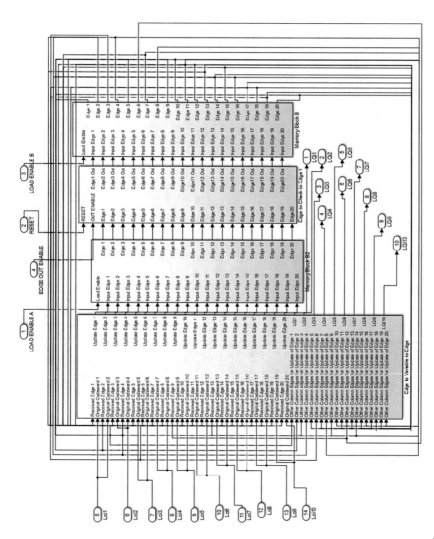

Fig. 5 LDPC decoder

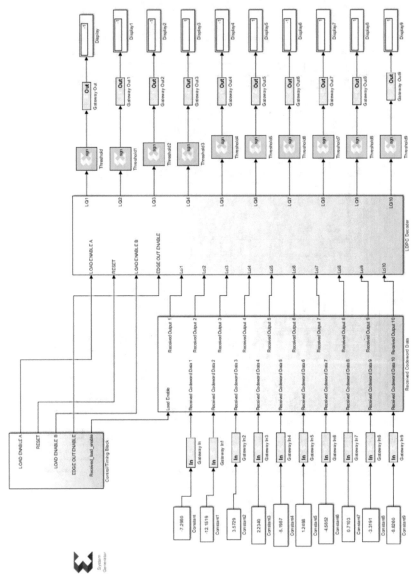

Fig. 6 Full LDPC decoder system

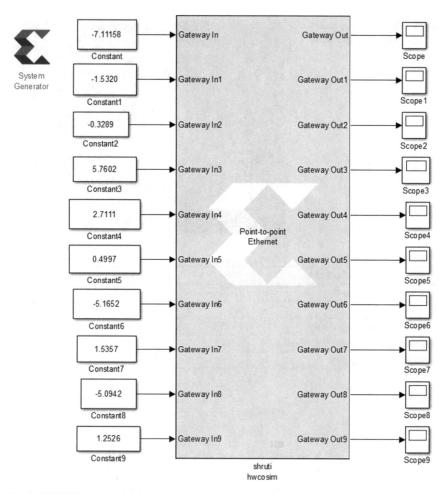

Fig. 7 LDPC Decoder implemented on Virtex-5 FPGA device

Technology (XST). This synthesizer converts HDL code into gate-level netlist. The table shows the synthesis report which contains the information about device utilization. The device utilization summary gives the information that how much total hardware resources are utilized while compiling the design. Table 1 shows the Xilinx device utilization summary.

Timing report was also generated by Xilinx timing analyzer which shows that the maximum clock frequency is 83.354 MHz which corresponds to minimum period of 11.997 ns.

Fig. 8 Virtex-5 board used to implement LDPC decoder

Table 1 Xilinx device utilization summary

Logic utilization and distribution	Used	Available	% used
Number of slice registers	7702	69,120	11
Number of slice LUTs	9060	69,120	13
Number of LUT FF pairs used	13,132	–	–
Number of fully used LUT FF pairs	3630	13,132	27
Number of bonded IOs	0	640	0

5 Conclusion

LDPC decoder was implemented in Simulink using Xilinx system generator blocks and was dumped on Virtex-5 FPGA board. Device utilization summary resources of FPGA board were shown. There is trade-off between size of lookup table (LUT) which is used for nonlinear function and accuracy of stored value. In future, we can use piecewise linear approximation to approximate the nonlinear phi function to get more accurate results, and the work can be extended for large block length.

Acknowledgements This project is funded by MHRD under IMPRINT 2.

References

1. Shannon, C.E.: A mathematical theory of communication. ACM SIGMOBILE Mobile Comput. Commun. Rev. **5**(1), 3–55 (2001)
2. Gallager, R.: Low-density parity-check codes. IRE Trans. Inf. Theory **8**(1), 21–28 (1962)
3. Ryan, W.E., et al.: An Introduction to LDPC Codes, pp. 1–23 (2004)
4. Sebastian, N., et al.: FPGA implementation of min-sum algorithm for LDPC decoder. In: Trends in Electronics and Informatics (ICEI), pp. 821–826. IEEE (2017)

5. Chithra, K., Ranjitha, C.R., Thomas, J.: A Brief Study on LDPC Codes
6. Remmanapudi, S., Bandaru, B.: An FPGA implementation of low density parity-check codes construction and decoding. In: 2012 International Conference on Devices, Circuits and Systems, pp. 216–220. IEEE (2012)
7. Pei, Y., Yin, L., Lu, J.: Design of irregular LDPC codec on a single chip FPGA. In: Proceedings of the IEEE 6th Circuits and Systems Symposium on Emerging Technologies: Frontiers of Mobile and Wireless Communication (IEEE Cat. No. 04EX710), May 2004, vol. 1, pp. 221–224
8. Bhavsar, N.P., Vala, B.: Design of hard and soft decision decoding algorithms of LDPC. Int. J. Comput. Appl. **90**(16) (2014)
9. Madi, A.A., Mansouri, A., Ahaitouf, A.: Design, simulation and hardware implementation of low density parity check decoders using min-sum algorithm. Int. J. Comput. Sci. Issues (IJCSI) **9**(2), 83 (2012)
10. Fanucci, L., Rossi, F.: A throughput/complexity analysis for the VLSI implementation of LDPC decoder. In: Proceedings of the Fourth IEEE International Symposium on Signal Processing and Information Technology, pp. 409–412 (2004)
11. Pei, Y., Yin, L., Lu, J.: Design of irregular LDPC codes on a single chip FPGA. In: Proceedings of the IEEE 6th Circuits and Systems Symposium on Emerging Technologies: Frontiers of Mobile and Wireless Communication, vol. 1, pp. 221–224. IEEE (2004)

A Multiclass Classification of Epileptic Activity in Patients Using Wavelet Decomposition

Daya Gupta, Divyashikha Sethia⊙, Abhra Gupta⊙ and Trideep Sharma⊙

Abstract Epilepsy is a chronic disorder caused in the brain where seizures occur multiple times unreliably causing unconsciousness or tremendous convulsions over the entire body. The identification of epileptic seizure activities in electroencephalography (EEG) signals by manual inspection is prone to errors and time-consuming. The proposed study suggests using Discrete Wavelet Transform to decompose the EEG signals into frequency sub-bands. A certain subset of the frequency sub-bands was chosen for feature selection. Following the DWT decomposition, the proposed method calculates the standard deviation for each sub-band present in the subset. Finally, it feeds the standard deviation values of the sub-bands to the classifiers. This work investigated the three-class classification problem focused on classifying an EEG signal into one of the three classes, which are (1) healthy patient with eyes closed, (2) patients in inter-ictal stage whose EEG recordings have been recorded from the hippocampal formation of the opposite hemisphere of the brain, and (3) patients experiencing seizure activities. The accuracy achieved in proposed work is 98.45% which beats the state-of-the-art accuracy in this three-class problem. Additionally, the proposed method achieves the highest accuracy of 100% in classifying normal EEG signals (eyes open and eyes closed) and seizure EEG signal in two separate experiments which is comparable with the existing state of the art EEG signal classification techniques. The proposed work uses six different classifiers in each of the three experiments conducted where every classifier has been used with 8 different Daubechies wavelets db1 to db8. The results obtained from these experiments provide valuable insights establishing that SVM performs the best in most of the experiments with the db4 wavelet among the 8 wavelets achieving the highest accuracy.

Keywords Epilepsy · Electroencephalogram (EEG) signals · Wavelet decomposition · Classification

---

D. Gupta · D. Sethia · A. Gupta (✉) · T. Sharma
Delhi Technological University, New Delhi 110042, India
e-mail: abhragupta2@gmail.com

© Springer Nature Singapore Pte Ltd. 2020
S. Agarwal et al. (eds.), *Machine Intelligence and Signal Processing*,
Advances in Intelligent Systems and Computing 1085,
https://doi.org/10.1007/978-981-15-1366-4_33

1   Introduction

Epilepsy is a chronic disorder caused in the brain where seizures occur multiple times unreliably. Electroencephalography (EEG) is the measure of the ionic current that flows from the neurons in our brain, and the electrical activity is recorded on the EEG. The clinical EEG equipment is used for patients suffering from attention-deficit/hyperactivity disorder (ADHD), Alzheimer disease, epilepsy. Usually, the device kept is called the scalp EEG. The electrodes are used to measure the electrical potential generated by the neurons. The EEG device has the electrodes arranged symmetrically. The task of distinguishing EEG signals from normal signals is done by professionals where characteristic patterns in the form of spike and slow waves are observed in EEG signals [9]. The detection of EEG signals visually is a difficult task because there is hardly any difference in EEG activity between epileptic seizures and non-epileptic seizures [20]. Therefore, automated systems for faster and more precise detection of epileptic seizures from EEG signals were developed. In developing countries like India, there is a lack of proper treatment and amenities to epilepsy. With the advancement of technology, automated epilepsy detection and reporting can assist in timely treatment and patient care.

Epileptic seizures are broadly classified into many types based on which part of the brain their origin is. Focal seizures arise from a localized region of the brain's cortex. There are generalized seizures which affect the entire cerebral cortex of the brain. The generalized seizure often results in loss of consciousness. The general sign in the patients is jerking or atonic seizure.

Most of the works are based on the binary classification problem using the datasets B-E, A-E, C-E and multiclass classification using the datasets ABCD-E. Thus, it is needed to work on a problem which is intermediary of the two classification problems because of very limited research on three-class classification using the datasets BC-E.

This paper focuses on three classification problems, one being a three-class classification problem and the other two being binary-class classification problems. The study uses Discrete Wavelet Transform to decompose the EEG signals into frequency sub-bands and using Support Vector Machine, classifies an EEG epoch into one of the three classes namely healthy, inter-ictal, and seizure. The accuracy achieved in this work is 98.45%. Additionally, the study achieves an accuracy of 100% in the binary classification problem in classifying EEG epoch into healthy and seizure classes.

The rest of the paper is organized as follows: Sect. 2 presents the related work.

2   Related Works

Most of the works have studied set A and set E for two-class classification of seizure detection, obtaining an accuracy of 90–100%; classification of seizure detection obtaining an accuracy of 90–100%. The details of the all the five datasets A, B, C, D, and E obtained from the University of Bonn [1] have been discussed in Sect. 4.

The proposed work resembles the most to the research work done by Sharmila and Geethanjali [19]. Their work has used DWT to decompose the EEG signals into sub-bands and applied naïve Bayes classifier (NB) and k-NN classifiers on the statistical features calculated on the sub-bands, concluding that the computation time using NB classifier is lesser than k-NN classifier and the former achieves greater accuracy.

Having worked on three-class classification problem on the healthy, inter-ictal and ictal dataset, our proposed work achieves an accuracy of 98.45%, which is greater than the accuracy in the Sharmila and Geethanjali's work [19].

The most related works on seizure detection considered to be a part with the state of the art have been compiled in Table 1 with the datasets used and the corresponding accuracy obtained. The related works included in Table 1 mostly used wavelet transform to decompose the mother wavelet in an attempt to design a framework which has optimal DWT settings to improve accuracy and reduce the computational cost of detection of epileptic activity.

Table 1 Results and features used in previous and proposed method using the same EEG dataset

Authors	Method used	Type of experiment	Accuracy (%)
Tzallas et al. [23]	Time frequency features using ANN	A-E	100
		ABCD-E	97.73
Guo et al. [10]	DWT and line length feature using ANN	ABCD-E	97.77
Orhan et al. [14]	DWT and clustering using multilayer perceptron ANN	A-E	100
		ABCD-E	99.60
Gandhi et al. [8]	DWT and energy, std and entropy features; using SVM and probabilistic NN	ABCD-E	95.44
Nicolau et al. [13]	Permutation entropy and SVM	A-E	93.55
		B-E	82.88
		C-E	88.00
		D-E	79.94
		ABCD-E	86.10
Kaya et al. [12]	1D LBP and functional tree 1D LBP and BayesNet	A-E	99.5
		D-E	95.5
		CD-E	97

(continued)

Table 1 (continued)

Authors	Method used	Type of experiment	Accuracy (%)
Samiee et al. [17]	Rational discrete short time Fourier transform using multilayer perceptron	A-E	99.80
		B-E	99.30
		C-E	98.50
		D-E	94.90
		ABCD-E	98.10
Peker et al. [15]	DTCWT using complex valued NN	A-E	100
		ABCD-E	99.15
Swami et al. [21]	DTCWT and energy, std, Shanon entropy features using GRNN	A-E	100
		B-E	98.89
		C-E	98.72
		D-E	93.33
		AB-E	99.18
		CD-E	95.15
		ABCD-E	95.24
Sharma et al. [18]	ATFFWT and FD feature using LS-SVM	A-E	100
		B-E	100
		C-E	99
		D-E	98.50
		AB-E	100
		CD-E	98.67
		AB-CD	92.50
		ABCD-E	99.20
Tiwari et al. [22]	Key point based LBP and SVM	ABCD-E	99.3
Chen [5]	DTCWT and Fourier features with NN classifier	A-E	100
		ABCD-E	100
Bajaj and Pachori [2]	Amplitude and frequency modulation bandwidths of IMFs and least squares-SVM	ABCD-E	99.5

(continued)

Table 1 (continued)

Authors	Method used	Type of experiment	Accuracy (%)
Yuan et al. [24]	ApEn, hurst exponent, scaling exponents of EEG and extreme learning machine algorithm	ABCD-E	99.5
		D-E	96.5
Bhattacharya et al .[3]	TQWT-based multi-scale K-NN entropy	A.E	100
		B-E	100
		C-E	99.5
		D-E	98.5
		A-BCDE	99
Das et al. [6]	SVM using NIG parameters as features in dual-tree complex wavelet transform domain	ABCDE	100
		A-E	100
		AD-E	100
		D-E	100
		C-E	100
Sharmila and Geethanjali [19]	Nave Bayes/k-NN classifiers	A-E	100
		B-E	99.25
		C-E	99.62
		AB-E	99.16
		AC-E	99.50
		BC-E	98.25
		CD-E	98.75
		ABC-E	98.68
Proposed work	Discrete wavelet transform, SVM classifier	A-E	100
		B-E	100
		BC-E	98.45
		AD-E	95
		BD-E	96.87

3 Dataset

The dataset [1] used in our work is obtained from the Department of Epileptology of the University of Bonn. Each dataset consists of 100 single-channel EEG epochs. It is useful for a variety of applications. The EEG recordings were recorded with a 128-channel electrode system. The sampling rate used was 173.61 Hz with a duration of 23.6 s. It has five sets of data consisting of 100 EEG recording for each set. Set A and B are for healthy patients with the eyes open and closed, respectively. Set C and D are for the inter-ictal case of epilepsy. The EEG signals in set C were recorded from a hippocampal formation of an opposite hemisphere of the brain while the EEG signals in set D were recorded from the hippocampal formation which was the epileptogenic area. Set E is for ictal or the patient having a seizure attack.

4 Proposed Method

The proposed work deals with three experiments, one among which is multiclass classification problem and the rest of the two are binary classifications. The pre-ictal state which the proposed work classifies the EEG signal is a very useful class for medical purposes. **If the start of the pre-ictal state is predicted correctly, the medication can be provided at the earliest so as to avoid the next seizure attack.** The **first stage** of our proposed method consists of gathering the sampled EEG signals and then perform pre-processing on them (Fig. 1).

Having done the pre-processing, the **second stage** is creating 3 matrices corresponding to the three classes used in multiclass classification and the two classes in cases of binary classification.

The **third stage** of our proposed method is mainly analysis of the EEG signal of the patient using Discrete Wavelet Transform(DWT). DWT is a convenient technique

Fig. 1 Sample signals

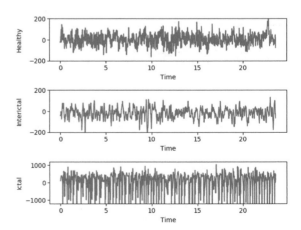

for this purpose because using DWT, original signal can be decomposed into its frequency components to obtain EEG sub-bands in various ranges of frequency. It is appropriate to mention here that the brain waves existing at different frequency ranges are most active during certain human activities. The main advantage behind using wavelets is that these are localized in both time and frequency, whereas, on the other hand, the Fourier transform is localized only in frequency. The wavelet transform strikes a balance between the frequency resolution and time resolution of the signal. In the proposed work, wavelet decomposition up to the eight levels is applied to the dataset considered over which the three-class classification is to be done. The structure of the eight-level wavelet decomposition is shown below. According to Haddad et al. [11], there is a high correlation between delta and gamma sub-bands with temporal seizures. Their research reveals that gamma spikes have more probability to originate on the same nodes after delta high voltages have been recorded and there is a high correlation in these events. According to Ren et al. [16], it was found out that the gamma waves often precede the inter-ictal epileptiform spike discharges(IED) in certain brain areas. Their research work shows a strong correlation between the seizure onset zone and gamma-IEDs. **So, the proposed work included the coefficients** D_3 **and** A_8.

Only few data exist where focal beta activity is manifested as an electroencephalographic seizure pattern. This pattern has only been observed in patients with intractable seizure disorders [4]. The reason for including the coefficients D_4 and D_5 is because the EEG signals of the healthy patients were recorded with their eyes open and with their eyes closed. It is likely that the alpha waves and the beta waves were the most active in these patients the entire time, considering that alpha waves are most dominant during wakeful alertness, calmness, and mental coordination while the beta waves are most active during waking state of consciousness when the patient is paying attention toward cognitive tasks and the outside world. Considered coefficients for all healthy, inter-ictal, ictal periods of seizure have been provided in Figs. 2, 4 and 3.

The coefficients A_8, D_3, D_4, and D_5 have been considered for the proposed method in the **fourth stage**. Having selected the four wavelet coefficients for every feature

Fig. 2 Healthy wavelet decomposition

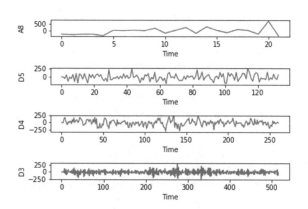

Fig. 3 Inter-ictal wavelet decomposition

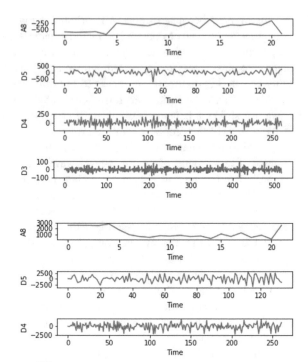

Fig. 4 Ictal wavelet decomposition

vector where a feature vector represents a patient of a class, the standard deviation of the considered coefficients has been computed for all the EEG epochs of a patient in this stage. Thus, dataset has 4097 discrete EEG readings for every patient to begin with after which wavelet decomposition was done up to eight levels creating nine coefficients. Four coefficients were selected out of them whose standard deviation was calculated. The feature matrix consists of five features as the standard deviation of approximation coefficient A_8 and detail coefficients D_3, D_4, and D_5 along with the class label for every feature vector.

In the **fifth stage**, a train-test split is made for every class and the training samples are fed to the classifiers. The proposed work uses 60% training data and 40% testing data.

The classifiers are imported and the classification is done in the **sixth stage**. This work consists of five types of experiments which are mainly classification problems, namely:

1. Binary-class classification of healthy patients, with eyes open and patients experiencing seizures
2. Binary-class classification of healthy patients, with eyes closed and patients experiencing seizures

3. Multiclass classification of healthy patients, with eyes closed and patients in inter-ictal state(whose EEG readings were recorded from hippocampal formation of an opposite hemisphere of the brain) and patients experiencing seizures.

This work compared the result obtained using eight different Daubechies wavelets namely db1, db2, db3, db4, db5, db6, db7, and db8. The results obtained after using each of these wavelets are tabulated in the next chapter and the insights obtained from the results are mentioned thereby. The features extracted which are essentially the statistical standard deviation of the selected coefficients in the previous stage are fed into the Machine Learning classifiers. The classifiers used in the proposed work are SVM, Decision Tree, Random Forest, Naive Bayes, Nearest Neighbors, and AdaBoost. The results obtained using each classifier in each of the three experiments are mentioned and tabulated in the next section.

5 Results and Comparison

The proposed work deals with the three-class classification problem where the classes are consisted of healthy patients (with eyes closed), patients in an inter-ictal stage whose EEG readings have been recorded from the opposite hemisphere of the brain, and patients experiencing epileptic activity.

5.1 Multiclass Classification

Table 2 shows the list of all the classifiers used in the work where the highest classification accuracy achieved was 98.45% (by Support Vector Machine). The accuracy is obtained while using the classifier for multiclass classification on datasets *BC-E*.

Table 2 Comparison between accuracy achieved by different classifiers in multiclass classification of BC-E

Classifier	Accuracy (%)
Support Vector Machine	98.45
Nearest Neighbors	95.34
Random Forest	95.34
Naive Bayes	94.57
Decision Tree	93.79
AdaBoost	74.41

5.2   Two-Class Classification

Table 3 shows the list of all the classifiers used in the work where the highest classification accuracy achieved was 100% (by Support Vector Machine). The accuracy is obtained while using the classifier for two-class classification on datasets *A-E* and *B-E*

5.3   Comparison of Accuracy with Different Daubechies Wavelets

The proposed work deals with the three-class classification problem where EEG epochs are classified into healthy patients with eyes closed, seizure-free intervals, and seizure (ictal periods) epochs along with two two-class classification problems where one problem was classification of EEG epochs into healthy patients(with eyes closed) and patients experiencing seizure. The another problem is being classification of EEG epochs into healthy patients(with eyes open) and patients experiencing seizure. Tables 4 and 5 show a comparison of accuracies obtained using different wavelets. The description of the datasets *A*, *B*, *C*, *D*, and *E* is summarized in Chap. 4. It is observed that the highest accuracy is obtained using the db4 wavelet which gives 98.45% in the three-class classification problem and 100% in both the binary classification problem. This might be due to the smoothing factor of db4 making it suitable to detect changes in the EEG Epoch.

5.4   Conclusions from Results

The results obtained from the various experiments conducted provide many insights to look into. Working with an automated system which accomplishes the task of multiclass classification instead of a system which does two-class classification is a

Table 3 Comparison between accuracy achieved by different classifiers in two-class classification

Classifier used	Accuracy for A-E (%)	Accuracy for B-E (%)
SVM	96.67	100
Nearest Neighbors	100	96.67
Random Forest	100	96.67
Naive Bayes	100	96.67
Decision Tree	96.67	100
AdaBoost	96.67	100

Table 4 Comparison between accuracy achieved by using different Daubechies wavelets using sets BC-E, B-E and A-E

Wavelet used	Accuracy for BC-E (%)	Classifier used	Accuracy for B-E (%)	Classifier used	Accuracy for A-E (%)	Classifier used
db1	97.67	SVM	100	SVM	100	SVM
db2	97.67	SVM	100	Decision Tree	100	SVM
db3	97.67	SVM	100	SVM	100	SVM
db4	98.45	SVM	100	SVM	100	SVM
db5	96.96	SVM	100	SVM	100	SVM
db6	96.96	SVM	100	Decision Tree	100	SVM
db7	96.96	SVM	100	Nearest Neighbors	100	SVM
db8	96.96	SVM	100	SVM	100	Nearest Neighbors

Table 5 Comparison between accuracy achieved by using different Daubechies wavelets using BD-E and AD-E

Wavelet used	Accuracy for BD-E (%)	Classifier used	Accuracy for AD-E (%)	Classifier used
db1	95.34	SVM	94.57	SVM
db2	95.34	SVM	94.57	AdaBoost
db3	96.89	SVM	95.34	SVM
db4	96.89	SVM	96.12	Random Forest
db5	94.57	SVM	96.12	Random Forest
db6	95.34	SVM	95.34	Random Forest
db7	96.89	SVM	94.57	SVM
db8	96.89	SVM	94.57	SVM

better choice because in the former, the inter-ictal activity is detected separately. This can help physicians to administer the vaccines and provide immediate medication to these patients in inter-ictal state before they experience the next epileptic activity. This is not possible to achieve in the binary classification scenario.

The results obtained from Table 3 show that both the sets *A-E* and *B-E* are linearly separable considering 100% accuracy using SVM in the case of *B-E*. SVM achieves 96.67% accuracy in set *A-E* which may be due to outliers. However, on using k-NN classifier where $k = 3$, achieved 100% accuracy in set *A-E*. This explains that the outliers are very few in population thereby leading the k-NN classifier to the right classification everytime the testing point is assigned the label of the majority count.

It is observed from the comparison Tables 4 and 5 that the highest accuracy is achieved using the fourth-order Daubechies (db4) wavelet. This can be attributed to their better localization performance in both time domain and frequency domain compared to other Daubechies wavelet. Furthermore, for signal processing in EEG, db4 wavelets are better suited due to their smoothing factor and shape [7].

6 Conclusion and Future Work

It is a very time-consuming and computationally expensive task to detect epileptic seizure activities from lengthy EEG signals. This process is usually done by trained professionals. In this work, an automated epileptic seizure detection system has been developed based on DWT followed by classification of the EEG epochs into one of the three classes namely healthy, inter-ictal, and ictal by an SVM classifier. The EEG recordings obtained from healthy patients with eyes closed and patients having seizures have been classified with an accuracy of 100%. The EEG recordings obtained from healthy patients with eyes open and patients having seizures have been classified with an accuracy of 100%. The EEG recordings obtained from the three classes during healthy, inter-ictal, and ictal dataset have been classified with an accuracy as high as 98.45%. The proposed work achieves a greater accuracy because of selecting a subset of the coefficients which helps avoid redundancy. The coefficients considered in the proposed work are the detail coefficients D3, D4, and D5 and the approximation coefficient A8. The reason for selecting these four coefficients is because the most important information related to epileptic activities are found to be present in these frequencies windows. It would be beneficial if the proposed method is used in real time for three-class classification. It is expected that with this system, subtle information which are usually hidden in the EEG data will be revealed to the clinicians with the help of which better decision can be taken. The proposed model must be validated using some other database than the one used here. Additionally, if the proposed model is to be used in real-time epilepsy seizure detection, the proposed model needs to be tested with long-duration EEG records with more patients.

As a part of future work, it is planned to deploy the method to test and predict seizure activity in epileptic patients in a clinical environment and apply deep learning methods.

Acknowledgements We would like to thank and offer sincere gratitude to Dr. Vibha Sharma and Dr. Puneet Talwar (IHBAS, Delhi) who have guided us with their patience and knowledge throughout the research work.

References

1. Andrzejak, R.G., Lehnertz, K., Mormann, F., Rieke, C., David, P., Elger, C.E.: Indications of nonlinear deterministic and finite-dimensional structures in time series of brain electrical activity: dependence on recording region and brain state. Phys. Rev. E **64**, 061907 (2001). https://doi.org/10.1103/PhysRevE.64.061907
2. Bajaj, V., Pachori, R.B.: Classification of seizure and nonseizure EEG signals using empirical mode decomposition. IEEE Trans. Inf. Technol. Biomed. **16**(6), 1135–1142 (2012). https://doi.org/10.1109/TITB.2011.2181403
3. Bhattacharyya, A., Pachori, R.B., Upadhyay, A., Acharya, U.R.: Tunable-q wavelet transform based multiscale entropy measure for automated classification of epileptic EEG signals. Appl. Sci. **7**(4) (2017). https://doi.org/10.3390/app7040385. http://www.mdpi.com/2076-3417/7/4/385
4. Bonati, L.H., Naegelin, Y., Wieser, H.G., Fuhr, P., Ruegg, S.: Beta activity in status epilepticus. Epilepsia **47**(1), 207–210 (2006)
5. Chen, W., Zhuang, J., Yu, W., Wang, Z.: Measuring complexity using FuzzyEn, ApEn, and SampEn. Med. Eng. Phys. **31**(1), 61–68 (2009)
6. Das, A.B., Bhuiyan, M.I.H., Alam, S.M.S.: Classification of EEG signals using normal inverse Gaussian parameters in the dual-tree complex wavelet transform domain for seizure detection. Signal Image Video Process. **10**(2), 259–266 (2016). https://doi.org/10.1007/s11760-014-0736-2
7. Gajic, D., Djurovic, Z., Gligorijevic, J., Di Gennaro, S., Savic-Gajic, I.: Detection of epileptiform activity in EEG signals based on time-frequency and non-linear analysis. Front. Comput. Neurosci. **9**, 38 (2015)
8. Gandhi, T., Panigrahi, B.K., Anand, S.: A comparative study of wavelet families for EEG signal classification. Neurocomputing **74**(17), 3051–3057 (2011). https://doi.org/10.1016/j.neucom.2011.04.029
9. Geva, A.B., Kerem, D.H.: Forecasting generalized epileptic seizures from the EEG signal by wavelet analysis and dynamic unsupervised fuzzy clustering. IEEE Trans. Biomed. Eng. **45**(10), 1205–1216 (1998). https://doi.org/10.1109/10.720198
10. Guo, L., Rivero, D., Pazos, A.: Epileptic seizure detection using multiwavelet transform based approximate entropy and artificial neural networks. J. Neurosci. Methods **193**(1), 156–163 (2010)
11. Haddad, T., Ben-Hamida, N., Talbi, L., Lakhssassi, A., Aouini, S.: Temporal epilepsy seizures monitoring and prediction using cross-correlation and chaos theory. Healthcare Technol. Lett. **1**(1), 45–50 (2014)
12. Kaya, Y., Uyar, M., Tekin, R., Yıldırım, S.: 1d-local binary pattern based feature extraction for classification of epileptic EEG signals. Appl. Math. Comput. **243**, 209–219 (2014)
13. Nicolaou, N., Georgiou, J.: Detection of epileptic electroencephalogram based on permutation entropy and support vector machines. Expert Syst. Appl. **39**(1), 202–209 (2012)
14. Orhan, U., Hekim, M., Ozer, M.: EEG signals classification using the k-means clustering and a multilayer perceptron neural network model. Expert Syst. Appl. **38**(10), 13475–13481 (2011). https://doi.org/10.1016/j.eswa.2011.04.149
15. Peker, M., Sen, B., Delen, D.: A novel method for automated diagnosis of epilepsy using complex-valued classifiers. IEEE J. Biomed. Health Inform. **20**(1), 108–118 (2016). https://doi.org/10.1109/JBHI.2014.2387795
16. Ren, L., Kucewicz, M.T., Cimbalnik, J., Matsumoto, J.Y., Brinkmann, B.H., Hu, W., Marsh, W.R., Meyer, F.B., Stead, S.M., Worrell, G.A.: Gamma oscillations precede interictal epileptiform spikes in the seizure onset zone. Neurology **84**(6), 602–608 (2015)
17. Samiee, K., Kovcs, P., Gabbouj, M.: Epileptic seizure classification of EEG time-series using rational discrete short-time Fourier transform. IEEE Trans. Biomed. Eng. **62**(2), 541–552 (2015). https://doi.org/10.1109/TBME.2014.2360101

18. Sharma, M., Pachori, R.B., Acharya, U.R.: A new approach to characterize epileptic seizures using analytic time-frequency flexible wavelet transform and fractal dimension. Pattern Recognit. Lett. **94**, 172–179 (2017). https://doi.org/10.1016/j.patrec.2017.03.023
19. Sharmila, A., Geethanjali, P.: DWT based detection of epileptic seizure from EEG signals using naive Bayes and k-NN classifiers. IEEE Access **4**, 7716–7727 (2016). https://doi.org/10.1109/ACCESS.2016.2585661
20. Subasi, A., Kevric, J., Abdullah Canbaz, M.: Epileptic seizure detection using hybrid machine learning methods. Neural Comput. Appl. **31**(1), 317–325 (2019). https://doi.org/10.1007/s00521-017-3003-y
21. Swami, P., Gandhi, T.K., Panigrahi, B.K., Tripathi, M., Anand, S.: A novel robust diagnostic model to detect seizures in electroencephalography. Expert Syst. Appl. **56**, 116–130 (2016). https://doi.org/10.1016/j.eswa.2016.02.040
22. Tiwari, A.K., Pachori, R.B., Kanhangad, V., Panigrahi, B.K.: Automated diagnosis of epilepsy using key-point-based local binary pattern of EEG signals. IEEE J. Biomed. Health Inform. **21**(4), 888–896 (2017). https://doi.org/10.1109/JBHI.2016.2589971
23. Tzallas, A.T., Tsipouras, M.G., Fotiadis, D.I.: Automatic seizure detection based on time-frequency analysis and artificial neural networks. Comput. Intell. Neurosci. **2007** (2007)
24. Yuan, Q., Zhou, W., Li, S., Cai, D.: Epileptic EEG classification based on extreme learning machine and nonlinear features. Epilepsy Res. **96**(1–2), 29–38 (2011)

Hexa-Directional Feature Extraction for Target-Specific Handwritten Digit Recognition

Sanjay Kumar Sonbhadra⬭, Sonali Agarwal⬭ and P. Nagabhushan⬭

Abstract Handwritten numeral recognition (HNR) is the most challenging task in the area of optical character recognition (OCR). OCR process involves both feature extraction and selection that is being generated by central or distributed sources. Presence of unimportant features in a feature set may lead to the "curse of dimensionality" and causes malfunctioning of the recognition system. A feature set is claimed as a good feature set if it contains only useful and discriminative features. In this research, the proposed model considers the projection distance of available black pixels from sharp boundary edges from all four directions (left, right, top and bottom) and directional longest run length of black pixels for rows, columns, principal diagonal and off-diagonal of any handwritten digit/character. This feature extraction algorithm is advantageous because it yields less number of features compared to zone and projection distance-based approach thus reduces computation cost without compromising the classification accuracy. Two levels of experiments have been performed to validate the authenticity of the proposed approach. In the first level, we consider the group of confusing classes, e.g. (0, 6, 8 and 9) and perform one-class classification for target-specific mining using support vector data description (SVDD); whereas, in the second level we consider all classes from 0 to 9 and perform one-class classification. Experiments are performed on own generated and MNIST data sets. For both data sets, the proposed model demonstrates better results as compared to zoning and directional-based approach of feature extraction. This paper considers classification accuracy, training time and feature set size as comparison parameters.

Keywords Handwritten numeral recognition (HNR) · Target specific · Feature extraction · Feature selection · Support vector data description (SVDD) · Digit recognition

S. K. Sonbhadra (✉) · S. Agarwal · P. Nagabhushan
Indian Institute of Information Technology, Allahabad, India
e-mail: rsi2017502@iiita.ac.in

S. Agarwal
e-mail: sonali@iiita.ac.in

P. Nagabhushan
e-mail: pnagabhushan@iiita.ac.in

© Springer Nature Singapore Pte Ltd. 2020
S. Agarwal et al. (eds.), *Machine Intelligence and Signal Processing*,
Advances in Intelligent Systems and Computing 1085,
https://doi.org/10.1007/978-981-15-1366-4_34

427

1 Introduction

Handwritten numeral recognition is a typical example of a decision-making problem where an image of a number is to be classified. HNR is a challenging task due to the diverse handwriting assets and styles that are substance to writer and time deviations. Every language has its implicit features, basic writing rules and unique writing complexity; hence, a powerful recognition approach is required for the classification of a handwritten character image. Several techniques have been hosted for the recognition of Chinese, Japanese, Persian and Latin language characters [1–7]. Trier et al. reviewed feature extraction methods for offline handwritten characters and digits [8]. In recent years, some milestone works have been done for Arabic character recognition [6, 7, 9]. The Arabic language is of cursive nature, and a number of profiles of the same character in a word based on its occurrence make the recognition problem classier. For different Indian scripts and English languages, many approaches have been published during the last three decades [10–14]. Many real-time applications, such as writer identification, zip code matching, cheque amount validation and vehicle number identification use HNR technique. An unconstrained handwritten character/digit recognition system [15–20] involves several phases: pre-processing (filtering, segmentation, thinning and normalization), feature extraction (followed by feature selection), classification and confirmation.

Feature extraction is the most crucial step to extract key features from an image because classifiers cannot process raw images. The performance of a classifier depends on the quality of the extracted features of an object. Frequently used features in handwritten script recognition are zoning [21], characteristic loci [22, 23], Fourier descriptors [24], directional distance [21, 25], crossing points, contours [26, 27], Walsh–Hadamard transform [28], Gabor transform [29], wavelets [30], strokes [31, 32], etc. In this research, a hybrid approach of feature extraction has been proposed that uses directional distance and maximum run length for simplicity and reduced computational cost. Euclidian distances of a digit's black pixels from boundary edges in all directions (left, right, top and bottom) along with directional longest run lengths of black pixels for rows, columns, principal diagonal and off-diagonal pixels as features. For the validation of the proposed method, experiments have been performed on our self-generated and MNIST [33] data sets. Experimental results show that this approach outperforms over zoning and directional distance approach [21].

Selection of discriminant features is the next challenging task after feature extraction, and considering the same objective, many classification algorithms have been published in recent years [30, 34–36] for HNR. Machine learning techniques like support vector machines (SVMs) [37, 38], Fisher's discriminant analysis (FDA) [39], principle component analysis (PCA) [30, 40] and neural networks [33, 35] have been applied to solve the problem of handwritten script recognition. Among all proposed classifiers, artificial neural network became very popular in the 1980s as revealed by the performance achieved by artificial neural networks [3, 10]; whereas, the introduction of support vector machines (SVMs) [37] has given a classical thrust to the classification of data in data sciences. Literature highlighted that artificial neural network aims at minimizing the empirical risk; whereas, SVM minimizes generalization risk and for handwritten digit recognition, SVM outperforms other classifiers [41]. Conventionally, classifiers are capable of classifying the test object to any of the well-defined classes, but the problem becomes worse if the test object belongs to a new class and classifier misclassifies it. This problem is also known as one-class classification or target-specific mining [37]. Due to the large applicability of OCC, in the proposed research work SVDD has been used for target-specific classification.

The organization of the rest of the paper is as follows: in Sect. 2, the background of the proposed work has been discussed. Section 3 describes the proposed approach. The experimental results and performance analysis are covered in Sect. 4. The last section contains the concluding remarks and future scope of the proposed work.

2 Background

In this section, we first define the notations, and then SVDD has been discussed for the proposed HNR model.

2.1 Notation

Let a two-dimensional array $I_{N \times M}$ represent an image I_x of dimensions $N \times M$ such that

$$I_{N \times M} = f(i, j) : 0 \leq i \leq N - 1 \text{ and } 0 \leq j \leq M - 1 \tag{1}$$

where $f(i, j)$ denotes the intensity of the pixel at position (i, j) and for a binary image $f(i, j)$ is either 0 or 1. For the English language, there are ten unique digits to form any number; hence, image classes be $\{y_i\}$ where $0 \leq i \leq 9$. Let n_i be the number of samples of class i, hence available sample images are

$$\bigcup_{i=0}^{9} [x_{i1}, x_{i2}, \ldots x_{in_i}] \tag{2}$$

Let d_{lm} represent lth directional distance with respect to direction m where m $\in \{$left, right, top, bottom$\}$ and r_{ab} represent ath maximum run length with respect to direction b where $b \in \{$rows, columns, principal diagonal, off diagonal$\}$. After preprocessing (scaling, normalization and padding) of images, the following is used as input to the classifier:

$$\left[\left\{ (\vec{x}_{01}, y_0)v(\vec{x}_{02}, y_0)v \ldots v(\vec{x}_{0n_0}, y_0) \right\} \otimes \left\{ (\vec{x}_{11}, y_1)v(\vec{x}_{12}, y_1)v \ldots v(\vec{x}_{1n_1}, y_1) \right\} \otimes \right.$$

$$\{(\vec{x}_{21}, y_2)v(\vec{x}_{22}, y_2)v \ldots v(\vec{x}_{2n_2}, y_2)\} \ldots \otimes \{(\vec{x}_{81}, y_8)v(\vec{x}_{82}, y_8)v \ldots v(\vec{x}_{8n_8}, y_8)\} \otimes$$
$$\{(\vec{x}_{91}, y_9)v(\vec{x}_{92}, y_9)v \ldots v(\vec{x}_{9n_9}, y_9)\}] \tag{3}$$

where $\vec{x}_{ij}$ is defined as follows:

$$\vec{x}_{ij} = [d_{1l}, d_{2l}, d_{3l} \ldots d_{Nl}, d_{1r}, d_{2r}, d_{3r} \ldots d_{Nr}, d_{1t}, d_{2t}, d_{3t} \ldots d_{Mt},$$
$$d_{1b}, d_{2b}, d_{3b} \ldots d_{Mb}, r_{1r}, r_{2r}, r_{3r} \ldots r_{Nr}, r_{1c}, r_{2c}, r_{3c} \ldots r_{Mc},$$
$$r_{1p}, r_{2p}, r_{3p} \ldots r_{(N+M-3)p}, r_{1o}, r_{2o}, r_{3o} \ldots r_{(N+M-3)o}] \tag{4}$$

$\vec{x}_{ij}$ has total $5 \times (N + M) - 6$ features where projection distance (Fig. 4a) of available pixel $y(i, j)$ from respective boundary point $z(i, j)$ is given by

$$\sqrt{(z_i - y_i)^2 + (z_j - y_j)^2} \tag{5}$$

and maximum run length r of black pixels from rows, columns, principal diagonal and minor diagonal is calculated as shown in Fig. 4b.

2.2 Support Vector Data Description (SVDD)

Conventionally, classification requires at least two well-defined classes. Classifier fails to classify a test object if it does not belong to any of the class for which the classifier has been trained. For such cases, test class will be classified for some well-defined class, e.g. if a classifier is trained with instances of cats and dogs, and the test object from a different domain (e.g. a tiger from the category of animals) is given, then the classifier will classify tiger as either cat or dog. Classification is not only the process to classify a test object as a pre-existing class, but most importantly it is supposed to check whether the test object belongs to a class or not. Above discussed problem becomes worst when data set suffers from data irregularity problems such as class imbalance, class distribution skew and absent features. When a class is under sampled or not well defined or absent, then the traditional classification models do not work as assumed. Minter [42] was the first to observe this problem and termed it "single-class classification". For the next few decades, this concept has not recovered due attention; Moya et al. [43] finally originated the term one-class classification(OCC) where the classifier was trained only by training samples of the majority class.

In a research work proposed by Tax and Duin [37, 38], one-class classification consists of similar problems as occurred in the conventional classification techniques especially with multi-classes. In this paper, the authors identified standard one-class classification problems such as measuring the complexity of a solution, the estimation of classification error, the generalization of classification methods and the curse of dimensionality. As a solution to the stated problems, a method called the support

Fig. 1 Support vector data
descriptor [37]

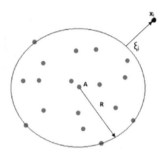

vector data description (SVDD) was proposed, where a hypersphere encloses all positive data objects (or say the class of interest), and the support vectors (the data points nearby the decision boundary) create the boundary of hypersphere. If a test object falls outside the hypersphere, then SVDD rejects it as an outlier or negative class otherwise it will accept the test object as a positive class (Fig. 1). The objecting function for SVDD is as follows:

$$L(R, a, \alpha_i, \gamma_i, \xi_i)$$

$$= R^2 + C \sum_i \xi_i - \sum_i \alpha_i \left\{ R^2 + \xi_i - \left(\|X_i\|^2 - 2a.X_i + \|a\|^2 \right) - \sum_i \gamma_i \xi_i \right. \quad (5)$$

subject to $\|X_i - a\|^2 \leq R^2 + \xi_i, \xi_i \geq 0$ i

where R is the radius of a hypersphere, $x_{i(s)}$ are outliers, and boundary dots are support vectors; whereas, to penalize outliers, there exist slack variables $\xi_i > 0$. The objective is to minimize the volume of a hypersphere (i.e. minimize R) so that it encompasses all training data objects with the penalty of the slack variable for the data points which are outliers thus there can be an increase in classification accuracy. By introducing kernel functions, the hypersphere model of SVDD can be made more flexible. Tax and Duin [44] found that generally the Gaussian kernel worked superior to the polynomial kernel and resulted in the tighter description. If Gaussian kernel width is high, then fewer support vectors selected that lead more spherical description and help to make the decision boundary lose or vice versa. It also demonstrates the effectiveness of the approach on handwritten digit recognition. In the proposed work, we use SVDD with Gaussian kernel as a classifier.

3 Proposed Algorithm

In this research work, a simple and novel approach has been proposed to extract directional features (distance and run length) from an image of a handwritten digit/character to train one-class classifier. Features are extracted concerning six directions: left, right, top, bottom, principal diagonal and off-diagonal. The proposed model consists of two-pass process to generate a feature set. In the first pass, each image must be cropped and covered by a sharp boundary along with some paddings to project the available black pixel in handwritten character and normalized to a specific size. In the second pass, feature extraction process is performed where it calculates Euclidian distances of character's black pixels from boundary edges in all four directions (left, right, top and bottom) followed by the maximum run length of black pixels from horizontal, vertical, principal diagonal and minor diagonal directions as feature set. Following are the proposed algorithms for image pre-processing and feature extraction.

Algorithm 1 (*Pre-processing*) Input: The Handwritten raw image $I(k)$ and Output: processed image $I^f(k)$.

1. Repeat step 2–5 for all k images.
2. Convert raw image $I(k)$ to binary image $I^b(k)$.
3. Crop image $I^b(k)$ sharply.
4. Add some paddings to all four directions.
5. Draw closed boundary to get processed image $I^f(k)$ of dimension $N \times M$.

Algorithm 2 (*Feature Extractor*) Input: Handwritten image $I^f(k)$ of dimension $N \times M$ and Output: $5 \times (N + M) - 6$ features.

1. Repeat step 2 and 3 for all images.
2. Calculate the Euclidian distance of the first available pixel (pixel value $= 1$) concerning all boundaries (left, right, top and bottom) Fig. 3a.
3. Calculate the maximum run-length of black pixels from four directions (horizontal, vertical, principal diagonal and minor diagonal) Fig 3b.

Figure 2 shows some pre-processed samples from different classes, and Fig. 3 describes the feature extraction mechanism.

4 Experiment and Results

Conventional methods use pixel density of an image as features, i.e. if an image is of dimension $N \times M$, then the number of features will be $N \times M$; whereas, in the proposed approach it reduces to $5 \times (N + M) - 6 <<< N \times M$ if N and M are too large. Zoning and directional-based approach, proposed by Rajashekararadhya and Vanaja Ranjan [21] generates $Z \times 2 \times (P + Q)$ features where Z is the number of zones, and each zone is of the dimension $P \times Q$. They normalized the image to 50×50 and divided the image into 25 zones of size 10×10, and then the total number of features generated is 1000; whereas, in the proposed model, the number of features is $5 \times (50 + 50) - 6 = 494$. Here, we have chosen $N = M = 50$ for the sake of minimum computational overhead. In this paper, two steps of validation have been performed on each data set. In the first phase, we make subsets of confusing

Fig. 2 Sample images

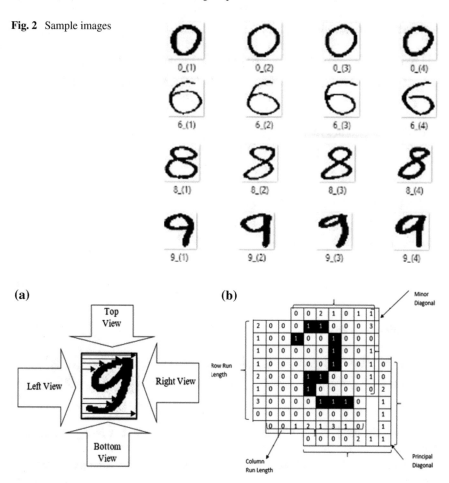

Fig. 3 Proposed feature extraction technique

handwritten digits, and in the second phase, we consider all samples associated with all available classes. HP ProLiant DL580 Gen9 Server is used to perform all the experiments. Following are the experiments and their results.

4.1 Validation for Confusing Digits

In this phase, we consider sets of confusing handwritten digits for the experiment, e.g. (0, 6, 8 and 9). Let the set of the class be $C = \{y_j\}$ where $0 \leq j \leq 9$ and let the confusing subset be $S1 = \{y_0, y_6, y_8, y_9\}$. Hence, available sample space is

$$[[x_{01}, x_{02}, \ldots x_{0n}], [x_{61}, x_{62}, \ldots x_{6n}], [x_{81}, x_{82}, \ldots x_{8n}], [x_{91}, x_{92}, \ldots x_{9n}]] \quad (6)$$

and training vectors are

$$
\begin{aligned}
&[\{(\vec{x}_{01}, y_0)v(\vec{x}_{02}, y_0)v \ldots v(\vec{x}_{0n_0}, y_0)\} \otimes \{(\vec{x}_{61}, y_6)v(\vec{x}_{62}, y_6)v \ldots v(\vec{x}_{6n_6}, y_6)\} \\
&\ldots \otimes \{(\vec{x}_{81}, y_8)v(\vec{x}_{82}, y_8) \ldots v(\vec{x}_{8n_8}, y_8)\} \otimes \{(\vec{x}_{91}, y_9)v(\vec{x}_{92}, y_9) \ldots v(\vec{x}_{9n_9}, y_9)\}]
\end{aligned}
\tag{7}
$$

and m distinct testing vectors are $[(\vec{x}_{j1}, y_j) \ldots (\vec{x}_{jm}, y_j)]$ such that $m >= 1$ where j is a unique class among 0, 6, 8 or 9.

4.2 Validation for All Classes

Hence, available sample space is

$$
\bigcup_{i=0}^{9}[x_{i1}, x_{i1}, \ldots x_{in}]
\tag{8}
$$

and training vectors are

$$
\begin{aligned}
&[\{(\vec{x}_{01}, y_0)v(\vec{x}_{02}, y_0)v \ldots v(\vec{x}_{0n_0}, y_0)\} \otimes \{(\vec{x}_{11}, y_1)v(\vec{x}_{12}, y_1)v \ldots v(\vec{x}_{1n_1}, y_1)\} \\
&\ldots \otimes \{(\vec{x}_{81}, y_8)v(\vec{x}_{82}, y_8) \ldots v(\vec{x}_{8n_8}, y_8)\} \otimes \{(\vec{x}_{91}, y_9)v(\vec{x}_{92}, y_9) \ldots v(\vec{x}_{9n_9}, y_9)\}]
\end{aligned}
\tag{10}
$$

testing vectors are $[(\vec{x}_{j1}, y_j) \ldots (\vec{x}_{jm}, y_j)]$ such that $m >= 1$, where j represents a unique class among 0–9.

Performance of the proposed algorithm is compared with zone and projection distance (ZPD)-based feature extraction mechanism [21]. For experiments, two data sets have been used. The first data set is generated by our own for which handwritten English digits have been collected from 100 individual writers. It contains 1500 samples from each class and a total of 15,000 samples. Pre-processing process has been performed as discussed in the first algorithm under Sect. 3 to all the images followed by the second algorithm to generate features to feed one-class classifier. For experiments, 400 samples from a class have been chosen randomly to train SVDD, and remaining 14,600 samples are available as testing data set; whereas, MNIST is used as the second data set for experiments. The MNIST repository is a collection of handwritten digits written by approximately 250 persons. It consists of 60,000 training samples and 10,000 testing images of size 28×28. All the images are pre-processed according to the first algorithm under Sect. 3 and resized to 50×50 followed by the second algorithm for feature set generation to feed one-class classifier. To calculate classification accuracy, 1000 samples from each class are chosen randomly to train classifier Z.

Table 1 Confusing class accuracy

Class	$Z \times 2 \times (P + Q)$ zone and projection distance [21]		$5 \times (N + M) - 6$ [proposed method]	
	Experiment 1	Experiment 2	Experiment 1	Experiment 2
0	99.12	88.32	99.66	89.36
6	96.37	87.77	99.26	87.26
8	95.12	90.12	98.66	91.66
9	94.66	91.36	99.12	93.12
Average accuracy	96.3175	96.32	99.18	99.18

Table 2 Samples versus average training time (seconds)

Number of training samples	$Z \times 2 \times (P + Q)$ zone and projection distance [21]	$5 \times (N + M) - 6$ [proposed method]
400	13.66	7.69
600	16.23	9.88
800	17.65	11.32
1000	24.36	12.79
1200	33.27	15.02
Average training time	21.03	11.34

4.3 Experiment 1: On Own Generated Data set

To perform the confusing class experiment, we consider classes 0, 6, 8 and 9. Training samples are 400 from each class to train SVDD, and remaining 5600 samples are available as testing samples in this experiment. Table 1 contains the accuracy rate; whereas, Table 2 shows the average training time.

To perform an overall class experiment, we consider all ten classes (0–9). Training samples are 400 from a class, and testing samples are 14,600 in this experiment. Table 3 shows the classification accuracy.

4.4 Experiment 2: On MNIST Data set

Thousand training samples from a particular class are randomly chosen to train the classifier, and 10,000 testing samples are used to test the performance of the classifier. For the confusing class experiment, 4000 testing samples are used, and Table 1 shows the classification accuracy rate; whereas, Table 4 shows training time with varying the number of training samples.

To perform an overall class experiment, we consider all ten classes (0–9), and Table 3 shows the overall classification accuracy.

Table 3 Overall class accuracy

Class	$Z \times 2 \times (P + Q)$ zone and projection distance [21]		$5 \times (N + M) - 6$ [proposed method]	
	Experiment 1	Experiment 2	Experiment 1	Experiment 2
0	99.12	89.17	99.67	89.17
1	99.37	90.37	99.37	93.37
2	97.36	92.26	98.96	95.76
3	96.37	94.37	99.26	95.46
4	96.36	94.36	98.66	94.66
5	95.69	95.69	99.12	98.32
6	95.12	94.32	98.66	97.66
7	97.56	91.16	98.96	92.96
8	96.37	94.77	99.16	96.26
9	94.66	93.66	99.12	94.12
Average accuracy	96.80	93.01	99.094	94.77

Table 4 Samples versus average training time (seconds)

Number of training samples	$Z \times 2 \times (P + Q)$ zone and projection distance [21]	$5 \times (N + M) - 6$ [proposed method]
800	17.65	11.32
1600	33.36	15.37
2400	48.73	23.89
3200	56.43	30.79
4000	74.91	37.12
Average training time	46.22	23.70

Table 5 shows the rapid reduction in number of features generated as compared to zone and projection distance-based feature extraction [21]. For this experiment,

Table 5 Dimension versus number of features generated

Dimension of image	$Z \times 2 \times (P + Q)$ zone and projection distance [21]	$5 \times (N + M) - 6$ [proposed method]
50×50	1000	494
100×100	2000	994
200×200	4000	1994
300×300	6000	2994
400×400	8000	3994
500×500	10,000	4994
1000×1000	20,000	9994

each image is divided into 25 zones. Results show that the proposed model generates 50% fewer features as compare to zone and projection distance-based approach.

5 Conclusion

Results of the above experiments show that the use of more number of features does not always assure better accuracy. Hence, to find a less complex way of feature extraction to reduce computational overhead and to increase classification accuracy is always an interesting and challenging task. It is also evident that discriminating features ensure good classification accuracy with reduced resource consumption.

In this research, hexa-directional projection distance and maximum run length-based feature extraction mechanism are proposed for handwritten numerals. It gives a reduced feature set and significant improvement in classification accuracy which are 2.30 and 1.76% as observed in Table 3. From Tables 2 and 4, it is observed that a reduction in computational cost is 46.09 and 48.72%, respectively. This paper highlights the importance of directional distance and run length to build a feature extractor for an enhanced character recognition system. However, in this research, the algorithm has been applied only to English numeric digits that can be further extended to other languages.

References

1. Chang, T., Chen, S.: Character segmentation using convex-hull techniques. Int. J. Pattern Recognit. Artif. Intell. **13**(6), 833–858 (1999)
2. Cheriet, M.: Extraction of handwritten data from noisy gray-level images using a multiscale approach. Pattern Recogn. Artif. Intell. **13**(5), 665–684 (1999)
3. Okamoto, M.: Online handwriting character recognition method using directional and direction-change features. Int. J. Pattern Recogn. Artif. Intell. **13**(7), 1041–1059 (1999)
4. Qian, K.: Gray image skeletonization with hollow preprocessing using distance transformation. Pattern Recogn. Artif. Intell. **13**(6), 881–892 (1999)
5. Su, T.: Chinese Handwriting Recognition: An Algorithmic Perspective, Springer (2013)
6. Alaei, A., Nagabhushan, P., Pal, U.: A new dataset of Persian handwritten documents and its segmentation. In: Proceedings 2011 7th Iranian Conference on Machine Vision Image Processing. MVIP, 2011
7. Alaei, U., Pal, U., Nagabhushan, P.: Using modified contour features and SVM based classifier for the recognition of Persian/Arabic handwritten numerals. In: Proceedings 7th International Conference on Advances in Pattern Recognition ICAPR, 391–394 2009
8. Trier, D., Jain, A.K., Taxt, T.: Feature extraction methods for character recognition-a survey. Pattern Recogn. **29**(4), 641–662 (1996)
9. Alaei, A., Nagabhushan, P., Pal, U.: Fine classification of unconstrained handwritten Persian/Arabic numerals by removing confusion amongst similar classes. In: Proceedings International Conference Document Analysis and Recognition ICDAR, 601–605 2009
10. Bhattacharya, U., Chaudhuri, B.B.: Handwritten numeral databases of indian scripts and multistage recognition of mixed numerals. IEEE Trans. on PAMI **31**(3), 444–457 (2009)

11. Pal, U., Wakabayashi, T., Sharma, N., Kimura, F.: Handwritten numeral recognition of six popular indian scripts. In: Proceedings of 9th ICDAR, 749–753 2007
12. Pal, U., Sharma, N., Wakabayashi, T., Kimura, F.: Off-line handwritten character recognition of Devnagari script. In: Proceedings of 9th ICDAR, 496–500 2007
13. Kulkarni, S.R., Rajendran, B.: Spiking neural networks for handwritten digit recognition—Supervised learning and network optimization. Neural Netw. **103**, 118–127 (2018). ISSN 0893-6080
14. Cilia, N.D., De Stefano, D., Fontanella, F., di Freca, A.S.: A ranking-based feature selection approach for handwritten character recognition. Pattern Recognit. Lett. (2018)
15. Alaei, A., Pal, U., Nagabhushan, P.: A new scheme for unconstrained handwritten text-line segmentation. Pattern Recognit. **44**(4), 917–928 (2011)
16. Gonzalez, R.C., Woods, R.E.: Digital Image Processing, p. 976. Nueva Jersey (2008)
17. Rajashekararadhya, S.V., Vanaja Ranjan, P., Manjunath, V.N.: Isolated handwritten Kannada and Tamil numeral recognition: zoning and a novel approach. In: First International Conference on Emerging Trends in Engineering and Technology ICETET 08, 1192–1195 2008
18. Das, N., et al.: A statistical-topological feature combination for recognition of handwritten numerals. Appl. Soft Comput. J. **12**(8), 2486–2495 (2012)
19. Dasgupta, J., Bhattacharya, K., Chanda, B.: A holistic approach for off-line handwritten cursive word recognition using directional feature based on Arnold transform. Pattern Recognit. Lett. **79**, 73–79 (2016)
20. Surinta, O., Karaaba, M.F., Schomaker, L.R., Wiering, M.A.: Recognition of handwritten characters using local gradient feature descriptors. Eng. Appl. Artif. Intell. **45** (2015)
21. Rajashekararadhya, S.V., Vanaja Ranjan, P.: Zone based feature extraction algorithm for handwritten numeral recognition of Kannada script. In: IEEE International Advance Computing Conference (IACC 2009), Patiala, India, 6–7 March 2009
22. Oliveira, L.S., Sabourin, R., Bortolozzi, F., Suen, C.Y.: Automatic recognition of handwritten numerical strings: a recognition and verification strategy. IEEE Trans. Pattern Recogn. Mach. Intell. **24**(11), 1438–1454 (2002)
23. Bao-Chang, P., Si-Chang, W., Guang-Yi, Y.: A method of recognizing handprinted characters. Plamondon, R.C., Suen, Y., Simner, M.L. (eds.) Computer Recognition and Human Production of Handwriting, World Scientific, pp. 37–60 (1989)
24. Manjunath Aradhya, V.N., Hemantha Kumar, G., Noushath, S.: Multilingual OCR system for South Indian scripts and English documents: an approach based on Fourier transform and principal component analysis. Eng. Appl. Artif. Intell. **21**(4), 658–668, ISSN 0952-1976 2008
25. Boukharouba, A., Bennia, A.: Novel feature extraction technique for the recognition of handwritten digits. Appl. Comput. Informatics **13**(1), 19–26 (2017)
26. Mukhopadhyay, P., Chaudhuri, B.B.: A survey of Hough transform. Pattern Recognit. **48**(3), 993–1010 (2015)
27. Shivakumara, P., Hemantha Kumar, G., Guru, D.S., Nagabhushan, P.: A new boundary growing and Hough transform based approach for accurate skew detection in binary document images. In: Proceedings of International Conference on Intelligent Sensing and Information Processing (ICISIP), IEEE, 140–146 2005
28. Bull, D.R.: Transforms for image and video coding, communicating pictures, pp. 133–169. Academic Press (2014). ISBN 9780124059061
29. Shustorovich, A.: A subspace projection approach to feature extraction: the two-dimensional gabor transform for character recognition. Neural Netw. **7**(8), 1295–1301 (1994). ISSN 0893-6080
30. Hadjadji, B., Chibani, Y., Nemmour, H.: An efficient open system for offline handwritten signature identification based on curvelet transform and one-class principal component analysis. Neurocomputing **265**, 66–77 (2017)
31. Arica, N., Yarman-Vural, F.T.: An Overview of character recognition focused on off-line handwriting. IEEE Trans. Syst. Man Cybern. Part C Appl. Rev. **31**(2), 216–233 (2001)
32. Chaudhuri, A., Mandaviya, K., Badelia, P., Ghosh, S.K.: Optical Character Recognition Systems for Different Languages with Soft Computing, vol. 352 (2017)

33. LeCun, Y., Bottou, L., Bengio, Y., Haffiner, P.: Gradient based learning applied to document recognition. Proc. IEEE **86**(11), 2278–2324 (1998)
34. Zhu, W., Zhong, P.: A new one-class SVM based on hidden information. Knowl. Based Syst. **60**, 35–43 (2014)
35. Zhang, J., et al.: Watch, attend and parse: an end-to-end neural network based approach to handwritten mathematical expression recognition. Pattern Recogn. **71**, 196–206 (2017)
36. LeCun, Y., et al.: Learning algorithms for classification: a comparison on handwritten digit recognition. Neural Networks Stat. Mech. Perspect. **2**, 261–276 (1995)
37. Tax, D.M.J., Duin, R.P.W.: Support vector domain description. Pattern Recognit. Lett. **20**(11–13), 1191–1199 (1999)
38. Tax, D.M.J., Duin, R.P.W.: Support vector data description. Mach. Learn. **54**, 45–66 (2004)
39. Ye, Q.L., Zhao, C.X., Zhang, H.F., Chen, X.B.: Recursive 'concave-convex' fisher linear discriminant with applications to face, handwritten digit and terrain recognition. Pattern Recognit. **45**(1), 54–65 (2012)
40. Jolliffe, T.: Principal Component Analysis. Springer-Verlag, New York (1986)
41. Surinta, O., Karaaba, M.F., Schomaker, L.R.B., Wiering, M.A.: Recognition of handwritten characters using local gradient feature descriptors. Eng. Appl. Artif. Intell. **45**, 405–414 (2015)
42. Minter, T.: Single-class classification (1975)
43. Koch, M.W., Moya, M.M., Hostetler, L.D., Fogler, R.J.: Cueing, feature discovery, and one-class learning for synthetic aperture radar automatic target recognition. Neural Networks **8**(7–8), 1081–1102 (1995)
44. Tax, D.M.J., Duin, R.P.W.: Uniform object generation for optimizing one-class classifiers. J. Mach. Learn. Res. **2**, 155–173 (2001)

Cardiovascular Disease Prediction Using Machine Learning Tools

Ashutosh Kumar, Rahul Gyawali and Sonali Agarwal

Abstract Cardiovascular disease deals with the disarray of blood vessels and heart. It is one of the leading causes of death globally. A doctor cannot be equally skilled in each and every domain and their availability is limited in most parts, especially in a country like India. Since the inception of machine learning, healthcare has been one of the most rewarding fields for its application. In this paper, we compared the accuracy of different machine learning algorithms like multilayer perceptron, k-nearest neighbour, logistic regression and support vector machine in approximating the severity level of cardiovascular disease in a person considering various physiological parameters. The heart disease dataset is taken from the UCI machine learning repository which is publicly available and is the most widely used dataset for heart disease prediction.

Keywords Cardiovascular disease · Machine learning algorithms · Heart disease prediction

1 Introduction and Motivation

Cardiovascular disease (CVD) is a disease that affects the cardiovascular system, particularly the blood vessels or heart. According to World Health Organization, approximately 17.9 million people in 2016 died from cardiovascular disease and making it the leading cause of death globally [1]. The risk of getting affected by CVD comes a decade earlier in Indians as compared to Europeans, mostly around their middle life years [2]. Thus, discovering the risk of CVD in early stages helps us to mitigate the possible casualties in future.

A. Kumar (✉) · R. Gyawali · S. Agarwal
Indian Institute of Information Technology, Allahabad, India
e-mail: pse2016003@iiita.ac.in

R. Gyawali
e-mail: iit2015005@iiita.ac.in

S. Agarwal
e-mail: sonali@iiita.ac.in

© Springer Nature Singapore Pte Ltd. 2020
S. Agarwal et al. (eds.), *Machine Intelligence and Signal Processing*,
Advances in Intelligent Systems and Computing 1085,
https://doi.org/10.1007/978-981-15-1366-4_35

India is the second-largest country in terms of population and lacks the provision of quality healthcare at affordable cost. India has a growing number of CVD patients. Therefore, automation in this field would be highly beneficial. Most of the work done in this field using machine learning techniques is based on different physiological factors like blood sugar level, chest pain and blood pressure. Our main aim is to analyse the efficiency of different algorithms and test the accuracy against the data from MySignals device [3].

2 Related Work

A momentous amount of research has been carried out in the healthcare sector over the past few decades with the availability of state-of-the-art algorithms and prodigious amount of medical data. By exploiting medical data, we discover significant information which helps to prevent possible dangers in future. Some astonishing work can also be witnessed particularly in predicting the risk of cardiovascular disease (CVD), which is considered to be one of the leading causes of mortality around the globe. This paper aims to explore different machine learning and deep learning methodologies in predicting cardiovascular diseases.

A review paper written by A. Hazra et al. explores minutely about cardiovascular disease (CVD) as well as legions of machine learning and data mining techniques used for prediction [4]. With some briefing about cardiovascular disease and its prevalence, the paper inquires into different data mining algorithms and tools. A comprehensive review of the past work has been presented in the literature review spanning over 35 papers.

Some of the techniques adopted were radial basis function-neural network (RBF-NN), support vector machine (SVM), fuzzy cognitive map (FCM), structural equation modelling (SEM), naive Bayes, decision tree, k-nearest neighbour (KNN), maximal frequent itemset algorithm (MAFIA), adaptive neuro-fuzzy inference system (ANFIS), random forest, weighted associative classifier (WAC), k-means, CART, ID3, a priori, etc. A contrasting dataset was used for prediction task which included UCI repository, Canadian Community Health Survey (2012), Sahara Hospital (Aurangabad), etc.

M. Singh et al. in their work [5] used a novel approach based on structural equation modelling (SEM) and fuzzy cognitive map (FCM). They used the Canadian Community Health Survey (2012) dataset where 80% of the dataset was used for training SEM model and 20% for testing the FCM model. Using 20 significant attributes, 74% accuracy was achieved.

S. K. Sen using WEKA Tool, an open-source data mining tool written in Java, predicted heart disease [6]. Using heart disease dataset from UCI learning repository, he took 14 attributes into consideration and used four different classification algorithms, i.e. naive Bayes classifier, support vector machine, decision tree and k-nearest neighbour. Upon prediction, it was observed that support vector machine had the greatest accuracy with 84.15%, while naive Bayes classifier with 83.49%, decision tree with 77.55% and k-nearest neighbour with 76.23%.

Kusuma and Divya Udayan [7] presented a complete analysis of different machine learning and deep learning algorithms over the year in order to predict heart disease. It was observed that approximately 30% of the researchers used deep learning approach for the problem while the rest 70% used machine learning tools. Looking over the past 4 years from 2014 to 2018, it was observed that the year 2016 had the highest number of articles related to the prediction of heart disease. It was also observed that Cleveland dataset is the most used dataset to the date (2018) for heart disease prediction. Authors claimed that out of all the algorithms used for prediction, support vector machine performed better than the rest of them.

Chitra [8] et al. integrated the results obtained from machine learning algorithms used for cardiovascular disease prediction. A novel approach of using cascaded neural network (CNN) was proposed in his paper. The dataset was taken from UCI machine learning repository which contained 76 attributes out of which 13 were used. To analyse high dimensional data in support vector machine, a kernel trick based on RBF kernel was used for classification purpose. The training and testing accuracy using CNN was 78.55% and 85%, respectively, while SVM classifier produced 75% and 82% training and testing accuracy. CNN classifier outperformed the rest of the algorithm for classification.

Bhuvaneswari Amma [9] writes about combining principal component analysis and adaptive neuro-fuzzy inference system (ANFIS). It will consist of two steps: first reducing the 13 attributes to 6 using PCA and in second, using ANFIS, predicting heart disease. The dataset used was obtained from UCI machine learning repository. It had a classification accuracy of 93.2.

Salamay et al. [10] predicted coronary artery disease using two machine learning algorithms which are fast decision tree and C4.5 prune tree. The dataset used was taken from the UCI machine learning repository and all the four databases including Cleveland heart disease database were used. The final accuracy was found out to be 78.06% which is higher than the average accuracy.

Ramalingam et al. [11] conducted research based on various machine learning algorithms such as k-nearest neighbour, support vector machine and random forest to predict heart disease. He found that reducing dimensionality using PCA improved accuracy and that the random forest algorithm could overcome some of the problems which other algorithms could not.

3 Method and Experiment

The main task is to classify a record containing physiological attributes of an individual for the severity of cardiovascular disease. The records are taken from heart disease dataset available at UCI machine learning repository. The dataset contains databases from four different regions, namely Switzerland, Hungary, Cleveland and the VA

Long beach. Each database contains 303 instances of patient's record containing a total of 76 attributes.

Firstly, all the dataset are integrated into a single one. Upon further preprocessing, the redundant attributes are removed and recorded with missing values in attributes are replaced with mean or median. Then, the data is visualized to get an insight about the values of different attributes. Different classification algorithms such as multilayer perceptron, k-nearest neighbour, logistic regression and support vector machine will be trained on the dataset.

The output from these algorithms will be valued and spanning across different classes which indicate the severity of cardiovascular disease. Accuracy of the algorithms will be tested against the data taken from MySignals and e-health platform. An individual record will be classified in a class by combining the output from different algorithms.

4 Implementation Plan

4.1 Exploratory Data Analysis

The heart disease dataset available at UCI machine learning repository contains four databases, namely, Cleveland, Hungary, Switzerland and the VA Long Beach [index]. The dataset contains 76 features per instance. Each one of the datasets has different instances where Cleveland contains 303, Hungarian contains 294, Switzerland contains 123 and Long Beach VA contains 200 instances, this can be visualized in the form of pie chart in Fig. 1.

4.2 Data Cleaning—Handling Corrupted and Missing Values

All the four databases were then integrated into single database containing 920 instances. Since few of the instances were corrupted in the original dataset, they were removed as a part of data cleaning. The final integrated dataset contains 899 instances of the record. As per our convenience with E-Health Kit (MySignals), we took 11 features into consideration for prediction. The features include the following:

- Age
- Sex
- Chest pain type
- Resting blood pressure
- Years of smoking
- Fasting blood sugar
- Diabetes history
- Family history of coronary artery disease

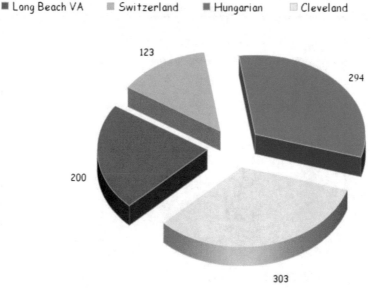

Fig. 1 Distribution of record instances across different databases

- Resting ECG
- Maximum heart rate achieved
- Degree of severity.

The selected features contained missing values after removing corrupted data. These missing values are required to be filled in order to acquire desired result. There is no better option to fill missing value than filling it with its original value. Due to the unavailability of the original value, the missing values were replaced by their mean or median values.

4.3   Data Statistics

The missing values in the dataset were replaced with mean and mode to handle missing data. The distribution of the classes among the instances in the dataset is given in Fig. 2. It is clearly visible that the majority of the instances (approximately 45%) belong to the 'Class 0', which denotes the absence of the disease. Rest of the instances in the dataset representing the presence of the disease are spread across four classes representing the severity.

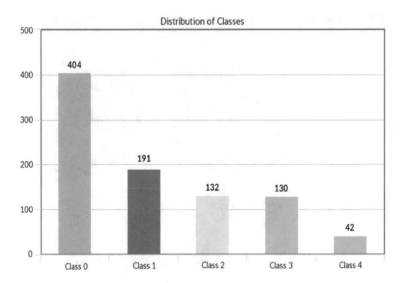

Fig. 2 Instances of different classes in dataset after cleaning

4.4 Algorithms

After cleaning and filling the missing values in the dataset, the next step was to train different models in order to make predictions. Before feeding the dataset into the model, they were normalized. Normalization adds great advantage to the prediction of the model. Normalization regulates the different numeric values across the columns in the dataset to a common range of values. This helps to converge the algorithm faster by attaining global minima more quickly.

Since we are dealing with categorical data, we need to convert it to a suitable representation for using classification model. One-hot encoding was adopted for representing categorical data. Instead of just giving class values or tags in target column, the target column is extended to multiple columns, one for each class. The presence of one class in an instance is marked with 1 in the instance of that class while the rest of the columns are filled with 0s (Fig. 3).

Target
2
1
3

➡

Target 1	Target 2	Target 3
0	1	0
1	0	0
0	0	1

Fig. 3 One-hot encoding method for categorical classification

Fig. 4 Graph of sigmoid function $\phi(z)$

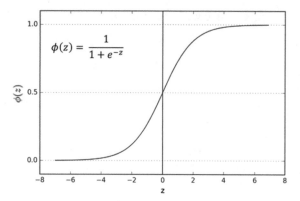

$$\phi(z) = \frac{1}{1 + e^{-z}}$$

Logistic Regression

Logistic regression is supervised learning-based classification algorithm, where we feed labelled data into the algorithm and try to predict the associated class for the new data. Given input 'X', Logistics regression outputs 'y', which denotes the class where given X belongs. It uses logistic function or commonly known as sigmoid function for its hypothesis, such that the output 'y' always lies between 0 and 1. This is mainly accompanied by the nature of sigmoid function given in Fig. 4.

Hypothesis:

$$f_\Theta(X) = y = g\left(\theta^T X\right) \tag{2}$$

$$g(z) = \frac{1}{1 + e - z} \tag{3}$$

where X is a feature vector of the form $(x_1, x_2, x_3, \ldots, x_n)$ containing 'n' features.

θ is a set of parameters to be learned for making predictions. y is the output which is either 0 or 1 denoting classes.

The hypothesis $f_\Theta(X)$ is often interpreted as estimated probability that $y = 1$ given X and θ. This can be expressed mathematically as:

$$f_\Theta(X) = P(y = 1|X; \theta) \tag{4}$$

Cost Function:

The cost function in logistic regression for prediction $f_\Theta(X)$ on given X where the correct label is 'y' can be expressed as:

$$\text{cost}\left(f_\Theta(X), y = -y^* \log(f_\Theta(X)) - (1 - y)^* \log(1 - f_\Theta(X))\right) \tag{5}$$

The overall cost function for 'n' training example is given as:

$$E(\theta) = \frac{1}{n} \sum_{i=1}^{n} \text{Cost}(f_{\Theta}(X_i), y_i) \tag{6}$$

The main aim is to find the minimum value of overall cost function $E(\theta)$, i.e. global minima to make better model. The value of parameters θ when $E(\theta)$ is minimum are the optimal values which is to be used for classifying new examples. The optimal value of parameters is obtained using gradient descent algorithm given as:

Repeat until convergence {

$$\theta_i = \theta_i - \mathrm{d}(E(\theta))/\mathrm{d}(\theta_i) \qquad \text{for } i = 0, 1, \ldots, n \tag{7}$$

}

K-Nearest Neighbours

K-nearest neighbour is a supervised classification algorithm which classifies a data point to a particular class based on the class of its neighbouring data points. 'K' is a parameter in the algorithm which denotes the number of neighbours to consider while classifying a data point. In figure 'h', we can observe that the data point 'X' is classified to square, square and triangle while taking $K = 3$, 5 and 7, respectively (Fig. 5).

The algorithm starts off with choosing suitable value of 'K'. Then we need to measure the distance between given data point and all other points in the dataset using suitable distance metric like euclidean, manhattan, etc. The 'K' nearest point from the given point is found based upon their distances. Once 'K' nearest neighbours are found, Then the given data point falls in the class where the majority of the neighbour fall. This algorithm falls in the category of instance-based learning.

Support Vector Machine

Fig. 5 Different values of K for classifying a point 'X'

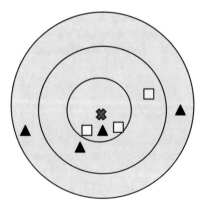

Fig. 6 Different supports with varied margin between two classes, i.e. squares and circles

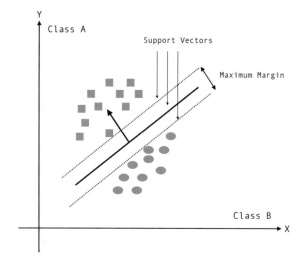

In machine learning paradigm, support vector machine, commonly known as 'SVM', is a supervised learning algorithm commonly used for classification task. The main aim of the algorithm is to find an optimal hyperplane which classifies a given feature vector to one of the classes. The optimal hyperplane is the one which has maximum margin from both of the classes, such that it is not biased to any of the classes and robust to noises (Fig. 6).

The equation of a hyperplane is given as:

$$g(X) = W^T X + b \qquad (8)$$

X is a d-dimensional feature vector. W is a vector perpendicular to the hyperplane decides the orientation of the hyperplane in d-dimensional space. b is a bias decides the position of the hyperplane in d-dimensional space.

The hyperplane divides the d-dimensional space into two half-spaces. For each feature vector 'X', we need to compute $g(X)$ to determine in which half-space does the given feature vector falls. For example, for a feature vector 'X_1':

If $g(X_1) > 0$ then X lies in the positive side of the hyperplane.
If $g(X_1) < 0$ then X lies in the negative side of the hyperplane.
If $g(X_1) = 0$ then X lies on the hyperplane.

The classifier should be able to classify any given feature vector correctly into its respective class. Therefore, classifier needs to be trained, for which values of W and b is calculated.

Artificial Neural Network

The artificial neural network model is adopted from the concept of biological neurons. In human brain, a neuron receives input from other neurons, do calculations and

Fig. 7 A single neuron takes inputs X is which gets multiplied to corresponding weights Wi s and are summed together. The bias X_0 is then added to the sum and then total sum is passed through an activation function ϕ to produce output 'Y'

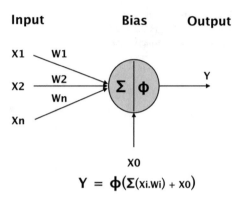

$$Y = \phi(\Sigma(x_i \cdot w_i) + x_0)$$

forward the result to other neurons. As shown in Fig. 9, A neuron receives a number of inputs $X_1, \ldots, X_n$ each of which are associated with corresponding weights w_0, $w_1, \ldots, w_n$, etc. The inputs and the weights are multiplied to each other and summed up. Then an external bias is added to the sum, which is then passed through an activation function. Here, an activation function behaves like a switch, which can turn on or off the flow of activation to other neurons. The choice of the activation function is very subjective. Some popular activation functions are: Sigmoid, ReLU, Softmax, Tanh, etc. (Fig. 7).

The concept of single neuron also called 'perceptron' can be extended to multi-layer perceptron to carry out different classification and regression tasks. Neurons are grouped into different successive layers; activations from previous layers are forwarded to the next layers. Input is fed into input layers; processing is done in hidden layers and the result is obtained in hidden layer.

In classification task, for a particular set of input 'X', the output 'Y' is expected. If there is an error in classification, the error gradient is backpropagated in previous layers which alters the weights and biases such that there is minimum error. This process is called backpropagation.

5   Results

The available dataset was split into 80% training data and 20% testing data, respectively. Different algorithms were trained with training data and accuracy were measured against testing data. The results from different algorithms are as follows (Table 1).

Even though there exist methods to diagnose CVD, its application and required effort to carry it out is huge, needless to say its affordability in a country like India. So the automation of predicting the severity of CVD not only will make it much easier and handy to use but also quite affordable for each.

Table 1 Accuracy obtained from different machine learning algorithms

Algorithm	Description	Remarks (%)
Logistic regression	Used as multiclass classifier	Accuracy of 88.88
K-nearest neighbours	For different values of 'k' from 1 to 10	Accuracy of 84.44
Support vector machine	Used RBF kernel	Accuracy of 88.88
ANN	Three-layer architecture was using different activation functions like ReLU, Sigmoid, Tanh, etc.	Accuracy of 88.89

Acknowledgements This research supported by "ASEAN—India Science & Technology Development Fund (AISTDF)", SERB, Sanction letter no.—IMRC/AISTDF/R&D/P-6/2017. Authors are thankful to AISTDF for providing us required fund for the research. Authors are also thankful to the authorities of "Indian Institute of Information Technology, Allahabad at Prayagraj", for providing us infrastructure and necessary support.

References

1. The World Health Organization, "Cardiovascular diseases (CVDs)", May 17, 2017 [Online]. Available https://www.who.int/en/news-room/fact-sheets/detail/cardiovascular-diseases-(cvds). Accessed 19 Feb, 2019
2. Ahajournals: Cardiovascular Diseases in India", Current Epidemiology and Future Directions, Circulation, vol. 133, pp. 1605–1620, Apr. 19, 2016 [Online]. Available https://www.ahajournals.org/doi/full/10.1161/CIRCULATIONAHA.114.008729. Accessed 19 Feb 2019
3. MySignals, "What is MySignals". Available http://www.my-signals.com/. Accessed 19 Feb 2019
4. Hazra, A, Mandal, S.K., Gupta, A., Mukherjee, A., Mukherjee, A.: Heart disease diagnosis and prediction using machine learning and data mining techniques: a review. Adv. Comput. Sci. Technol. vol. **10**, 2137–2159 (2017). ISSN 0973-6107
5. Singh, M., Martins, L.M., Joanis, P., Mago, V.: Building a cardiovascular disease predictive model using structural equation model & fuzzy cognitive map. In: IEEE International Conference on Fuzzy Systems (FUZZ-IEEE), pp. 1377–1382, Vancouver, 24–29 July 2016
6. Sen, S.K.: Predicting and diagnosing of heart disease using machine learning. Int. J. Eng. Comput. Sci. **6**(6), 21623–21631 (2017). ISSN 2319-7242
7. Kusuma, S., Divya Udayan, J.: Machine learning and deep learning methods in heart disease (HD) research. Int. J. Pure Appl. Math. **119**(18), 1483–1496 (2018)
8. Chitra, R.: Heart disease prediction system using supervised learning classifier. Bonfring Int. J. Softw. Eng. Soft Comput. **3**(1) (2013)
9. Bhuvaneswari Amma, N.G.: An intelligent approach based on principal component analysis and adaptive neuro fuzzy inference system for predicting the risk of cardiovascular diseases. In: Fifth International Conference on Advanced Computing (ICoAC), pp. 241–245 (2013)
10. El-Bialy, R., Salamay, M.A., Karam, O.H., Essam Khalifa, M.: Feature analysis of coronary artery heart disease data sets. In: International Conference on Communication, Management and Information Technology (ICCMIT), vol. 65, pp. 459–468 (2015)
11. Ramalingm, V.V., Dandapath, A., Karthik Raja, M: Heart disease prediction using machine learning techniques: a survey. Int. J. Eng. Technol. 684–687 (2018)

Analysis of Global Motion Compensation and Polar Vector Median for Object Tracking Using ST-MRF in Video Surveillance

Ashwini and T. Kusuma

Abstract Paper presents tracking a single moving region in the compressed domain using a spatiotemporal Markov random field (ST-MRF) model. An ST-MRF model naturally integrates the spatial and temporal aspects of the region's motion. Built upon such a model, the proposed method uses only the motion vectors (MVs) and block coding modes (BCMs) from the compressed bitstream to perform tracking. First, the MVs are pre-processed through intra-coded block motion approximation and global motion compensation (GMC). It is for region tracking in H.264/AVC-compressed video, which could be used to enhance important region analysis and, more generally, scene interpretation, in the compressed domain. It is well known that motion has a strong impact on saliency in dynamic scenes.

Keywords Block coding modes · Global motion compensation (GMC) · H.264/AVC-compressed video · Motion vectors · Spatiotemporal Markov random field

1 Introduction

Video-based object tracking is one of the challenging problems with a variety of applications, such as video surveillance, video indexing, and retrieval, video editing, communication, compression, etc. As the first step, object detection aims to locate and segment interesting objects in a video. Then, such objects can be tracked from frame to frame, and the tracks can be analyzed to recognize object behavior. Thus, object detection plays a critical role in practical applications. There are two major groups of approaches to segment and track moving objects in a video sequence, distinguished by the domain in which they operate: pixel domain and compressed domain. The former has the decoding is required to generate pixel-domain information. On the other hand,

Ashwini (✉) · T. Kusuma
Global Academy of Technology, Bangaluru, India
e-mail: dr.ashwini.k@gat.ac.in

T. Kusuma
e-mail: kusuma_t@yahoo.com

© Springer Nature Singapore Pte Ltd. 2020
S. Agarwal et al. (eds.), *Machine Intelligence and Signal Processing*,
Advances in Intelligent Systems and Computing 1085,
https://doi.org/10.1007/978-981-15-1366-4_36

"compressed-domain" approaches make use of the data from the compressed video bitstream, such as motion vectors (MVs), block coding modes, motion-compensated prediction residuals or their transform coefficients, etc. Despite the recent progress in both pixel-domain and compressed-domain video object tracking, the need for a tracking framework with both reasonable accuracy and reasonable complexity still exists. We propose a method for tracking moving objects in H.264/AVC-compressed video sequences using a spatiotemporal Markov random field (ST-MRF) model.

2 Literature Survey

Object Detection and Tracking in H.264/AVC Bitstreams: Many methods have been developed for moving object detection and tracking in H.264/AVC bitstream domain. A method based on partial decoding and initial object position determination is proposed in. Although additional information such as colors and the initial position of the objects provide reliable detection results, it cannot be applied for automatic monitoring systems because this is a semi-automatic approach which requires initial object position information by human intervention... It performs automatic object detection by observing the bitstream in MB level.

As a result, this method might fail to detect small objects, especially those of size smaller than an MB size of 16 pixels. Another method using partial decoding is proposed by Huang et al. [1] to detect moving vehicles in traffic surveillance recording. Since the method assumes that the objects always move in two opposite directions in two different lanes of roads, it may not be used for more general applications. Moreover, the use of partial decoding and background segmentation may require high computational complexity.

Compressed-Domain Segmentation and Tracking: An iterative scheme that combines global motion estimation (GME) and macroblock (MB) rejection is exploited in [2] to identify moving object blocks, which are then tracked via MB-level tracking. This scheme, however, is not able to segment and track the moving objects whose motion is not sufficiently distinct from the background motion. Käs and Nicolas [12] [16] estimate the trajectories of moving objects from H.264-AVC to SVC MVs. Foreground objects are identified by applying the background subtraction technique followed by temporal filtering to remove the noise. Afterward, motion segmentation is performed by timed-motion history images approach, and finally, the trajectory is estimated by object correspondence processing. Mean shift clustering is used in [4] to segment moving objects from MVs and partition size in H.264 bitstream.

After obtaining salient MVs by applying spatial–temporal median filter and global motion compensation (GMC), this method applies spatial-range mean shift to find motion-homogenous regions, and then smoothes the regions by temporal-range mean shift.

You et al. [14] presented an algorithm to track multiple moving objects in H.264/AVC-compressed video based on probabilistic spatiotemporal MB filtering and partial decoding. Their work assumes a stationary background and relatively slow-moving objects. Liu et al. [11] [14] perform iterative backward projection for accumulating over time to obtain the salient MVs.

The spatial homogenous moving regions are formed by statistical region growing. In [2], the segmented regions are further classified temporally using the block residuals of GMC. Another line of research [6–12] addresses segmentation and tracking problems using Markov random field (MRF) models, which provide a meaningful framework for imposing spatial constraints. Treetasanatavorn et al. [6] proposed an algorithm for motion segmentation and tracking from MV fields through the Gibbs–Markov random field theory and Bayesian estimation framework. The segmentation of the first frame is carried out by using the stochastic motion coherence model [8]. For the subsequent frames, the algorithm predicts a partition hypothesis by projecting previous partitions to the current frame based on the affine displacement model and then relaxes the partition labels [7]. The optimal configuration of this partition is found in the relaxation process so that it minimizes the maximum a posteriori (MAP) cost, described by hypothesis incongruity and the motion coherence model.

Zing et al. [19] apply MRF classification to segment and track moving objects in H.264/AVC-compressed video. First, the MVs are classified into multiple types, such as background, edge, foreground, and noise. In the second stage, moving blocks are extracted by maximizing MAP probability. In these methods, moving regions are coarsely segmented from MV fields before boundary refinement, which uses color and edge information. The authors employ GMC and MV quantization to extract a preliminary segmentation map, which is used later in initializing their spatial MRF model.

3 Objectives of the Work

Our work presents a method for tracking moving objects in H.264/AVC-compressed video sequences using a spatiotemporal Markov random field (ST-MRF) model. An ST-MRF model naturally integrates the spatial and temporal aspects of the object's motion. The proposed method evaluates the labeling dependence of consecutive frames through MVs as well as the similarity of context and neighboring MVs within the frame, and then assigns MVs into the most probable class (object vs. non-object, motion vs. no-motion), as measured by the posterior probability.

One important difference between and the proposed method is the calculation of motion coherence in which first classifies MVs to multiple types under the assumption of a static camera and then estimates motion coherence of a block according to the MV classification. On the contrary, our proposed method directly uses the MV observation in computing motion coherence. Our work aims to provide a framework for tracking moving objects in a compressed domain based on MVs and associated block coding modes alone.

The proposed method is tested on several standard sequences, and the results demonstrate its advantages over some of the recent state-of-the-art methods. In each frame, the proposed method first approximates MVs of intra-coded blocks, estimates global motion (GM) parameters, and removes GM from the motion vector field (MVF). The estimated GM parameters are used to initialize a rough position of the region in the current frame by projecting its previous position into the current frame. Eventually, the procedure of iterated conditional modes (ICM) updates and refines the predicted position according to spatial and temporal coherence under the maximum a posteriori (MAP) criterion, defined by the ST-MRF model.

3.1 Global Motion Estimation

Global motions in a video sequence are caused by camera motion, which can be demonstrated by parametric transforms. The process of estimating the transform parameters is called global motion estimation. Information is very important for video content analysis. In surveillance video, usually the camera is stationary, and the motions of the video frames are often caused by local motion objects. Thus, detecting motions in the video sequences can be utilized in anomalous events detection; in sports video, the heavy motions are also related to highlights. Motion estimation and compensation are the fundamental of video coding. Coding the residual component after motion compensated can save bit-rates significantly.

Algorithm

Step 1. Define constants (smallest block size in H.264)
Step 2. Center coordinates in image size only keep large size blocks and discard small size blocks and skip/intra-coded blocks
Step 3. Remove the region of object (imagine the object in the previous frame has the same position in the current frame)
Step 4. Reduce weights for those blocks which are different from their neighbors
Step 5. If there are not enough blocks to estimate GME then use all blocks
Step 6. Discard those blocks which we ignored it before
Step 7. Only use those blocks which might have background motion
Step 8. If there are not enough blocks to estimate GME stop iterative estimation

3.2 Preprocessing

Our proposed tracking algorithm makes use of two types of information from the H.264/AVC-compressed bitstream: BCM (partition) information and MVs. Texture data does not need to be decoded in the proposed method. H.264/AVC defines four basic macroblocks (MB) modes 16×16, 16×8, 8×16, and 8×8, where the 8×8 mode can be further split into 8×4, 4×8 and 4×4 modes. Since the smallest coding mode (partition) in H.264/AVC is 4 × 4, to have a uniformly sampled MVF, we map all MVs to 4 × 4 blocks. This is straightforward in inter-coded blocks, as well as skip blocks where the MV is simply set to zero. However, interpreting the motion in the center-coded blocks is more involved. These steps are the management of intra-coded blocks and eliminating GM.

3.2.1 Polar Vector Median for Intra-coded Blocks

Intra-coded blocks have no associated MVs. However, to run the ST-MRF optimization in the previous section, it is useful to assign MVs to these blocks. We propose to do this based on the neighboring MVs using a new method called polar vector median (PVM). For this purpose, we employ MVs of the first-order neighboring MBs (North, West, South, and East) that are not intra-coded. Figure 1 shows a sample intra-coded MB along with its first-order neighboring MBs. In this example, the goal is to find v (b) for all blocks b in the intra-coded MB. We collect MVs of the 4 × 4 blocks from the neighboring MBs that are closest to the current intra-coded MB and store them in the list V. The next step is to assign a representative vector from this collection of vectors. The standard vector median [5] would be the vector from the list with the minimum total distance to other vectors in the list. The existence of outliers, however, harms the vector median. We, therefore, propose another method called PVM, which has proved less cost demanding and more robust in our project.

Figure 2 shows the effect of assigning PVM of neighboring blocks to intra-coded blocks on the frame of the hall monitor sequence.

Algorithm

Fig. 1 Intra-coded blocks indicated by yellow square

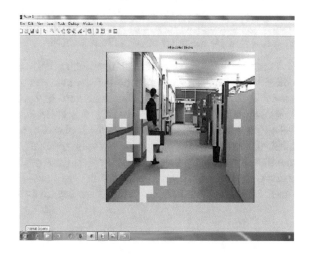

Fig. 2 Tracking result with
PVM assignment

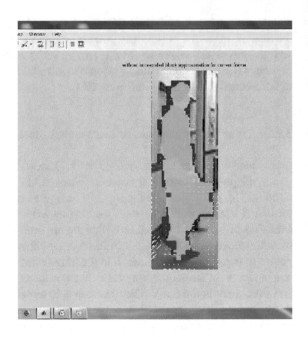

Input x: (array) x-component of a set of vectors y: (array) y-component of a set of vectors
Output: med_x: (real value) x-component of output vector med_y: (real value) y-component of output vector
Step 1: define constants
Step 2: compute the length of polar vector median
Step 3: compute the angle of polar vector median
Step 4: discard vectors with zero (or close to zero) length
Step 5: calculate difference angles in a clockwise direction
Step 6: calculate the polar median filtered vector

Figures 3 and 4 show the effect of assigning PVM of neighboring blocks to intra-coded blocks on the frame of the hall monitor sequence. The intra-coded blocks are indicated in Fig. 3, the labeling with zero MV assignment to intra-coded MBs is shown in Fig. 4, while the labeling with PVM assignment to intra-coded MBs. When the block labeling is computed, all pixels in a given block are assigned the label of that block. By comparing with the manually segmented ground truth, one can identify correctly labeled region pixels (true positives—TP), non-region pixels incorrectly labeled as region pixels (false positives—FP), and missed region pixels that are labeled as non-region pixels (false negatives—FN).

Fig. 3 Frame #70 of coastguard

Fig. 4 Target superimposed by scaled MVF after GMC

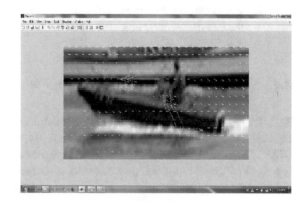

3.3   Global Motion Compensation

Global motion compensation (GMC) is a new coding technology for video compression in H.264/AVC standard.

By extracting camera motion, H.264/AVC coder can remove the global motion redundancy from the video. In H.264/AVC 4, each macroblock (MB) can be selected to be coded use GMC or local motion compensation (LMC) adaptively during mode determination [24]. Intuitively, some types of motion, e.g., panning, zooming, or rotation, could be described using one set of motion parameters for the entire video object plane (VOP). For example, each MB could potentially have the same MV for the panning.

GMC allows the encoder to pass one set of motion parameters in the VOP header to describe the motion of all MBs. Additionally, H.264/AVC allows each MB to specify its MV to be used in place of the global MV. In H.264/AVC advanced simple contour, the main target of global motion compensation (GMC) is to encode the global motion in a video object plane (VOP) using a small number of parameters.

A two-dimensional vector used for inter prediction that provides an offset from the coordinates in the decoded picture to the coordinates in a reference picture measured by posterior probability is a conditional probability conditioned on randomly observed data. Global motion (GM), caused by camera movement, affects all pixels in the frame. Since GM adds to the object's native motion, for accurate tracking, it is important to remove GM from the MV field before further processing. Given the parameters of the affine model, $[m_1 \cdots m_6]$, a block centered at (x, y) in the current frame will be transformed into a quadrangle centered at (x', y') in the reference frame,

$$
\begin{aligned}
x', y' \\
x^1 &= m_{1+}m_2 x + m_3 y \\
y^1 &= m_{4+}m_5 + m_6 y
\end{aligned}
\tag{1}
$$

This method reduces the influence of outliers in a reweighted iteration procedure based on the estimation error obtained using least squares estimation.

3.4 Spatiotemporal Markov Random Fields

ST-MRF model has been effectively used in pixel-domain applications such as moving object detection [13], object segmentation [1, 14, 15], unmanned aerial vehicles (UAV) [16], etc. The ST-MRF models for compressed-domain video object tracking. A moving rigid object is generally characterized by spatial compactness (i.e., not dispersed across different parts of the frame), relative similarity of motion within the region occupied by the object, and a continuous motion trajectory. The motion of flexible objects is much more difficult to characterize but, at least in principle, these objects can be treated in a divide and conquer manner as a collection of smaller sufficiently rigid objects.

A moving rigid object/region is generally characterized by spatial compactness (i.e., not dispersed across deferent parts of the frame), relative similarity of motion within the region, and a continuous motion trajectory. Our ST-MRF model is based on rigid object/region motion characteristics. We treat moving region tracking as a problem of inferring the MAP solution of the ST-MRF model. More specially, we consider the frame to be divided into small blocks (4×4 in our experiments). Region blocks will be labeled 1, non-region blocks 0. We want to infer the block labels formation. Here, the motion information consists of the MVs from the compressed bitstream, denoted $v^t(n)$, and the BCM and partition size $k^t(n)$, where $n = (x; y)$ indicates the position of the block within the frame. Therefore, our ST-MRF model is based on rigid object motion characteristics. We treat moving object tracking as a problem of inferring the MAP solution of the ST-MRF model.

4 Results and Discussions

Object-tracking data is the most typical ground truth for visual models [17]. To evaluate saliency models, each model's saliency map is compared with recorded locations of the subjects. Some recent publicly available object-tracking datasets are used in this work. The reader is referred to [12] for an overview of other existing datasets in the field.

Experimental setup
Several standard test sequences were used to evaluate the performance of our proposed approach. Sequences were in the YUV 4:2:0 format, at two resolutions, CIF (Common Intermediate Format, 352,288 pixels) and SIF (Source Input Format, 352,240 pixels), all at the frame rate of 30 fps. All sequences were encoded using the H.264/AVC JMv.18.0 encoder, at various bitrates, with the GOP (Group-of-Pictures) structure IPPP, i.e., the first frame is coded as intra (I), and the subsequent frames are coded predicatively (P). Motion and partition information was extracted from the compressed bitstream, and MVs were remapped to 4 × 4 blocks. Some of the characteristics of the proposed approach are its robustness and stability. To show this, we use the same parameters throughout all the experiments (Fig. 3).

As seen from Fig. 4, the MVF around the target (small boat) is somewhat erratic, due to fast camera motion in this part of the sequence.

The proposed ST-MRF-based tracking algorithm computes the energy function for the chosen MRF model, which is shown in Figs. 5 and 6.

Pixels shaded with deferent colors indicate TP, FP, and FN, as explained in the caption of Fig. 3. As seen here, some erroneous decisions are made around the boundary of the target object, but the object interior is detected well. The proposed method has produced a much better result compared to other methods in the face of sudden and fast camera movement. As seen here, some erroneous decisions are made around the boundary of the target object, but the object interior is detected well. The proposed method has produced a much better result compared to other methods in the face of sudden and fast camera movement.

Fig. 5 The heatmap visualization of MRF energy

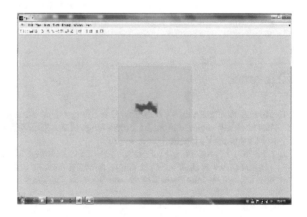

Fig. 6 Tracking result by
the proposed method

Some of the characteristics of the proposed approach are its robustness and stability. To show this, we use the same parameters throughout all the experiments. We found that the average performance does not change much if some of these parameter values are changed, especially the parameters $c1$, $c2$, and $c3$ that represent the effect of camera motion on energy function, and only affect a few frames. Figure 5 illustrates a few intermediate results from the tracking process for a sample frame from coastguard.

5 Conclusion

We studied important video processing problem—tracking in the compressed domain. We have presented a novel approach to track a moving object/region in an H.264/AVC-compressed video. The only data from the compressed stream used in the proposed method are the MVs and BCMs. After the preprocessing stage, which consists of intra-coded block approximation and global motion compensation (GMC), we employed an MRF model to detect and track a moving target. Using Morrkov Random model estimate the labeling of the current frame was formed based on the previous frame labeling and current motion information.

References

1. Huang, R., Pavlovic, V., Metaxas, D.N.: A new spatio-temporal MRF framework for video-based object segmentation. In: Proceedings of the MLVMA Conjunct. European Conference on Computer Vision, 2008, pp. 1–12
2. Cherian, A., Anders, J., Morella's, V., Papanikolopoulos, N., Mettler, B.: Autonomous altitude estimation of a UAV using a single onboard camera. In: Proceedings of the International Conference on Intelligent Robots and Systems, Oct 2009

3. Liu, Z., Lu, Y., Zhang, Z.: Real-time spatiotemporal segmentation of video objects in the H.264 compressed domain. J. Vis. Commun. Image Represent. **18**(3), 275–290 (2007)
4. Fei, W., Zhu, S.: Mean shift clustering-based moving object segmentation in the H.264 compressed domain. IET Image Process. **4**(1), 11–18 (2010)
5. You, W., Sabirin, M.S.H., Kim, M.: Real-time detection and tracking of multiple objects with partial decoding in H.264/AVC bitstream domain. In: Proceedings of the SPIE, Real-Time Image and Video Processing, Mar 2009, vol. 7244, pp. 72440D-1–72440D-12
6. Treetasanatavorn, S., Rauschenbach, U., Heuer, J., Kaup, A.: Model-based segmentation of motion fields in compressed video sequences using partition projection and relaxation. In: Proceedings of the Visual Communications and Image Processing, Beijing, China, Jul 2005
7. Treetasanatavorn, S., Rauschenbach, U., Heuer, J., Kaup, A.: Stochastic motion coherency analysis for motion vector field segmentation on compressed video sequences. In: Proceedings of the IEEE Workshop on Image Analysis for Multimedia Interactive Services, Apr 2005, pp. 1–4
8. Zeng, W., Du, J., Gao, W., Huang, Q.: Robust moving object segmentation on H.264/AVC compressed video using the block-based MRF model. Real-Time Imaging **11**(4), 290–299 (2005)
9. Chen, Y.-M., Bajic, I.V., Saeedi, P.: Motion segmentation in compressed video using Markov random fields. In: Proceedings of the IEEE International Conference on Multimedia and Expo, Singapore, Jul 2010, pp. 760–765
10. Winkler, S., Subramanian, R.: Overview tracking datasets. In: 5th International Workshop on Quality of Multimedia Experience (QoMEX), pp. 212–217. IEEE (2013)
11. Engelke, U., Kaprykowsky, H., Zepernick, H.J., Ndjiki-Nya, P.: Visual attention inequality assessment. IEEE Signal Process. Mag. **28**(6), 50–59 (2011)
12. Hadizadeh, H., Enriquez, M.J., Bajic, I.V.: Eye-tracking database for a set of standard video sequences. IEEE Trans. Image Process. **21**(2), 898–903 (2012)
13. Yin, Z., Collins, R.: Belief propagation in a 3D spatio-temporal MRF for moving object detection. In: Proceedings of the IEEE Computer Vision and Pattern Recognition, Jun 2007, pp. 1–8
14. Wang, Y., Loe, K.-F., Tan, T., Wu, J.-K.: Spatiotemporal video segmentation based on graphical models. IEEE Trans. Image Process. **14**(7), 937–947 (2005)
15. Wang, Y.: A dynamic conditional random field model for object segmentation in image sequences. In: Proceedings of the IEEE Computer Vision and Pattern Recognition, vol. 1, Jun 2005, pp. 264–270
16. Käs, C., Nicolas, H.: An approach to trajectory estimation of moving objects in the H.264 compressed domain. In: Proceedings of the 3rd Pacific-Rim Symposium on Image and Video Technology. LNCS, vol. 5414, pp. 318–329 (2009)
17. H.264/AVC Reference Software JM18.0 (Aug 2012) [Online]. Available at: http://iphome.hh
18. Arvanitidou, M.G., Tok, M., Krutz, A., Sikora, T.: Short-term motion-based object segmentation. In: Proceedings of the IEEE International Conference on Multimedia and Expo, Barcelona, Spain, Jul 2011
19. Aeschliman, C., Park, J., Kak, A.C.: A probabilistic framework for joint segmentation and tracking. In: Proceedings of the IEEE Computer Vision and Pattern Recognition, San Francisco, CA, Jun 2010
20. Comaniciu, D., Ramesh, V., Meer, P.: Kernel-based object tracking. IEEE Trans. Pattern Anal. Mach. Intell. **25**(5), 564–575 (2003)
21. Drummond, T., Cipolla, R.: Real-time visual tracking of complex structures. IEEE Trans. Pattern Anal. Mach. Intell. **24**(7), 932–946 (2002)
22. Kato, Z., Pong, T.-C., Lee, J.C.-M.: Color image segmentation and parameter estimation in a Markovian framework. Pattern Recognit. Lett. **22**(3–4), 309–321 (2001)
23. Treetasanatavorn, S., Rauschenbach, U., Heuer, J., Kaup, A.: Bayesian method for motion segmentation and tracking in compressed videos. In: Proceedings of the 27th DAGM Conference on Pattern Recognition. LNCS, vol. 3663, pp. 277–284 (2005)

24. Khatoonabadi, S.H., Bajic, I.V.: Compressed-domain global motion estimation based on the normalized direct linear transform algorithm. In: Proceedings of the International Technical Conference on Circuits/Systems, Computers, and Communications

Author Index

© Springer Nature Singapore Pte Ltd. 2020
S. Agarwal et al. (eds.), *Machine Intelligence and Signal Processing*,
Advances in Intelligent Systems and Computing 1085,
https://doi.org/10.1007/978-981-15-1366-4

Printed in the United States
By Bookmasters